Guide To

Natural Gas

Cogeneration

Guide to

Natural Gas

Cogeneration

NELSON E. HAY
Editor

Published by
THE FAIRMONT PRESS, INC.
700 Indian Trail
Lilburn, GA 30247

Library of Congress Cataloging-in-Publication Data

Guide to natural gas cogeneration.

1. Cogeneration of electric power and heat.   2. Gas, Natural.
I. Hay, Nelson E.
TK1061.G85 1988   621.1'9   86-46133
ISBN 0-88173-041-6

Published by The Fairmont Press, Inc.
700 Indian Trail
Lilburn, GA 30247
ISBN   0-13-370388-6   PH
ISBN   0-88173-041-6   FP

Printed in the United States of America.

Distributed by Prentice Hall
A division of Simon & Schuster
Englewood Cliffs, NJ 07632

Prentice-Hall International (UK) Limited, *London*
Prentice-Hall of Australia Pty. Limited, *Sydney*
Prentice-Hall International (UK) Limited, *London*
Prentice-Hall Hispanoamericana, S.A., *Mexico*
Prentice-Hall of India Private Limited, *New Delhi*
Prentice-Hall of Japan, Inc., *Tokyo*
Simon & Schuster Asia Pte. Ltd., *Singapore*
Editora Prentice-Hall do Brasil, Ltda., *Rio de Janeiro*

Foreword

"Is America Running Out of Steam?"

As I write this foreword, I have on my desk a recent *Washington Post* page 1 article entitled, "Is America Running Out of Steam?" Coincidentally, President Ronald Reagan is expected to sign legislation today which effectively repeals the constraints on natural gas use embodied in the Powerplant and Industrial Fuel Use Act and in the NGPA Incremental Pricing provisions. These events seem very appropriate to this book on natural gas-fired cogeneration. In fact, they are right to the point.

Productivity, competitiveness, profitability, employment, balance of trade—these are the key words of the central economic issue facing our country today. They translate into "standard of living." Can we maintain the standard of living which we have? Can we improve upon it and spread its benefits more broadly throughout our society?

If we are ultimately to answer these questions in the affirmative, we must overcome misconceptions and special interests where they deter us from choosing the most economically and environmentally desirable means of achieving our goals. Clearly, technological improvements such as cogeneration must be one of our tools. Cogeneration can allow us to recover and put to use energy which would

---

[1]Behr, Peter, "Is America Running Out of Steam?" Washington, D.C., *Washington Post*, April 17, 1987, p. 1.

v

otherwise be wasted. In the process, the need for new energy projects can be reduced, resources can be preserved, and energy costs and environmental damage can be minimized. In short, productivity can be increased. In most cases, these benefits can be greatest when clean and abundant natural gas is the cogeneration fuel.

While hailing these benefits of cogeneration, it is important that we also keep our feet on the ground. Cogeneration is one of many worthy concepts. It is not a panacea. It is not always economically desirable. When thorough analysis demonstrates its economic and environmental benefits, however, it should be pursued. We can little afford to do otherwise.

Nelson E. Hay

Acknowledgments

The editor wishes to express his thanks to the authors and copyright holders for permission to publish the works which are collected in this volume. Particular recognition should also be given to Mary Wosoogh, Paul Wilkinson, Bruce McDowell, David Sgrignoli, Eben Fodor, and Mike German for their contributions and support. The book was indexed by Mary Wosoogh, and typeset by Wanna Walker.

Contents

Section III — Natural Gas Prime Movers

Section IV — Natural Gas for Efficient Electric Generation

Section V — Regulatory Considerations

Appendices

Section I

Advantages of
Natural Gas-Fired Cogeneration

PAUL L. WILKINSON
Manager, Policy Analysis

BRUCE P. McDOWELL
Policy Analyst

American Gas Association

Chapter 1

Gas-Fired Cogeneration: What and Why?

"Cogeneration" refers to the sequential production within one system of both electrical (or mechanical) energy and thermal energy. Two-thirds of the energy consumed by conventional electric power plants is normally lost to the environment. Cogeneration systems, in contrast, recapture much of the otherwise wasted thermal energy and use this energy for a variety of purposes—e.g., space or water heating and industrial process needs. Cogeneration systems may be based on any of a number of fuels and commercially available technologies.

Cogeneration systems offer inherent efficiency advantages relative to conventional means of power production and, in addition, they may offer environmental benefits. A primary advantage of cogeneration systems is that they are inherently energy efficient. That is, they minimize the waste of primary energy resources—oil, gas, and coal. However, the net impact of cogeneration systems on the environment relative to conventional power production is not clear-cut. In spite of the fact that they consume less fuel, small, dispersed internal combustion or coal-based cogeneration systems may be less environmentally attractive than large central powerplants with tall stacks and pollution control equipment. For example, the production of NO_x via a diesel fuel-fired internal combustion cogeneration system would be over twice the NO_x level produced by the combina-

3

Exhibit 1-1

Comparison of Emissions from Gas Turbine and Diesel-Powered Internal Combustion Systems versus a Conventional Coal- and Oil-Fired System for Production of Steam and Electricity

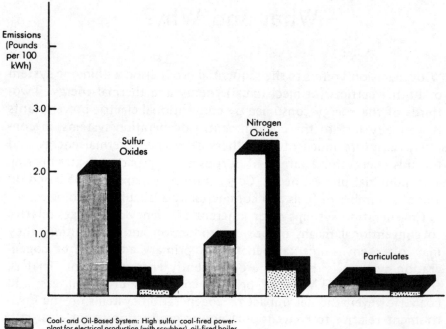

Emissions (Pounds per 100 kWh)

Sulfur Oxides

Nitrogen Oxides

Particulates

Coal- and Oil-Based System: High sulfur coal-fired power-plant for electrical production (with scrubber), oil-fired boiler for steam production.

Gas Turbine Cogeneration System.

Internal Combustion Cogeneration System, diesel-fuel powered.

Note: Emissions from each system based on production of 100 kWh of electricity and 610,000 Btu of useable steam; all equipment assumed to meet new sources performance standards.

Source: Congress of the United States, Office of Technology Assessment, *Industrial and Commercial Cogeneration*, (Washington, DC: February, 1982) pp. 286, 287.

tion of a scrubber-equipped coal-fired powerplant and a coal-fired boiler (with electrical and steam production capacities equivalent to the cogeneration system).

Gas-Fired cogeneration systems are an attractive option from both an environmental and an energy efficiency standpoint. Gas-fired cogeneration systems, such as gas turbines or gas boiler/steam turbine combinations, offer a clean and efficient means of electrical power and thermal energy production. According to the Office of Technology Assessment, a gas turbine cogeneration system would require approximately 25 percent less input energy than the combination of a new coal-fired powerplant and an oil-fired boiler producing steam (assuming optimal system sizing). In addition, the gas-fired cogenerator would emit less than 1 percent of the SO_x, 27 percent of the particulates, and 50 percent of NO_x produced by the comparably sized conventional coal- and oil-fired system (with pollution control equipment).

Chapter 2

An Economic and Environmental Comparison of Natural Gas and Coal Use For Large-Scale Industrial Cogeneration

Cogeneration can allow energy consumers to lower their purchased energy costs through use of the energy normally wasted in conventional systems. Industrial energy consumers are best suited to take advantage of this technology, primarily as a result of their large and relatively constant energy demand. Therefore, most existing cogeneration capacity is currently concentrated in the industrial sector.

Natural gas and coal are the two predominant fuels consumed in industrial cogenerators. In small-scale projects with fuel inputs below 50 million Btu per hour (MMBtu/hr), natural gas is much more widely used than coal for both technical and economic reasons. For larger capacity applications, natural gas and coal are more competitive.

This chapter examines the relative economics and environmental impacts of coal-fueled and natural gas-fueled large-scale industrial cogeneration systems. Two steam turbine cogeneration systems are analyzed, one fueled by gas and one by coal, as well as one gas-fueled combustion turbine system.

EXECUTIVE SUMMARY

Natural gas is economically and environmentally preferable to coal for use in most large-scale industrial cogeneration applications.

Economic Results

Gas-fueled cogeneration is economically preferable to coal-fueled cogeneration in large-scale industrial applications, in spite of the fact that coal is generally less costly than gas, due to the lower costs of installing and operating gas-fueled systems.

- The total levelized annual costs of "typical" gas turbine ($4.6 million per year) and gas-fueled steam turbine ($7.5 million per year) cogeneration systems are only 46 percent and 75 percent, respectively, of the cost of a comparable coal-fueled system ($10.0 million per year). [All costs in constant 1986 dollars, based on 30 year projected national average fuel prices, a 100,000 pound per hour steam output, a 10.2 percent real after-tax cost of capital, a 70 percent capacity utilization factor, and a 4.4 cent per kilowatt hour buy-back rate for *excess* electricity—produced only by the gas turbine system. See Table 2-1.]

  — For this "typical" system, the annual installed capital charge is lowest for the gas-fueled steam turbine—$1.3 million, per year—versus $2.5 million for the gas turbine, and only 36 percent of the $3.6 million capital charge of the coal system.

  — The annual operation and maintenance expenditure for the gas turbine cogeneration system, $0.4 million per year, is approximately one-third of the annual O&M requirement of the gas-fueled steam turbine ($1.3 million), and only 13 percent of the significantly higher O&M requirement of the coal-fueled system—$3.0 million annually.

  — Due to the additional electricity required to operate the necessary pollution control equipment of the coal-fueled system, the cost of purchased electricity for the coal system would exceed purchased electricity costs of the gas-fueled steam

Table 2-1

Baseline Cogeneration Systems' Total Levelized[1] Annual Costs
(Thousands of 1986 $)

| | Process Steam Requirement (lbs/hr) | | | | |
	50,000	100,000	200,000	400,000	800,000
Gas Turbine:					
Capital	$1,394	$2,467	$ 3,956	$ 6,275	$12,550
Fuel	2,854	5,709	11,418	22,835	45,670
O&M	225	449	899	1,797	3,596
Electricity[2]	(2,001)	(4,003)	(8,004)	(16,010)	(32,020)
Total	$2,472	$4,622	$ 8,269	$14,897	$29,796
Steam Turbine, Gas:					
Capital Charge	$ 880	$1,301	$ 3,388	$ 5,603	$ 9,344
Fuel	1,808	3,616	7,207	14,438	28,876
O&M	956	1,250	2,115	3,199	4,837
Purchased Electricity	690	1,381	2,736	5,499	10,998
Total	$4,334	$7,548	$15,447	$28,740	$54,056
Steam Turbine, Coal:					
Capital Charge	$1,962	$3,603	$ 6,362	$11,161	$19,608
Fuel	939	1,877	3,742	7,497	14,994
O&M	1,896	2,970	4,565	7,248	11,554
Purchased Electricity	768	1,536	3,046	6,120	12,239
Total	$5,565	$9,987	$17,715	$32,027	$58,395

Sources: Based on data from R. W. Hess, J. J. Turner, W. H. Krase, and R. Y. Pei for the U.S. Department of Energy, *Factors Affecting Industry's Decision To Cogenerate* (Rand Corp., Santa Monica, CA: January 1983), Office of Technology Assessment, Congress of the United States, *Industrial and Commercial Cogeneration* (Washington, DC, February, 1983), and the fuel prices as projected by the American Gas Association, *The A.G.A.-TERA Reference Case 1986-1* (Arlington, VA: 1986).

[1] Levelized annual cost refers to a method by which total project expenditures, which fluctuate from year to year, are converted to a constant annual expenditure figure over the life of the project.

[2] Negative indicates production of electricity in excess of plant's requirements—excess sold to grid at rate of 4.4 cents/kWh.

turbine by 11 percent—$155,000 per year. The coal to gas turbine contrast is appreciably more striking—an annual electricity cost of $1.5 million for the coal system versus an annual *income* of $4.0 million for the gas turbine—attributable to the higher electricity to steam output of the gas turbine system which allows substantial sales of excess electricity to the grid.

Environmental Results

- Large-scale natural gas-fueled cogeneration systems are environmentally preferable to coal-based systems with pollution control equipment regardless of the system size. For example, when compared to a coal-fueled system with a scrubber and an electrostatic precipitator, a gas-fueled steam turbine emits only 6 percent of the sulfur oxides, 10 percent of the particulates, and 28 percent of the nitrogen oxides.

  - In order to comply with the Clean Air Act, even the smaller coal systems examined herein must include expensive and cumbersome pollution control equipment due to their potential to emit more than 250 tons annually of a regulated pollutant.

Because many site-specific factors affect the feasibility of industrial cogeneration, it is imperative to perform a site-specific analysis when considering such an investment. In no way does this analysis preclude the need for detailed feasibility studies for actual industrial (or other) applications.

BACKGROUND

In the United States today, the combined electrical output of commercial and industrial cogeneration facilities that either are operating or are in the design/construction phase capacity totals approximately 24,700 MW. Over 10,700 MW (about 43 percent) of the total is believed to be fueled by natural gas, with the remainder being powered by other fuels, such as coal, oil, biomass, wood, and wood waste. About 39 percent of the total capacity, and over 75 percent of the natural gas-based capacity, is located in the West South Central region of the United States.[1]

Although the number of these systems is growing, the potential has barely been tapped. The fuel input to cogeneration facilities currently accounts for less than one percent of the total fuel input to the approximately 1.8 million commercial and industrial boilers. Of this total, less than one percent (about 10,700) are large industrial boilers (fuel input greater than 50 MMBtu/hr). However, these large industrial boilers, which have steam outputs similar to those in the cogeneration systems examined in this analysis, account for over 37 percent of the total U.S. commercial and industrial boiler capacity. Furthermore, almost 90 percent of those large industrial boilers have capacities less than 250 MMBtu/hr[2]. (See Table 2-2 for the size distribution of large industrial boilers.)

Table 2-2

Size Distribution of the Industrial Large Boiler Population

Size (MMBtu/hr)	Number of Units	Total Capacity (MMBtu/hr)
50 — 100	5,640	414,390
100 — 250	3,833	564,220
250 — 500	919	300,600
500 — 1500	237	179,020
>1500	61	214,920
Total	10,690	1,673,150

Source: PEDCo Environmental Inc. for the U.S. Environmental Protection Agency, *Population and Characteristics of Industrial/Commercial Boilers in the U.S.* (Washington, DC: August, 1979).

Two types of cogeneration technologies are examined in this analysis: Steam turbines and gas-fueled combustion turbines. The steam turbine cogeneration system analyzed herein is comprised of a boiler (fueled by gas or coal) used in combination with a steam turbine to produce both electricity and process steam. The input fuel is combusted in the boiler, which produces high pressure steam. The high pressure steam is channeled to the turbine to turn a rotor which generates electricity. Low pressure steam exits the turbine and is used to meet the industry's process steam requirement.

The gas turbine system is comprised of a compressor, a combustion chamber, a turbine, a generator, and a waste heat boiler. Air is compressed and moves to the combustion chamber where the gas is burned to heat the air. The heated air drives the turbine, which turns a rotor in the generator to produce electricity. The waste heat boiler recovers heat from the hot gases that are not used in the turbine and, with the infusion of water, uses the heat to produce steam for use in the industrial process. The ratio of electricity to process steam produced for this type of system is about three and one-half times that of the steam turbine systems.

METHODOLOGY

Data Sources and Assumptions

Capital costs, operation and maintenance (O&M,—excluding fuel) costs, and system characteristics (size, capacity, etc.) for the steam turbine systems were obtained from the Rand Corporation's *Factors Affecting Industry's Decision to Cogenerate.*[3] For the gas turbine cogeneration system, O&M costs and system characteristics were obtained from the Office of Technology Assessment's (OTA) *Industrial and Commercial Cogeneration.*[4] The ratio of gas turbine to gas-fueled steam turbine capital costs was assumed to be 1:2, based on data obtained in the Rand, OTA, and other published reports (assuming equal electric output).

The sizes or capacities of the systems examined ranged from a process saturated steam requirement of 50,000 lbs/hr to 800,000 lbs/hr. Each of the three cogeneration systems is sized to meet the same process steam requirement. Since the gas turbine system has a higher electric to thermal ratio and a lower overall energy efficiency factor, the fuel inputs and the electric outputs of the gas turbine and steam turbine cogeneration systems differ. (The gas-fueled steam turbine and the coal-fueled steam turbine systems, however, are similar in terms of fuel input, steam output, and gross power production.) For example, a process steam requirement of 200,000 lbs/hr would require a fuel input of 303 MMBtu/hr and gross electric output of 10.90 MW for the steam turbine, while the gas turbine would require a fuel input of 480 MMBtu/hr and a gross electric output of 38.7 MW (see Table 2-3).

Capital costs (expressed in 1986 dollars) include the combustion unit, power generation and conditioning equipment, architect and engineering services, a contingency fund, interest during construction, and pollution control equipment. Flue gas desulfurization systems (scrubbers) are required on all of the coal-powered systems examined in this analysis, including the two with fuel inputs less than 250 MMBtu/hr. To comply with the Clean Air Act, these smaller systems must have scrubbers since they have the potential to emit more than 250 tons annually of a regulated pollutant, which would subject them to prevention of significant deterioration (PSD) review.[5] O&M costs include expenses arising from the use of the heat source, cogeneration, and pollution control equipment.

Baseline energy prices and real (net of inflation) escalation rates were taken from the *A.G.A.-TERA Reference Case 1986-1*[6]. [The American Gas Association's Total Energy Resource Analysis (A.G.A.-TERA) system of computer models is a detailed long-run simulation of the gas industry in terms of supply and demand, with consideration of the environment in which the industry operates.] These fuel prices (expressed in 1986 dollars) represent national average prices as projected for the industrial sector for coal, natural gas, and purchased electricity. See Appendix 2-1 for the 1986 energy prices and projected real escalation rates employed.

Table 2-3

Cogeneration System Costs and Characteristics

	Process Saturated Steam Requirement (lbs/hr)				
	50,000	100,000	200,000	400,000	800,000
Characteristics					
Required input heat content (MMBtu/hr)					
Steam Turbine	76	152	303	607	1,214
Gas Turbine	120	240	480	960	1,920
Gross power production (kW)					
Steam Turbine	2,722	5,444	10,888	21,775	43,550
Gas Turbine	9,700	19,400	38,700	77,400	154,800
Parasitic power requirement (kW)					
Steam Turbine, Coal	210	419	836	1,675	3,349
Steam Turbine, Gas	42	85	169	339	678
Gas Turbine	150	300	595	1,190	2,390
Costs (Thousands of 1984 $)					
Capital					
Steam Turbine, Coal	10,830	20,000	35,310	61,948	108,830
Steam Turbine, Gas	4,877	7,222	18,807	31,100	51,865
Gas Turbine	8,128	14,381	22,451	36,587	73,174
Annual O&M (@70% CU)					
Steam Turbine, Coal	1,896	2,970	4,565	7,248	11,554
Steam Turbine, Gas	956	1,250	2,115	3,199	4,837
Gas Turbine	225	449	899	1,797	3,596

Source: R. W. Hess, J. J. Turner, W. H. Krase, and P. Y. Pei for the U.S. Department of Energy, *Factors Affecting Industry's Decision To Cogenerate* (Rand Corp., Santa Monica, CA: January 1983); Office of Technology Assessment, Congress of the United States, *Industrial and Commercial Cogeneration* (Washington, DC, February 1983).

A 30 year operational life was assumed, beginning in 1986 and ending in 2015. Since a gas turbine's expected lifetime is only 20 years, it was assumed that this system was replaced in the 21st year, and half of its costs were applied to this investment analysis. The steam turbine is assumed to be in operation the full 30 years. The steam turbine system provides 65 percent of the facility's required electricity, necessitating the purchase of the remaining power requirement from an electric utility and not allowing for any sales from the cogenerator to the grid. (Because the pollution control equipment of the coal-based systems require electricity to operate—known as a parasitic power requirement—purchased electricity costs are higher for these coal-based systems.) The gas turbine system provides 233 percent of the facility's required electricity, and it is assumed that the excess is sold to the grid for a net cost of 80 percent of the industry's cost of purchased electricity on a dollar per kilowatt basis. For the baseline analysis, a 70 percent capacity factor was used.

Total Levelized Annual Cost Calculation

The economics of the coal and natural gas cogeneration options were determined by comparing their total levelized annual costs. Through the use of levelization, the actual stream of fixed and variable annual expenditures for the cogeneration investments is modified to show an equivalent constant annual expenditure figure for the entire plant's lifetime so that investment options can be more easily compared. The sum of the present values of these constant, levelized costs is equal to the sum of the present values of the actual annual expenditures. Therefore, levelized cost can also be used to calculate the cost of a project in terms of present value, so that the importance of the timing of the cash flows can be taken into account (i.e., a dollar in the hand today has greater "value" than a dollar in the hand ten years from now, even absent inflation).[7]

Only the total cogeneration costs were examined in this analysis. The more conventional option (operating a boiler to produce steam and purchasing all required electricity) was not examined. This analysis compares only the costs of natural gas-fueled and coal-fired cogeneration.

Environmental Impact

The estimated emissions of sulfur dioxide (SO_2), particulates (TSP), carbon monoxide (CO) and nitric oxides (NO_x) in tons per year resulting from burning coal and natural gas cogeneration systems were based on emission factors derived from Appendix B of *Industrial and Commercial Cogeneration* published by the Office of Technology Assessment (OTA).[8] The OTA document presents emission factors from steam and gas turbine cogeneration units in terms of pounds of pollutants per 100 kWh. Since the emission factors are based on an equivalent electric output, the gas turbine system, for the environmental comparison, was sized to match the electric output of the steam turbine system. In order to reach an equivalent process steam output, a back-up gas boiler was included, as well as its resultant emissions.

Annual electrical generation was calculated by multiplying the number of operating hours times the plant power rating. This result was then multiplied by OTA's emission factor to estimate annual emissions. The median plant size, 200,000 lbs/hr of steam, was selected for illustrative purposes, and emissions for systems of other sizes can be determined by multiplying the emissions of the selected plant by the percentage increase or decrease in plant size when compared to the selected plant (assuming the same capacity utilization factor).

RESULTS

Under the assumptions of this analysis, natural gas-fueled cogeneration is more economical than coal-fired cogeneration for large-scale industrial installations. This natural gas total cogeneration systems cost advantage over coal occurs, despite coal's fuel cost advantage over natural gas, because of the relatively larger capital, O&M, and purchased electricity costs for coal-fired systems. Large coal-fired cogeneration systems average from one and one-third to over twice the capital costs of natural gas-based systems (depending on the system technology employed) due to such items as pollution control equipment, larger land area needed (larger boiler size, fuel storage, etc.), and extra fuel handling equipment. O&M costs are higher because coal is more difficult to handle and burn than natural gas, requiring more personnel, training, fuel stockpiling, and maintenance. Purchased electricity costs are higher for coal-fired cogeneration because the operation of a coal boiler, particularly one with pollution control equipment, requires more electricity (parasitic power requirement) than a natural gas-fired system.

The gas turbine cogeneration system exhibits the most favorable economics of the three systems included in the analysis in all scenarios examined. The gas-fueled steam turbine system was found to be more economical than the coal-fueled steam turbine for industrial processes requiring 800,000 lbs/hr of steam or less (see Table 2-1). The overall economics of gas turbine cogeneration systems are more favorable than gas-fueled steam turbines partially because of their lower capital and O&M costs on a per energy output basis. While Tables 2-1 and 2-3 show that the gas turbines have higher absolute and levelized capital costs than its gas-fueled steam turbine counterpart, the gas turbine produces (and consumes) much more total energy in order to produce the equivalent amount of process steam. The cost of a gas turbine system in terms of dollars per kilowatt of *electric* capacity is about one-half that of the gas-fueled steam turbine, and in terms of dollars per MMBtu of *total* energy produced (steam and electricity), the cost of the gas turbine is about 85 percent that of the gas-fueled steam turbine.

In addition, the economics of the gas turbine are more favorable than the gas-fueled steam turbine because it produces a relatively higher amount of electricity. On a per MMBtu basis, electricity is more costly to industry than is steam. Thus, the gas turbine, because it produces a relatively greater proportion of electricity, produces a more costly output per dollar of input (assuming that the industry can utilize all of the electricity or sell its excess at a reasonable rate). This relatively higher cost per output more than compensates for the relatively lower overall thermal efficiency of the gas turbine cogeneration system (70 percent efficient) when compared to the steam turbine (79 percent efficient).

Natural gas-fueled steam turbines may be preferred over gas turbines in certain situations, although not indicated in this analysis.

For example, gas-fueled steam turbines may be favored in smaller applications, or in situations when excess electrical production is not desired. In addition, those companies that are capital-constrained may prefer the gas-fueled steam turbine cogeneration option because of its lower initial cost. On the whole, however, gas turbines have been favored over gas-fueled steam turbines in recent years for large-scale industrial cogeneration.[9]

While the turbine system exhibits more favorable economics than coal-fueled steam turbines for all sizes examined, the economic advantage for natural gas-fueled steam turbine cogeneration over a comparable coal-based steam turbine cogeneration system generally declines as the process steam capacity of the unit increases. As plant capacity increases, the ratio of capital and O&M to total costs decreases, while the fuel and purchased electricity costs rise in proportion to the plant's capacity (assuming a constant capacity utilization factor). Thus, fuel costs become a more significant factor in the total levelized annual costs as capacity increases, and the economic advantage for natural gas-fueled steam turbines declines under the national average pricing assumptions contained in the baseline case.

Natural gas-based cogeneration is also environmentally preferable to coal-fired cogeneration, regardless of plant size, location, or operating conditions. For the cogeneration systems examined in this analysis, the use of natural gas results in only a fraction of the total air pollutants emitted from coal-burning systems (14 percent for the gas-fueled steam turbine and 23 percent for the gas turbine). For example, when burning natural gas, a steam turbine cogeneration unit capable of providing 200,000 lbs/hr of steam emits only 6 percent of the sulfur dioxide, 10 percent of the particulate matter, 28 percent of the nitrogen oxide, and less than 1 percent of the carbon monoxide when compared to a coal-burning cogenerator equipped with pollution control equipment (see Table 2-4).

Since many factors have an impact on the feasibility of industrial cogeneration, it is imperative to perform a site-specific analysis when considering such an investment. In no way does this analysis preclude the need for detailed feasibility studies for actual industrial (or other) applications. Rather, this analysis is meant to provide a general indication of the economics and environmental impacts of gas versus coal use in large-scale industrial cogeneration applications.

Table 2-4

Annual Emissions from Cogeneration System[1] Operation[2]
(tons/year)

Pollutant	Steam Turbine		Gas Turbine with Back-Up Boiler[3]		
	Coal-Based[4]	Natural Gas-Based	Gas Turbine	Back-Up Gas Boiler	Total
Sulfur Oxides	1188	70	3	Neg.[5]	3
Particulates	100	10	10	5	15
Carbon Monoxide	40	Neg.[5]	50	9	59
Nitrogen Oxide	694	197	267	118	385

[1] Based on a steam requirement of 200,000 lbs/hr.

[2] Assumes a 70 percent capacity utilization.

[3] Since emission factors are based on an equivalent electric output, a back-up boiler is required for the gas turbine in order to reach an equivalent process steam output.

[4] Assumes the use of flue gas desulfurization and an electrostatic precipitator in order to meet new source performance standards.

[5] Negligible.

Source: Office of Technology Assessment, Congress of the United States, *Industrial and Commercial Cogeneration* (Washington, DC: February, 1983).

REFERENCES

1.   American Gas Association, "Estimated Current Natural Gas-Fueled Cogeneration Capacity: 1985 Update" (Arlington, VA: December 20, 1985).
2.   PEDCo Environmental Inc. for the U.S. Environmental Protection Agency, *Population Characteristics of Industrial/Commercial Boilers in the U.S.* (Washington, DC: August, 1979).
3.   R. W. Hess, J. J. Turner, W. H. Krase, and R. Y. Pei for the U.S. Department of Energy, *Factors Affecting Industry's Decision To Cogenerate* (Rand Corp., Santa Monica, CA: January, 1983).
4.   Office of Technology Assessment, Congress of the United States, *Industrial and Commercial Cogeneration* (Washington, DC: February, 1983).
5.   National Committee on Air Quality, *To Breathe Clean Air* (Washington, DC: March 1981).
6.   American Gas Association, *The A.G.A.-TERA Reference Case 1986-1* (Arlington, VA: 1986).
7.   Electric Power Research Institute, *Technical Assessment Guide* (Palo Alto, CA: July, 1979).
8.   Office of Technology Assessment, *Industrial and Commercial Cogeneration* (Washington, DC: February, 1983).
9.   Federal Energy Regulatory Commission, *The Qualifying Facilities Report* (Washington, DC: January 1, 1986).

Appendix 2-1

Data Assumptions

Startup Year (1986) Fuel Costs (in 1986 $/MMBtu)[1]
- Natural Gas: $ 3.78
- Coal: $ 1.75
- Electricity: $16.25

Fuel Cost Annual Real Escalation Rate[1] (%)
- Natural Gas: 0.28
- Coal: 1.48
- Electricity: 3.17

Capacity Utilization 70%

Real, After-Tax Cost of Capital 10.2%

Federal and State Income Tax Rate (%):	40
Property Tax and Insurance (% of capital costs):	2
Investment Tax Credit (%):	0
Book Life (years)	
Steam Turbine[2]	30
Gas Turbine[3]	20
Tax Life (years)	
Steam Turbine[2]	15
Gas Turbine[3]	20
Salvage Value	0
Ratio of required electrical energy to required heat energy:[2]	0.26

---

[1] Source: American Gas Association, *The A.G.A.-TERA Reference Case 1986-1* (Arlington, VA: 1986).

[2] Source: R. W. Hess, J. J. Turner, W. H. Krase, and R. Y. Pei for the U.S. Department of Energy, *Factors Affecting Industry's Decision to Cogenerate* (Rand Corp., Santa Monica, CA: January, 1983).

[3] Source: Office of Technology Assessment, Congress of the United States, *Industrial and Commercial Cogeneration* (Washington, DC: February, 1983).

[4] Source: American Gas Association, *Tax Reform Act of 1986; Effect of H. R. 3838 on Corporations and Individuals* (Arlington, VA: August 2, 1986).

Chapter 3

Benefits and Opportunities in Residential, Commercial and Industrial Natural Gas-Fired Cogeneration

One study[1] estimates that, for all sectors, less than one-third of the future gas-driven cogeneration capacity will be owned and operated by the actual cogeneration energy consumers. (The remaining capacity will be owned by outside investors, including gas utilities.) For the residential and commercial sectors, energy consumer ownership of cogenerators is expected to be less than one-fourth of the total gas-fueled capacity.

These projections reflect the fact that many obstacles exist that deter energy consumers from investing in cogeneration technologies. This is particularly true for the residential sector and to a lesser extent for the commercial sector. These obstacles include:

- Lack of capital—nonessential investments such as cogeneration may have a low priority for many firms with limited capital, despite a possible high expected return;

- High required rate of return—unregulated businesses have historically had a higher return requirement for investments not directly related to their primary business;

- Unfamiliarity with energy investments—while some large industrial firms are knowledgeable about investments in energy-related equipment, many smaller companies are not. This unfamiliarity with such investments may cause firms to reject cogeneration. Concerns over equipment maintenance and operation are such obstacles. A gas utility's expertise in such matters could help overcome this problem;

• Lack of familiarity with cogeneration as an energy option—not all of those who could employ cogeneration are aware that this option is feasible or that it even exists;

• Confusing government regulations—a myriad of federal, state, and local government regulations affect the cogeneration investment decision. Natural gas utilities are more familiar with these regulations and can select the proper cogeneration option within the regulatory framework;

• Uncertainty about future energy prices—gas utilities are usually more knowledgeable about long-term energy price projections than are most firms. Without this knowledge, firms could make an imprudent fuel choice based on short-term energy price trends; and,

• Inability to utilize tax benefits—nonprofit organizations and firms with low tax liabilities may not be able to take advantage of investment tax credits and depreciation schedules which make cogeneration investments more attractive. Gas utilities, either through subsidiaries or through their knowledge of different financial options, could maximize potential tax benefits.

Therefore, in order for cogeneration gas demand to reach its full potential (especially in the residential and commercial sectors), gas utilities will have to become involved in marketing cogeneration facilities. Some gas distribution companies are already participating in these activities. One utility offers funding assistance to its potential cogeneration customers for feasibility studies, design, and installation of the facility.[2] A few gas utilities are developing subsidiary companies that develop, own, and operate cogeneration facilities for specific customers, either as a partner in a joint venture or as sole owner.[3]

In order to overcome the largest obstacle facing potential cogeneration owners—capital availability—gas utilities may need to participate in financing cogeneration projects. Financial involvement by utilities can take several forms:

- Lending source—the utility can set up a subsidiary to act as a lending institution for gas-fired cogeneration projects;

- Joint partnerships—the energy consumer and the utility (or subsidiary) could, with or without the participation of other investors, invest jointly in the cogeneration project. This spreads out the risks and benefits of the project;

- Leasing—the utility could finance the entire investment, then lease the equipment to the cogenerator with an option to purchase the equipment at the end of the lease. A sale and leaseback arrangement could have certain tax advantages for certain cogenerators and utilities; and,

- Outright ownership—the utility could own the equipment outright and charge the consumer for the energy provided by the facility. Alternatively, the utility could form a subsidiary to own the equipment and charge the energy consumer for the thermal energy plus a lease payment on the equipment.

In this chapter, the outright ownership option by a subsidiary of a gas utility is used in the quantitative analysis.

METHODOLOGY

This chapter incorporates and modifies previous cogeneration research to illustrate the benefits and feasibility of cogeneration at the consumer, utility, and national levels. This analysis does not attempt to make new forecasts of cogeneration capacity or cogeneration gas demand by sector. Neither does this analysis attempt to develop new case studies proving the technical feasibility and economic viability of cogeneration in specific applications.

Four previously published case studies were examined and modified slightly: a 51-unit high rise condominium in the Pacific region employing a 100 kW cogenerator with a gas piston engine;[4] a 40- by 90-foot, fast-food restaurant in the East North Central region employing a 70 kW turbocharged engine with a 30-ton chiller-heater;[5] a

190-bed hospital in the West South Central region employing a 300 kW continuous-duty engine with a 60-ton absorbtion chiller,[6] and a textile plant in the South Atlantic region employing a 42 MW gas turbine with a supplementary fired recovery boiler.[7]

The cogeneration capital, electric standby, and operation and maintenance (O&M) costs (adjusted to reflect 1983 constant dollars —all costs and revenues are expressed in constant 1983 dollars), conventional energy use, and cogeneration energy use data were supplied by the original case study analyses. The *A.G.A.-TERA Fall 1983 Base Case*[8] supplied future regional energy prices and inflation factors. The first year of operation was assumed to be 1985 and operating life was assumed to be 15 years. (Actual equipment life would probably be longer.)

First, conventional and cogeneration energy costs were compared for each case to determine cost savings resulting from the switch to cogeneration. Conventional energy costs include electricity and gas or oil. Cogeneration costs include electricity, gas, electric standby, and O&M.

The cogeneration gas price was assumed to be at about five percent lower than the existing gas price (depending on the original case study's estimate) due to either higher demand resulting in a lower average price (rate schedule impact) or a special cogeneration or electric generation rate as is already found in some gas rates today. Also, only in the fast-food restaurant case was electricity assumed to be sold back to the grid, as that was the only case study in which such a sellback was considered in the source document.

Second, the tax effect of the cogeneration option was calculated. Taxable income was derived by deducting property taxes (assumed to be 2.5% of updepreciated capital costs), loan interest [condominium rate=13.0% (2.0% above prime rate), fast-food rate=11% (prime rate), hospital rate=12.5%, and textile plant rate=11.9% (corporate bond rate)],[9] and depreciation (accelerated depreciable life of 5 years for cogeneration equipment, 15 years for boiler equipment)[10] from the energy cost savings. Then the tax was calculated by applying the assumed tax rate [condominium=0% (nonprofit), fast-food restaurant and textile plant=50%, and hospital=41%] to the taxable income. A tax credit of 10% was applied to the cogeneration equipment capital costs (allowing for a 95% depreciable base, as prescribed in recent tax laws), which was taken in the first year.

The cash flow was calculated next. This entailed subtracting income taxes, property taxes, and insurance costs (2.2% of undepreciated capital cost) from energy cost savings.

From the cash flow estimates and the cogeneration equipment capital costs, two indices of profitability were calculated—real, after-tax internal rate of return (IRR) and payback. The real, after-tax IRR shows the yield for the investment, and is calculated by determining the discount rate at which the present value of the expected cash outflows (capital costs) equals the present value of the expected cash inflows.

The payback period shows the number of years required to recover the initial cash investment, and is calculated by comparing the first few years' cash inflows with the capital costs. (In two of the cases, the hospital and fast-food restaurant, the capital costs from the original case studies represent the costs of acquiring and assembling the cogeneration equipment in today's market. If these types of projects become successful, prepackaging economies of scale would reduce these capital costs substantially.)

Total energy consumption of the various systems was compared, including consideration of energy used or lost in the conversion and transmission of electricity provided by central powerplants (average conversion losses equal 67 percent; seven percent for transmission and distribution). The total energy use for cogeneration, measured in Btu, was expressed as a percent of the conventional system energy usage.

Alternate cases were derived as well, based on these case studies. A non-gas baseline was developed for three of the cases; oil/electric baselines for the condominium and hospital examples (calculated by assuming the amount of oil consumed for thermal use would equal the gas used in Btu terms) and an all electric baseline, with data provided by the original case study, for the fast-food restaurant. (No such alternative case was developed for the textile plant example.) The methodology for determining the IRR and payback for the non-gas baselines was identical to the methodology outlined above.

In another set of alternative cases, the IRR and the payback to a gas utility or its subsidiary were derived (under the assumption that a gas utility or its subsidiary owned and operated all the cogeneration facilities). It was assumed that the cost to the consumer for the service provided by a utility-owned and operated cogeneration system would allow the energy consumer to reduce his total energy bill by 10 percent (e.g., if the consumer's previous conventional total energy bill were $100, the utility subsidiary would charge him $90 for its service of providing an equal amount of energy).

This rate is calculated to cover the subsidiary's costs for gas, back-up electricity, and maintenance of the equipment. The rate also allows the subsidiary a reasonable return. (This is a simplifying assumption. In reality, the energy purchase contract between the energy consumer and the utility subsidiary would be more complex and would require negotiated terms.)

This return to the subsidiary provided the basis by which taxes and cash flow were determined. Those calculations varied from the methodology outlined above only in that debt costs were assumed to be 11% and the tax rate was assumed to be 50% in all instances. No extra discount on the gas price was assumed despite the subsidiary's relationship with the gas utility.

In addition, two different assumptions were changed in the utility subsidiary owned, fast-food, all-electric baseline example to estimate the cogeneration economic's sensitivity to changes in gas/electricity price differentials and capital costs. For the first sensitivity case, the original case study's 1985 estimated energy prices for the Chicago area were substituted for the East North Central region prices (which increased the price differential by 50 percent), and these 1985 Chicago energy prices were assumed to follow the same rate trajectory as the regional prices from 1986 to 1999.

The second sensitivity case lowered the cogeneration capital costs 15 percent to reflect the prepackaging cost reduction goal of GRI. In each of these sensitivity cases, only one assumption per case was changed, and all other assumptions were held constant.

RESULTS

Natural gas-fueled cogeneration offers substantial economic and non-economic benefits not only to industrial energy consumers, but to energy consumers in the residential and commercial sectors, to the nation, and to natural gas utilities as well.

The most important cogeneration benefit to energy consumers is reduced costs. Not only can many industrial companies reduce their energy costs through cogeneration, but a significant number of commercial firms can employ cogeneration to cut their energy costs as well. Residential cogeneration opportunities are limited, but in particular applications residential cogeneration is technically and economically feasible.

The four case studies presented illustrate the economic benefits of gas-driven cogeneration to individual energy consumers of these sectors—residential (a condominium), commercial (a fast-food restaurant and a hospital), and industrial (a large textile fiber-manufacturing plant)—and to natural gas utilities. The cases assume that the energy consumer owns and operates the equipment. These cases are later adapted to show the economics of utility ownership.

Results for the condominium case study (Table 3-1) show a real, after-tax IRR of seven percent when the cogenerator is offsetting a gas/electric baseline (i.e., thermal energy is provided by gas combustion and an electric utility supplies electrical power) with a payback of almost 10 years. For an oil/electric baseline, the IRR improves to 10.6 percent and the payback falls to 7.6 years. While these after-tax, real IRRs are acceptable, the payback periods are relatively long. One of the major impediments to economic success in this example is that the assumed owner, a nonprofit homeowner's association, cannot take advantage of the tax benefits associated with this investment.

The economics of the fast-food restaurant example are more favorable. For a gas/electric baseline, the real, after-tax IRR is about nine percent with a payback of just under 6 years. For an all electric baseline (electricity supplies all thermal and electrical energy for the conventional energy system), the IRR more than doubles to 18.4 percent and the payback decreases to 4 years. This major improvement over the gas/electric baseline results from the cogeneration

Table 3-1
Cogeneration Case Study Analysis Summary—Cogenerator Owned[1]

	Cogeneration System			
	Size (kW)	Cost (10^3 1983$)	IRR[2,3] (%)	Payback[3] (Years)
Individual Customer's Base System				
Condominium[4]				
Oil/Electric	100	249	10.61	7.6
Gas/Electric	100	249	7.01	9.6
Fast Food Restaurant[5]				
All Electric	70	150	18.36	4.1
Gas/Electric	70	150	8.89	5.9
Hospital[6]				
Oil/Electric	300	410	22.10	3.8
Gas/Electric	300	410	21.76	3.9
Textile Plant[7]				
Oil/Electric	42,000	36,202	27.32	3.9

---

[1] Energy user assumed to own, operate, and maintain cogeneration and associated equipment.

[2] Real, after-tax internal rate of return earned by the owners of the cogenerators. Nominal IRR can be calculated by using the equation $[(IRR/100)+1] \times 1.04$ (4.1 percent is the assumed annual inflation factor through 2000).

[3] Energy prices and inflation factors for these calculations provided by American Gas Association, *A.G.A.-TERA-Fall 1983 Base Case* (Arlington, VA: 1983).

[4] Cogenerator cost and energy use for baseline and cogeneration scenarios obtained from GKCO Consultants for the American Gas Association, *A.G.A. Manual: Cogeneration Feasibility Analysis* (Arlington, VA: April, 1982).

[5] Cogenerator cost and energy use for baseline and cogeneration scenarios obtained from the Sievert Corp. for the Gas Research Institute, *Packaged Gas-Fired Cogeneration Systems Developed for Fast Food Restaurants* (Chicago, IL: June, 1983).

[6] Cogenerator cost and energy use for baseline and cogeneration scenarios obtained from United Enertec, Inc. for the Gas Research Institute, *Packaged Gas-Fueled Cogeneration Systems Developed for Hospitals: Phase 1, Final Report* (Chicago, IL: June, 1983).

[7] Cogenerator cost and energy use for baseline and cogeneration scenarios obtained from TRW Energy Engineering Division for Department of Energy, *Handbook of Industrial Cogeneration* (Washington, DC: October, 1981).

system's thermal energy production displacing a more expensive conventional thermal energy source in the all electric baseline.

The economic results for the hospital case study prove very favorable for cogeneration. The real, after-tax IRR for both the oil/electric and the gas/electric baselines are about 22 percent while the paybacks are just under 4 years.

Results for the textile plant are also very favorable. The real, after-tax IRR is over 27 percent. This compares very favorably with the national average real, after-tax industrial IRR, estimated to be between eight and ten percent.[10] Payback is just under 4 years. The industrial example shows the highest return mainly because of its high, constant thermal energy demand.

The results for all the case studies show that cogeneration can be, in certain applications, economical to organizations from all three market sectors. The economics are better for those who can take advantage of the tax benefits and in those applications where a more expensive thermal fuel, such as electricity, is offset.

In addition, those energy consumers that exhibit a more constant energy consumption over time, such as manufacturing plants and hospitals, would be better candidates for cogeneration in terms of economics because the cogeneration facility can operate longer and displace a high proportion of the energy purchased annually. Facilities whose energy consumption patterns show more seasonal and daily fluctuations are less likely candidates for cogeneration, but this does not automatically preclude success for these facilities.

Not all energy consumers would want, or would be able, to make a cogeneration investment on their own. A gas utility, or its subsidiary, could assist in the investment process in many ways. The following quantitative results illustrate the economic benefits to a utility or its subsidiary if it owned and operated the previously described cogeneration facilities and sold the energy to the consumers.

Because the utility shares the economic benefit from cogeneration with the energy consumer as an inducement to the consumer to become a cogenerator, the real, after-tax IRRs that the utility earns on these projects are necessarily smaller than if the energy consumers owned and operated the cogeneration equipment (except for the condominium example, since the utility can utilize tax benefits which the homeowners association cannot).

These IRRs for the utility subsidiary range from more than six percent to 15.4 percent for the gas/electric conventional baselines and from over 11 percent to 19.7 percent for the non-gas baselines. All of these returns are higher than the real, after-tax cost of capital of 5.6 percent for the average natural gas utility.[11] The average payback to the utility's subsidiary was calculated to be under 5 years. These profitability indices would improve if other benefits, such as increased gas sales, were considered.

These case studies used regional average energy prices, which tended to keep the profitability indices lower than what can be obtained. Cogeneration economics are very sensitive to site-specific energy prices, and there are numerous areas within a region with a greater gas price advantage over electricity than does the regional average price.

To illustrate the price sensitivity, an alternate case was developed in which Chicago-specific rates were used in the utility owned, fast-food study which originally used only electricity as an energy source. The Chicago prices exhibited a gas price advantage over electricity that was 50 percent greater than the regional prices.

This difference caused the IRR to almost double, from 15.1 percent to 29.9 percent (see Appendix 3-1). For this case, then, a one percent change in the energy price differential resulted in almost a two percent change in the IRR. In addition, the payback improved from 4.4 years to just under 3.0 years.

Another factor that could improve cogeneration economics is a reduction in system capital costs. GRI projects that the fast-food cogeneration system capital costs will drop by 15 percent if this system can be prepackaged successfully.[12] Such a drop in capital costs would cause an 18.4 percent increase in the IRR to the gas company in the fast-food/all electric example. In addition, the payback period would fall from 4.4 years to 4.1 years.

The nation as a whole is a beneficiary of the use of gas-fueled cogeneration. Of primary importance is the more efficient use of the nation's energy resources. On average, two-thirds of the energy consumed by conventional electric utility powerplants is lost to the environment. This lost power is heat, or thermal energy, that is created when electricity is produced and is usually vented into the atmosphere or discharged into waterways.

Cogeneration technologies minimize this waste by utilizing the thermal energy. For the case studies, the net energy savings due to cogeneration from source of electrical production range from 12 to 32 percent (Table 3-2). Energy savings are greatest when gas-fueled cogeneration displaces electrically produced thermal energy.

The nation also benefits from gas-fueled cogeneration since the use of cogeneration technologies mitigates the need for large, expensive central electric powerplants. Larger cogenerators, such as those used in industry, are much more cost effective than new coal ($1,300/kW for capital costs) and new nuclear ($2,600/kW)[13] baseload powerplants.

Even smaller scale cogenerators, which are more expensive per kilowatt than larger cogenerators, are price competitive with these new baseload plants. For the case studies, capital costs range from $2,490/kW to $1,370/kW for the smaller, R/C sector cogenerators, and the large industrial cogenerator costs $860/kW.

Cogeneration capital costs are continuing to decline. In addition, cogenerators can be built in a relatively short time (average of less than 2 years),[14] while conventional baseload powerplants take approximately 5 to 10 years to construct.[15] Thus, cogeneration facilities can help meet capacity shortfalls of electric utilities forecast for the late 1980's and early 1990's.[16] In fact, a number of electric utilities are encouraging cogeneration in order to lower their need for future capacity additions.

In addition, gas-fueled cogeneration may be beneficial to the nation by helping to reduce air pollutant emissions. By mitigating the need for new, conventional powerplants and reducing the use of present facilities, cogeneration helps to reduce this potential to emit air pollutants. At the same time, gas-fueled cogeneration systems (e.g., gas turbines or gas boiler/steam turbine combinations) ". . . emit less than one percent of the SO_x, 27 percent of the particulates, and 50 percent of the NO_x produced by the comparably sized conventional coal- and oil-based system (with pollution control equipment)."[17]

It must be recognized that the engineering, economics, and energy savings of actual cogeneration systems must be evaluated on a case-by-case, site-specific basis. In no way does this analysis preclude the need for detailed feasibility studies for actual condominiums, fast-

Table 3-2

Cogeneration Case Study Energy Use Comparison[1]
(10⁹ Btu)

	Electric Powerplant Fuel Requirement	On Site Fuel Requirement	Total Fuel Requirement	Cogeneration Energy Use as a Percent of Conventional Energy Use (%)
SYSTEM				
Condominium				
Oil/ or Gas/Electric[2]	7.1	8.7	15.8	—
Cogeneration	0.2	12.1	12.3	78
Fast Food Restaurant[3]				
All Electric	8.5	—	8.5	—
Cogeneration	1.2	4.6	5.8	68
Gas/Electric	4.6	1.1	5.7	—
Cogeneration	0.1	4.7	4.8	84
Hospital				
Oil/ or Gas/Electric[2]	38.6	14.4	53.0	—
Cogeneration	10.8	35.7	46.5	88
Textile Plant				
Oil/Electric	5,224	4,840	10,064	—
Cogeneration	1,441	6,328	7,769	77

[1] Energy use calculated at point of fuel consumption to generate electricity and thermal power, thus accounting for all conversion, transmission, and distribution energy losses for purchased electricity.

[2] Oil/electric and gas/electric baselines' and cogeneration scenarios' energy use are the same.

[3] Electricity total for cogeneration adjusted for electric sellback.

Sources: GKCO consultants for the American Gas Association, *A.G.A. Manual: Cogeneration Feasibility Analysis* (Arlington, VA: April 1982); The Sievert Corp. for the Gas Research Institute, *Packaged Gas-Fired Cogeneration Systems for Fast Food Restaurants* (Chicago, IL: June 1983); United Enertec, Inc. for the Gas Research Institute, *Packaged Gas-Fired Cogeneration Systems for Hospitals: Phase 1 Final Report* (Chicago, IL: June 1983); TRW Energy Engineering Division for the Department of Energy, *Handbook of Industrial Cogeneration* (Washington, DC: October 1981).

food restaurants, hospitals, or textile plants (or other applications). Rather, this analysis is meant to provide a general indication of the magnitude of potential energy savings provided by gas-fired cogeneration, the types of markets in which gas-fired cogeneration can favorably compete, and those factors affecting the success of cogeneration.

REFERENCES

1. GKCO Consultants for the Gas Research Institute, *Natural Gas-Based Cogeneration Market Potential;* GRI-80/0102; PB 82-220567; (Chicago, IL: March 1982).
2. Office of Technology Assessment, Congress of The United States, *Industrial and Commercial Cogeneration* (Washington, DC: December 14, 1982) p. 108.
3. Brian Plants, "Key to Future of Cogeneration: Energy Joint Ventures?," *Cogeneration World* (Washington, DC: May/June 1982).
4. Based on data from GKCO Consultants for the American Gas Association, *A.G.A. Manual: Cogeneration Feasibility Analysis* (Arlington, VA: April 1982).
5. Based on data from the Sievert Corporation for the Gas Research Institute, *Packaged Gas-Fired Cogeneration Systems Developed for Fast Food Restaurants;* GRI-82/0059; PB 83-25768 (Chicago, IL: June 1983).
6. Based on data from United Enertec, Inc. for the Gas Research Institute, *Packaged Gas-Fueled Cogeneration Systems for Hospitals: Phase 1 Final Report;* GRI-82/0062; PB 83-249540 (Chicago, IL: June 1983).
7. Based on data from TRW Energy Engineering Division for the Department of Energy, *Handbook of Industrial Cogeneration;* (Washington, DC: October 1981).
8. American Gas Association, *A.G.A.-TERA Fall 1983 Base Case;* (Arlington, VA: 1983).
9. Rates derived from data in "Business Week Index," *Business Week* (McGraw-Hill Inc., New York, NY: October 17, 1983).
10. Based on General Energy Associates for the Department of Energy, *Industrial Cogeneration Potential (1980-2000) Targeting of Opportunities at the Plant Site (TOPS),* Volume II; (Washington, DC: May 1983).
11. Derived from data found in A.G.A.'s *Gas Facts 1981;* Moody's Investor Service, Dunn & Bradstreet Corporation, *Moody's Public Utility Manual 1983;* (New York, NY: 1983), and Value Line Inc., *The Value Line Investment Survey,* Pt. 3; (New York, NY: July 15, 1983). Nominal,

before tax costs were: Debt, 13.20%; preferred, 10.84%; common equity,
12.95%.

12. Sievert Corp. for GRI, *Prepackaged Gas-Fired Cogeneration System Developed for Fast Food Restaurants...*

13. American Gas Association, *Comparison of Initial Capital Investment Requirements for New Domestic Energy Supplies: 1983 Update;* (Arlington, VA: December 1983).

14. OTA, *Industrial and Commercial Cogeneration,* p. 121.

15. Energy Information Administration, Department of Energy, *Projected Costs of Electricity from Nuclear and Coal-Fired Power Plants,* Volume 2; (Washington, DC: November 1982).

16. See Department of Energy, *The Future of Electric Power in America: Economic Supply for Economic Growth;* DOE/PE-0045; (Washington, DC: June 1983).

17. American Gas Association, *Fact Book: Energy, The Environment, and Natural Gas;* (Arlington, VA: October 1983), p. 24.

Appendix 3-1

Cogeneration Economic Sensitivity to Price and Capital Cost Changes

	Lower Capital Cost By 15 Percent[1]		Increase Gas/Electric Price Differential By 50 Percent[2]	
	IRR	Payback	IRR	Payback
Original Case[3]	15.09%	4.4 yrs	15.09%	4.4 yrs
Alternate Case	17.87%	4.1 yrs	29.8%	3.0 yrs
Sensitivity Index	1.23	1.07	1.96	1.47

[1] The total capital cost falls from $150,000 to $127,700.

[2] Gas price advantage over electricity price in 1985 increases by 50 percent.

[3] Gas company-owned cogeneration system in a fast-food restaurant which originally used electricity only in its conventional system.

Chapter 4

Existing and Planned Cogeneration Capacity

Natural gas is the predominant fuel used in cogeneration systems. Environmentally, natural gas is the preferred fossil fuel in these facilities—when compared to a similarly-sized conventional coal- or oil-based system with pollution control equipment, a gas-fueled turbine cogeneration system emits less sulfur oxides, particulates, and nitrogen oxides. Moreover, natural gas is economically preferable for use in most industrial and commercial cogeneration applications, primarily due to the lower capital and operating costs associated with gas systems.

The amount of operating and planned cogeneration capacity has risen dramatically since the passage of the Public Utility Regulatory Policies Act of 1978 (PURPA), which provides qualifying cogenerators access to the electric utility grid for the purposes of selling and purchasing electricity.

- The amount of cogeneration capacity that is operating, under construction, or on order ("existing") is estimated to total about 24.7 gigawatts (see Table 4-1).* The amount of natural gas-fueled existing capacity is estimated to total 10.8 gigawatts, or 44 percent of the total.

---

*Detailed data for 1986 were not yet available as this chapter went to print. Consequently, data presented are for 1985. Total cogeneration capacity operating, under construction or on order is estimated to have risen to 27.0 gigawatts by year-end 1986.

Table 4-1
Estimated Cogeneration Capacity
(MW)

Census Region	All Fuels			Gas-Fueled		
	Existing[1]	Planned	Total	Existing[1]	Planned	Total
New England	872	1,578	2,450	13	104	117
Middle Atlantic	1,576	2,222	3,798	64	774	838
South Atlantic	3,048	1,980	5,028	267	68	335
East North Central	4,256	391	4,647	145	102	247
East South Central	1,110	210	1,320	77	38	115
West North Central	834	167	1,001	282	15	297
West South Central	9,693	5,374	15,067	8,466	4,007	12,473
Mountain	1,004	491	1,495	218	274	492
Pacific	2,321	5,163	7,484	1,234	3,604	4,838
Total	24,714	17,576	42,290	10,766	8,986	19,752

[1] Includes plants built, under construction, and ordered.

- Almost 17.6 gigawatts of additional cogeneration capacity is in the planning stage, nine gigawatts (51 percent) of which will be gas-fueled.

- When the amounts of existing and planned capacity are combined, the potential amount of cogeneration capacity expected in the near future totals 42.3 gigawatts, of which 19.8 gigawatts (47 percent) will be gas-fueled.

- Existing gas-fueled cogeneration systems consume 711 billion cubic feet (Bcf) of natural gas annually. If all the planned capacity is built, an *additional* 594 Bcf of gas will be consumed in cogeneration units annually in the near future (see Table 4-2).

- On a regional basis, the West South Central region has almost 40 percent of the existing capacity, followed by the East North Central (17 percent), the South Atlantic (12 percent) and the Pacific (9 percent) regions.

- Recent growth in cogeneration capacity has been greatest in the Pacific and the West South Central regions, due to growth in electric demand, favorable regulatory climates, and regional industrial and economic growth. This is particularly true for the Pacific region—if all the planned capacity is built, the region's share would double to almost 18 percent of the total market.

- These two regions also account for the vast majority of existing and planned gas-fueled capacity. The West South Central has 79 percent of the existing, and 45 percent of the planned, gas-fueled capacity. The Pacific region has 11 percent of the existing, and 40 percent of the planned, gas-fueled capacity. Therefore, 85 percent of new gas-fueled capacity is focused in these two regions.

Table 4-2

Estimated Cogeneration Gas Use[1]
(MMcf/yr)

Census Region	Existing[2]	Planned	Total
New England	859	6,874	7,733
Middle Atlantic	4,230	51,157	55,387
South Atlantic	17,449	4,494	21,943
East North Central	9,584	6,742	16,326
East South Central	5,089	2,511	7,600
West North Central	18,638	911	19,629
West South Central	559,553	264,839	824,392
Mountain	14,409	18,110	32,519
Pacific	81,560	238,203	319,763
Total	711,371	593,921	1,305,292

[1]Based on 65 percent capacity utilization and a 30 percent average conversion efficiency factor (for electrical production alone).

[2]Includes plants built, under construction, and ordered.

METHODOLOGY

Two sources provided the bulk of the data for this report: An Electric Power Research Institute (EPRI) report that estimates cogeneration capacity as of December, 1981[2] and a FERC report on filings by cogenerators seeking qualifying status under the provisions of PURPA.[3] The EPRI report was compiled after an extensive literature search and contacts with utilities and regulatory agencies. The report classifies cogeneration facilities as "known," "probable," "possible," and "planned/proposed." Facilities both connected and not connected to the utility grid were inventoried for this report. The report failed, however, to identify cogeneration capacity by fuel type.

The FERC regularly issues reports on those facilities filing for qualifying status under PURPA. These reports identify the applicant, location, prime mover, fuel type, and capacity of each facility. The FERC classifies these facilities as "new" or "existing." Those new facilities are either cogeneration units newly built, under construction, or planned or proposed. Some of the new facilities, however, may not ever be built—receiving qualifying status is just a first step in the planning stages for new units. The FERC publishes a summary report of these filings every year, and the *Energy Daily*[4] publishes periodic updates of those filings in between FERC publication dates.

For this chapter, the EPRI data for "known" and "probable" cogeneration capacity were used as the estimate for existing capacity prior to 1982. The FERC data for "new" filings since fiscal year 1980 were used as an estimate for post-1981 capacity. In an attempt to identify those FERC filings that have actually been built, are under construction, or have been ordered, and to capture those cogeneration facilities built or planned to be built after 1981 which did not apply to FERC for PURPA qualifying status, utilities and regulatory agencies were contacted and a literature search was undertaken. Those facilities which have been built or are in some phase of construction were grouped in the existing category. The others were left in the planned category.

Calculating the amount of cogeneration capacity built or proposed to be built after 1981 entailed totaling the capacity identified by FERC or by other sources. The EPRI data, which was used for the existing (pre-1982) capacity estimate, did not present a capacity breakout by fuel type, and the gas share had to be estimated. A natural gas market share, based on the FERC data and a less exhaustive survey of existing capacity done by the Edison Electric Institute in 1980 which did identify fuel types,[13] was calculated for each Census region and applied to the EPRI regional numbers. The average fuel share used for natural gas was 34 percent.

Annual gas use by cogeneration facilities was calculated by assuming a 30 percent average conversion efficiency factor (for electrical production alone) and a capacity utilization factor of 65 percent. This capacity factor may be somewhat low, thereby understating the required gas demand volume, since cogenerators often maximize their operational economics by employing relatively high utilization rates.

REFERENCES

1. American Gas Association, "Fact Sheet: Estimated Current Natural Gas-Fueled Cogeneration Capacity" (Arlington, VA: May 25, 1985).
2. Synergic Resources Corp. for the Electric Power Research Institute, *Evaluation of Dual Energy-Use Systems* (Palo Alto, CA: October 1982).
3. Federal Energy Regulatory Commission, *The Qualifying Facilities Report* (Washington, DC: January 1, 1985).
4. *The Energy Daily*, King Publishing Group (Washington, DC: various 1985 issues).
5. *California Energy Manual*, Frank Kester Associates (Belmont, CA: June 1985).
6. *Cogeneration*, Pequot Publishing (Fairchild, CT: various issues).
7. *Cogeneration World*, International Cogeneration Society and Research Institute (Washington, DC: various issues).
8. *Energy Business*, Ebasco Services Inc. (New York, NY: various issues).
9. *Energy User News*, Fairchild Publications (New York, NY: various issues).
10. *Gas Industries*, Gas Industries E&A News Inc. (Park Ridge, IL: various issues).
11. *Gas Turbine World & Cogeneration*, Pequot Publishing Inc. (Fairchild, CT: various issues).
12. *Oil & Gas Journal*, PennWell Publishing Co. (Tulsa, OK: various issues).
13. Edison Electric Institute, *Cogeneration and Small Power Production: Facilities and Rates* (Washington, DC: February 1980).

Section II

Economic and Engineering Feasibility Analysis

J. A. ORLANDO, P.E.
GKCO, Incorporated

41

Chapter 5

Evaluating Cogeneration Feasibility

Cogeneration systems have proved to be economically attractive investments. In general, older installations are located at larger energy intensive facilities, such as industrial plants, hospitals, shopping malls, large residential complexes and universities. These systems have been limited to applications where the energy end user, who was frequently the owner of the cogeneration system, had the expertise—either on staff or through consulting engineers—to adequately evaluate a cogeneration system, without risking a significant financial expenditure. Because cogeneration technical and economic viability is so site specific, detailed analyses performed for one application have had limited applicability to other facilities.

The final decision as to whether or not a cogeneration system is appropriate is usually made after a detailed, thorough engineering analysis. Among the factors that must be considered are the end user's energy requirements, the mechanical/electrical systems, purchased power rates, fuel costs, space requirements, staffing, insurance, taxes and site operational requirements. In addition, the study must utilize historic or simulated energy use data as a basis for modeling future energy consumption, potential variations in that usage and the amount of purchased energy that can be displaced by cogenerated power and heat. It includes a capital and operating cost analysis for alternative conventional and cogeneration systems, and the selection and optimization of the most attractive cogeneration configuration.

While the cost of a final feasibility analysis can be quite significant, that study should be viewed as a final step in a multistep planning process leading to the selection of the most cost-effective utility system for any specific site. This step is undertaken only after previous simpler, less comprehensive investigations have shown that the cost of the detailed analysis is justified.

This Section first describes an overall framework by which cogeneration feasibility can be tested. Key cost parameters, various economic measures, site data needs and collection techniques, procedures for the calculation of the cost of conventional and cogeneration-based utilities, capital cost estimation, cash flow analysis, tax implications and third-party financing are reviewed.

In evaluating the feasibility of a cogeneration application, it is important to utilize as much site-specific information as possible. However, the development and analysis of such information can be costly, with no assurance that the project will produce overall reductions in utility costs. The following procedure allows the potential cogenerator to explore the economics of cogeneration with a series of resource investments, each greater than the previous and each producing the information required to determine whether the costs of the next step are warranted. The three initial steps described below also are illustrated in Figure 5-1.

- Site Walkthrough and Technical Screening;
- Preliminary Economic Screening; and
- Detailed Engineering Study.

WALKTHROUGH

The objectives of an initial site inspection or walkthrough are to determine whether the site is technically compatible with cogeneration and whether a cogeneration system has the potential for economic viability. Technical feasibility is based on technical compatibility between the cogeneration system and the site's mechancal/electrical systems; a determination as to whether there is adequate

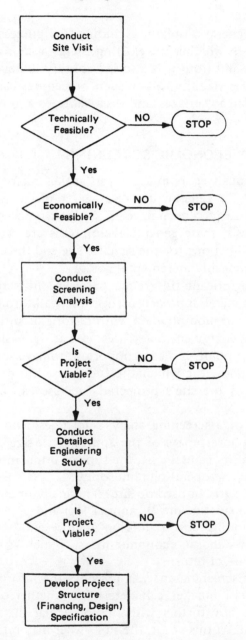

Figure 5-1. Overall Decision Making Process

space for a cogeneration plant, including fuel processing equipment; and an analysis of whether the existing energy systems are adequate.

During the walkthrough, it also is necessary to develop some measure of economic viability. Many such procedures can be used, each having particular advantages and disadvantages and each having differing information needs.

PRELIMINARY ECONOMIC SCREENING

The walkthrough economic analysis makes many simplifying assumptions and, in general, uses average annual or monthly utility costs rather than incremental costs. As one proceeds to the more thorough analysis, these general assumptions are replaced by more realistic approximations for the specific site and the economic analysis is based on monthly incremental costs.

The most significant difference between the screening analysis and the more general approach used in the walkthrough is the development and use of monthly data, with historic or anticipated monthly conventional energy characteristics used as the basis for any actual analysis. At this point, it also is important to have some understanding of the purposes for which the fuel is used in order to estimate what fraction of the site's projected fuel use can be displaced by recovered heat.

The output of a screening analysis includes a number of parameters, including an estimate of the approximate size of the cogeneration system and its capital cost, the type of prime mover likely to be used, the design and operating philosophy (e.g., Is cogenerated power to be sold to the grid or used on site? Thermally or electrically sized? Baseloaded or load tracking?), annual fuel and power requirements and costs, operating cost reduction (before and after taxes), projected cash flows and an economic measure such as simple payback or an internal rate of return.

In short, the screening analysis provides the information required so that the decisionmaker can determine whether or not he or she should proceed any further with this particular site and, if so, how to best proceed. At this point, the site owner may retain a consulting engineer to prepare a more detailed analysis, or bring in a qualified third-party developer.

ENGINEERING STUDY

It is in the engineering feasibility analysis where conceptual designs are fully detailed, systems optimized and costs more fully developed. Data requirements for an existing facility include specific expansion, construction, renovation and demolition plans; as-built plans and drawings; and information on recently implemented or anticipated energy conservation plans. This latter information is required to ensure that the feasibility analysis considers probable future conditions, rather than historic patterns which may be no longer relevant.

It also is important to consider the impact of recent or planned energy conservation activities because an effective conservation program can significantly change cogeneration economics. These energy use data are used to develop a site energy model which is the basis for the more extensive analyses of cogeneration options. Various system sizes, operating modes, equipment combinations and heat recovery options can be examined with this model.

Capital and operating costs can be estimated as a basis for an economic analysis, with various ownership and financing assumptions. Among the information available at the completion of this step are:

- Existing and projected energy use patterns;

- Fuel availability;

- Zoning requirements;

- Project requirements (space, service corridors, noise suppression, location, fuel handling);

- Plant footprints and layouts;

- Grid intertie requirements;

- Permitting requirements and costs;

- Schedule;

- Total budget; and

- Cash flow requirements.

As described above, a feasibility analysis consists of a number of successive analyses, each providing additional detail and each used as a basis for determining whether the cost of the next analysis is justified. The approach utilizes more general cost and performance data during the early studies, thus reducing owner costs and risks.

Chapter 6

Conventional and Cogeneration Costs

All economic analyses, regardless of the depth to which they are taken, follow the basic outline of Figure 6-1. The costs that the potential cogenerator would incur for utility service without cogeneration are computed as a base. The alternative cost of energy that a cogenerator would incur, including the cost of both the cogeneration system and of any supplemental conventional utility purchases, are then computed. The difference between these costs, or the operating cost reduction, is the project revenue. The capital costs of the cogeneration system are estimated to determine the required incremental investment and the economic merit of the project.

Two areas must be addressed in the preparation of any feasibility analysis. First is the identification of all cost components, and second is the development of actual cost estimates. In this chapter, the individual cost components are identified and reviewed. Each of them should be considered in the conduct of any feasibility study, for it is the depth to which they are analyzed that is the most significant contributor to the cost of the analysis. The following discussion reviews both the operating and the capital cost considerations of cogeneration economics.

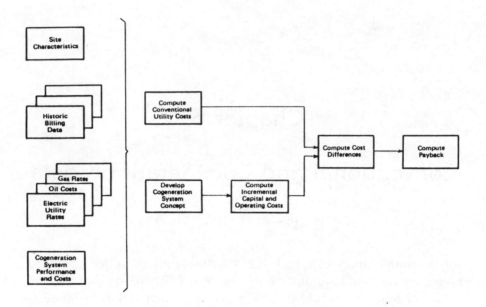

Figure 6-1. Feasibility Analysis Procedure

OPERATING COSTS

Key operating cost and, as appropriate, revenue parameters include:

- the conventional cost of purchased power;
- the cost of utility-supplied supplemental, standby, interruptible and maintenance power;
- the rate that an electric utility will pay for power purchased from a cogenerator;
- the cost of conventional fuels;
- the cost of the cogeneration system fuel, supplemental fuels and, if appropriate, alternate fuels;
- insurance costs;
- federal, state and local income taxes;
- state sales and gross receipt taxes;

- local property taxes;
- staffing;
- space requirements;
- maintenance, both scheduled and unscheduled; and
- administrative and management costs.

PURCHASED POWER COSTS

Investor-owned electric utility charges and the terms and conditions for the purchase of power are specified in a tariff which is usually subject to approval by a state regulatory commission. The most notable exceptions to this procedure are the municipal utilities whose rates are set by the local government and which are not usually regulated by a state commission.

In general, a utility rate consists of one or more of the following components:

Customer Charge—This item is a fixed charge applicable to all customers.

Demand Charge—This item is the amount charged by the utility for the capacity to deliver power. A ratchet refers to a demand billing which is based on some historic peak demand rather than the peak incurred in the current month. The billing demand, as contrasted to the actual or measured demand, consists of some fraction (usually from 50 percent to 100 percent) of the peak demand incurred during the ratchet period—usually eleven months. Ratchets can have a significant effect on operating cost savings produced by the cogeneration system, both during the first year of operation as a consequence of previous conventional power purchases and in subsequent years as a result of system shutdowns.

Energy Charges—This item is the amount charged by the utility for actual energy delivered. It is based on the electric utility's variable costs of power production, the most significant of which are fuel and labor.

Fuel Adjustment Charges—This item is the amount charged by the utility for recovery of the actual costs for fuel or purchased power. It is computed on a monthly, quarterly or annual basis. The Fuel Adjustment Charge (FAC) allows the utility to recover rapidly changing fuel-based expenses without the delay and cost of a more formal rate case. The purchased energy adjustment usually is applied on a per-kilowatt-hour basis for actual energy delivered.

Taxes—This item is the amount charged by the utility on behalf of some governmental body, such as the state or a local municipality. They usually take the form of a Sales or Gross Receipts Tax (GRT) or a Franchise Fee.

Generally, a utility bill will contain one or more of the above items, although some utilities also may include such miscellaneous charges as a conservation adjustment, an environmental surcharge, a nuclear power plant adjustment or economic development charges.

COGENERATOR POWER RATES

Qualified Facilities (QFs)[1] can both purchase power from the electric utility and/or sell power to it. Sales to a QF fall into any of the following categories.

1. *Standby*—Sales of power to the cogenerator which are required due to unscheduled shutdown of the cogenerator's power plant.

2. *Supplemental*—Routine sales of power to a cogenerator which are required to make up the difference between the power produced by the cogenerator and the site demand.

3. *Interruptible*—Sales of power to a cogenerator which are scheduled at the convenience of the electric utility. These sales can be suspended unilaterally by the utility in accordance with a pre-established set of conditions specified by the utility tariff or by contract.

---

[1] In order to be eligible for Public Utility Regulatory Policies Act (PURPA) benefits, a cogenerator must become a Qualified Facility (QF) as defined by the Federal Energy Regulatory Commission (FERC). To do so, the cogenerator must meet FERC-specified operating efficiency and output requirements and have no more than 50 percent ownership by an electric utility.

4. *Maintenance*—Sales of power to a cogenerator that are made according to a time schedule developed by the utility and the cogenerator.

QFs also may sell power to the electric utility and according to regulations developed by FERC. These sales, which may be based either on a published or negotiated rate, may be at 100 percent of the utility's avoided cost, although in some cases slightly more or less than the avoided cost may be paid. PURPA has defined avoided cost as the cost to an electric utility of energy or capacity or both, which, but for the purchase from a QF, the electric utility would incur if it were to construct the capacity itself or purchase the capacity from another source. The billing for a power sale to a utility may take the same form as a sale by the utility, with demand and energy charges.

Many utilities have developed avoided cost rates, with some utilities differentiating in the rate depending on the cogenerator's size. In addition, utilities have established different values for the capacity component depending on the date at which the cogeneration system becomes operational, the degree of control the utility has over the operation of the facility and the term of the contract between the cogenerator and the utility. To the extent that the cogenerator can provide long-term technical and economic guarantees of power availability to the utility, the capacity payment for this power usually is increased.

In some areas where avoided costs exceed purchased power costs the cogenerator can, at his or her option, sell all cogenerated power to the utility while simultaneously purchasing all required power from that same utility.

FUEL COSTS

Small and medium-sized cogeneration systems are fueled primarily by gaseous and liquid fuels, with liquid fuel usage limited to lighter and midweight oils. Gas distribution utilities generally offer firm gas, which is available on a year-round basis, and interruptible gas, with deliveries suspended for conditions specified by the distribution utility. The interruptible gas rate may be offered at a significant discount off the firm gas cost. Use of interruptible gas for cogeneration requires the cogenerator to use an alternate fuel, and/or the end user must use conventional purchased power and boilers during periods of that interruption.

Firm natural gas rates usually are based on a declining-cost, block structure, with smaller, high-cost initial blocks. The cost of larger tail-end blocks approaches the utility's own variable costs for gas. When natural gas is used for cogeneration at a facility that had been burning natural gas for heating or process use, the power plant fuel is an incremental load which should be priced accordingly. This pricing, which can provide a user all fuel at a lower average cost, may not be available when the cogeneration system is owned by a third party, which will be a separate utility account.

Baseloaded cogeneration systems operating at rated capacities exhibit fairly uniform monthly fuel requirements. In recognition of this high annual load factor, gas utilities increasingly are offering demand-commodity-based rates that provide cost savings. These rates, which are similar to the electric utility's demand-energy rate structure, may provide natural gas at rates that are significantly lower than heating gas rates. Finally, a number of gas utilities also are offering special incentive rates designed for cogeneration customers.

A cogeneration system may utilize a mixture of natural gas and a supplemental fuel, such as sewer or landfill gas, which might previously have been flared or otherwise disposed. Fuel mixing provides a constant calorific value while reducing the average fuel cost. Such by-product fuel has no value because it has not been previously used.

INSURANCE

In some cases, cogeneration equipment may be included under the site's overall insurance coverage. However, the more likely case is that the cogeneration system must be insured separately. The cost of such insurance will vary depending on equipment performance history, site and design characteristics. The annual charge may range from 0.25 percent to over one percent of the equipment capital cost.

TAXES

Just as with any business venture, it is necessary to examine cogeneration economics on both a before-tax and an after-tax basis considering income, property and sales taxes.

A viable cogeneration system will produce reduced utility costs for the end user. Depending on the end user's accounting policies, this cost reduction may contribute to increased profit through lower production cost or it may generate a profit directly through internal energy sales. In either case, the increased profit generally will be subject to federal, state and local income taxes. In addition, the cogeneration system itself will incur local property tax liabilities.

In any evaluation of cogeneration economics, it is important to utilize tax rates applicable to the particular end user in the specific locality and organization. A cogeneration system sited at a local manufacturing facility would be charged federal taxes at the overall corporate rate, regardless of the profitability of the project or the local facility. For example, a subsidiary of a larger corporation that has no federal tax liability may not pay any federal income taxes, even though the subsidiary itself is profitable and might otherwise be subject to taxes. This tax liability can be either advantageous or not, depending on the specific cogenerator.

Cogeneration systems also may qualify for selected tax benefits. Tax laws are constantly changing, and it is wise to consult with qualified experts to determine liabilities and benefits prior to a firm commitment to the project.

When a third party owns and operates the cogeneration property, retail energy sales from the cogenerator to the end user may be subject to local sales or utility Gross Receipts Taxes. This cost element should be explored in any third-party project.

Many facilities, such as hospitals and nursing homes, which have characteristics favorable to cogeneration, also enjoy tax-exempt status. A third party, owning a cogeneration system serving a tax-exempt end user, is not itself a tax-exempt entity and tax liability must be considered.

STAFFING

Cogeneration systems usually are capable of unattended operation, and their economic success frequently depends on an ability to operate without a full-time staff dedicated to any individual installation. Most systems delivering hot water or low-pressure steam do not require an operating engineer; however local codes should be consulted to verify that an on-site operator is not required.

Turbine-based systems delivering high-pressure steam may require an operating engineer.

MAINTENANCE COSTS

All prime movers require periodic inspection, servicing and replacement of parts; however, the time period for each scheduled maintenance procedure will depend on the type of prime mover, engine operating conditions and site conditions. Computerized monitoring and telecommunications systems can provide daily and hourly monitoring of system performance and allow planned maintenance programs to be based on an as-needed approach.

Manufacturer information can be used to estimate the cost for scheduled maintenance; however, such estimates must be tempered with information on local labor costs, service personnel's travel distance and time to the cogeneration facility, ease of access to the prime movers, and site ambient conditions. Long-term service contracts are becoming increasingly important as devices for assuring that adequate maintenance is available at predictable costs.

SPACE REQUIREMENTS

A cogeneration system—particularly one that is sited at an existing facility that had been serviced with conventional utilities—may require additional space, thus decreasing the amount of space that is available for the end user's primary business objectives. In such cases, it is necessary to charge the cogeneration plant with the "rental cost" of that space. In cases where the cogeneration system can reduce space requirements, it should receive a credit.

Because cogeneration systems are central plant utility systems, they often are able to trade off less expensive mechanical room space located in a subbasement, in a parking structure, on a roof or at a remote location for more expensive prime business space.

ADMINISTRATIVE AND MANAGEMENT COSTS

One additional cost item consists of clerical and management costs, incurred in any business activity. When a cogeneration system is owned and operated by a third party, these costs can be as high as 5 percent of total operating costs.

CAPITAL COSTS

First cost items include equipment costs; costs for remodeling the existing mechanical/electrical systems and building architectural modifications; and engineering, financing, legal and licensing fees.

EQUIPMENT

The cost of a cogeneration system will be most sensitive to the type and size of the prime mover, the selected fuel, environmental control requirements, and the intricacies of the installation and site. Some fees, such as engineering fees, are variable, with the fee being determined as a percentage of the project cost and often decreasing with increased project size. Other fees, such as for legal and financing services, do not vary significantly with the size of the project. In general, the larger the system, the greater the potential to realize the benefits of economies of scale.

This is particularly true with regard to smaller packaged or factory-assembled cogeneration systems. While the retail cost of the various modules is fairly uniform in terms of dollars per installed kilowatt, the fixed costs—including engineering, legal, financing, construction management and even installation costs—are somewhat insensitive to size. Thus, increasing module size results in a lower cost per installed kilowatt. However, it is important to recognize that at any specific site, larger modules producing more heat than can be used may not be able to develop the same savings per kilowatt as smaller systems matched to the site's thermal energy requirements.

ELECTRIC UTILITY INTERCONNECTION

The cost of interconnecting a cogenerator to the electric utility grid will depend on a number of factors, including the protective relaying requirements of the utility, whether or not the cogenerator will operate in parallel with the utility grid, whether an induction or a synchronous generator is used, whether or not an isolation transformer or power factor correction is required, and the voltage of the utility service and the cogenerator's generator. Utility interconnection costs will vary from as low as $20 per kVA, for simpler induction generator systems, to as high as $200 per kVA, for more complex systems.

BUILDING MODIFICATION COSTS

Interfacing a cogenerator to a building's mechanical/electrical systems may result in modifications to the existing systems at the point where the heat and power are fed into the building system, modifications to the fuel supply piping, and architectural and structural modifications to the building itself. Each of these cost elements must be identified separately and charged to the project as appropriate.

Chapter 7

Economic Measures of Performance

Cogeneration systems can reduce the total cost of utility service, and, in some instances where power is sold to an electric utility, can even produce a positive net revenue stream. That is, the total cogeneration revenue is greater than the cogeneration system's operating cost plus the cost of supplemental fuel and power. Whether it is sited at an existing facility or new construction, cogeneration systems do require an incremental investment over and above that which would be required if the end user were to utilize more conventional utility services. While the decision as to whether or not one should invest in cogeneration may consider such "intangibles" as predictability of future utility costs, reliability of electrical supply and the quality of that supply, the decision ultimately becomes one of basic economics. The owner frequently must quantify these intangibles to determine whether the added investment is justified. In performing such analyses, a reduction in the end user utility costs is treated as a revenue.

Each investment decision is unique and, usually, each potential investor has developed an economic measure that is best suited to his or her needs. These measures vary with regard to ease of use, accuracy and financial objective. Several approximate methods, such as simple payback, return on investment and return on average investment, are easy to apply and understand, and, therefore, are frequently used. The basic shortcoming of these measures is that they do not consider the time value of money—a significant problem when individual cost components escalate at different rates; when the system's economic life differs from the debt term; or when significant expenditures for equipment overhauls may be required at two-to-four-year intervals.

Internal rate of return, net present value, annual cost, cost benefit ratio, return on equity and the break-even period are exact measures in that they consider the discount rate. These measures are equivalent and will result in the same decision given the same set of alternatives. Each of these measures is capable of accurately reflecting tax impacts, which will vary over the life of the project.

Because of its wide acceptance, simple payback period, which is defined as the total capital investment divided by the annual revenue, is used as one basic measure of economic viability. For investors who explicitly consider the opportunity cost, or time value of money, the internal rate of return is more appropriate.

The internal rate of return is the discount rate which will produce zero as the present value of the sum of all project costs and revenues. The calculation of the internal rate of return, therefore, requires the development of cash flow projections for the economic life of the project.

Third-party relationships provide the end user with an opportunity to obtain some of the operating cost reductions of cogeneration without a capital investment and, therefore, some economic performance measures such as simple payback and internal rate of return cannot be computed. In this case, the net present value of future utility costs over the economic life of the project is an appropriate comparative economic measure.

Chapter 8

Site Data Collection

A cogeneration feasibility study consists of a determination as to whether any of the commercially available cogeneration alternatives are compatible with the end user's technical and financial requirements. It requires the collection of information about the patterns of energy use and cost at each specific potential cogeneration facility. As described in this chapter, these data requirements include information on the facility's mechanical/electrical systems, the electric utility's interconnection requirements, the uses of energy, the efficiency with which the conventional systems convert energy, the seasonal and daily variations in the pattern of energy use, and the current and projected costs of that energy.

MECHANICAL/ELECTRICAL SYSTEM CHARACTERISTICS

Gas-fueled cogeneration systems are economically viable because they are able to usefully and economically apply thermal energy that otherwise would go unused, either in total or in part. If cogeneration is to be feasible at any specific location, the energy systems at that location must be capable of utilizing the recoverable thermal energy. Thus, the first question to be answered in any cogeneration analysis is simply, "Can the heat recoverable from a prime mover be economically applied at the site?"

In a new or existing industrial facility or in an existing building, the answer to the above question is usually straightforward. Industrial thermal processes are rather well-defined as to the required temperature, whether that thermal energy is in the form of hot water, chilled water or steam. Table 8-1 is a listing of those energy intensive industries often considered primary cogeneration candidates, with summary data as to the temperature at which process heat is required. As a general rule, while most industrial processes that require thermal energy as water or steam are compatible with cogeneration, the selection of the prime mover will be influenced heavily by the temperature and/or pressures at which this heat is required.

Table 8-2 provides a set of guidelines that can be used to determine whether the mechanical systems in commercial buildings or building complexes are compatible with heat recovered from a cogeneration system. Caution is required in using these general guidelines, for if one is willing to spend enough money, almost any HVAC system can be made compatible with cogeneration. Table 8-2 assumes that there are no unique economic or technical considerations that would justify extraordinary retrofit costs.

The guidelines of Table 8-2 are applicable to new construction. However, in this case, they also can be viewed as design criteria. If cogeneration is economically viable based on overall conventional energy cost considerations and anticipated energy usage patterns and budgets, Table 8-2 can be used to identify those HVAC systems that should be considered for the building. For example, if cogeneration is viable in a hospital based on local electric and gas costs, then a central plant or water heating system would be designed instead of a unitary electric heating/cooling system.

Special performance requirements, such as a need for low-temperature coolants for food freezing or ice rinks, or extremely low-temperature differences between the supply and return for an end user thermal system, will have an impact on the ability to utilize recovered heat. The generic data included in this text may not apply to such atypical requirements and the end user should consult an engineer to determine their potential impact. Table 8-3 lists some of of the factors that should be considered when analyzing and designing

Table 8-1. Industrial Process Heat Requirements
(Percent Heat Requirements by Temperature)

SIC	Industry	Less Than 212°F	212°F to 350°F	Greater Than 350°F
1211	Bituminous Coal and Lignite	100.0		
2013	Sausages and Prepared Meats	97.6		2.4
2022	Natural Cheese	100.0		
2023	Condensed and Evaporated Milk	66.4	4.1	29.5
2046	Wet Corn Milling	21.7	62.5	15.8
2062	Cane Sugar Refining	13.6	75.0	11.4
2063	Beet Sugar	22.4	73.0	4.6
2075	Soybean Oil Mills	26.7	73.3	
2085	Distilled Liquors	15.0	85.0	
2261	Finishing Plants—Cotton	47.3	52.7	
2262	Finishing Plants—Synthetic	68.8	31.2	
2421	Sawmills and Planing Mills		100.0	
2435	Plywood		100.0	
2436	Veneer	100.0		
2511	Wooden Furniture	100.0		
2600	Pulp and Paper Mills	16.3	51.1	32.6
2812	Alkali and Chlorine		100.0	
2822	Synthetic Rubber	100.0		
2823	Cellulosic Man-Made Fibers		19.9	80.1
2824	Noncellulose Fibers	100.0		
2865	Cyclic Crudes and Intermediate		100.0	
2869	Organic Chemicals—NEC		100.0	
2911	Petroleum Refining	2.4	2.5	95.1
2951	Paving Mixtures		100.0	
3271	Concrete Block	34.7		65.3

Table 8-2. Commercial Building Heating System—
Cogeneration Compatibility

System	Interface Characteristics
2-Pipe Fan-Coil, Heating or Cooling	Generally a good match for cogeneration heat
Gas or Electric Furnaces—both interior and rooftop	Feasible, requires installation of heating coil in furnace
Electric Baseboard	Not feasible
Hot Water Baseboard	High-temperature baseboard radiators have insufficient heat transfer surface for lower-temperature cogeneration system heat
Unit Heaters—gas or electric	Not feasible
Unit Heaters—steam	Insufficient heat transfer surface if cogeneration system produces hot water
Unitary Electric Resistance Heating-Cooling and Air Source Heat Pump Heating-Cooling (through-the-wall units)	Costs tend to be prohibitive
Unitary Electric-Water Source Heat Pump Heating-Cooling	Excellent match and use for even low-temperature recovered heat
Air Handling or Built-Up Air Conditioning System with heating coils and electric cooling	May require supplemental heating system
Radiant Panel Heating—electric	Not feasible
Radiant Panel Heating—hot water	Excellent match

Table 8-3. Typical Cogeneration System Characteristics

System	Characteristic	Performance Levels
Electrical	Voltage	120/240, 120/208, 480, 480/277, 4800; ±5%
	Frequency	50 Hz, 60 Hz; ±.4 Hz
	Harmonic Content	Voltage: 2% total, 1% any single harmonic Current: 3% total, 2% any single harmonic
Thermal	Available Temperature	Up to 250°F, piston engines are acceptable 250°F to 350°F, gas turbines or piston engines may be acceptable Above 350°F, gas turbines
Architectural	Structural	Loadings up to 300 lb/sq ft required
	Space	5000 sq ft/engine plus access space
	Noise	Acceptable for commercial installations
	Accessibility	Access to mechanical/electrical systems access to engine-generator set for maintenance

a cogeneration system, and provides performance bounds that are considered as typical.

ELECTRICAL USAGE DATA

White data on an end user's electrical requirements are most reliably obtained directly from historic energy bills, from the electric utility and from on-site meters and submeters, it should not be assumed that all such data are correct. One objective of the walkthrough is to determine what data are available and whether such data are likely to be correct.

In general, electric utility bills will provide both the billing demand and the total energy requirements for the facility. Depending on the specific utility, that bill also may include such additional details as the actual demand, the site power factor, the time during which the actual peak occurred, the degree days during the billing period and the amount of energy in kilowatt hours used during each time-of-day billing period—on peak, intermediate and off peak. In some cases, peak demands for each period are provided. Figure 8-1 is an illustration of a utility electric bill.

Utilities which have instituted time-of-day rates will install recording meters that provide the demand and energy usage at specified intervals for a complete billing period. In general, the demand interval is specified as either 15, 30 or 60 minutes depending on the utility demand rate. Figure 8-2 is a sample printout from this type of meter; this detailed documentation is sometimes included with the utility bill. When these printouts are not routinely provided, they usually can be obtained from the utility for a nominal charge.

Data such as that shown in Figure 8-2 are extremely useful in sizing the cogeneration system and, in particular, in determining how much purchased power can be displaced by a baseloaded cogeneration system. Demand data are used to develop load-duration curves such as that shown in Figure 8-3.

Figure 8-3 can be developed by tabulating the number of hours that the facility's electrical or thermal demand exceeds specified levels. In practice, the curve is developed using demand increments or bins, the size of which may range from a few percent to

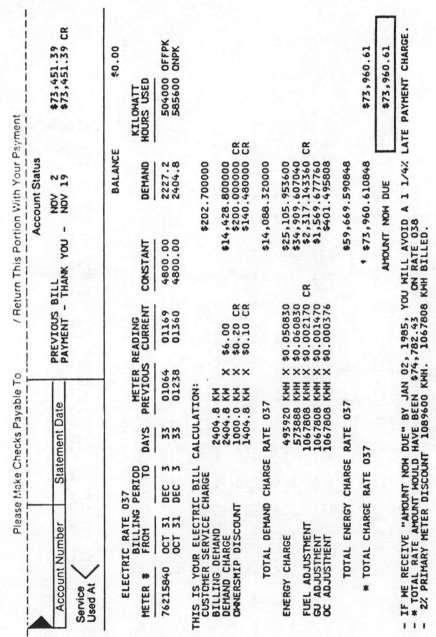

Figure 8-1. Sample Electric Bill

09/01/84 - 09/30/84 DEMANDS

MONTHLY ONE HALF HOUR DEMANDS IN KILOWATTS
(NOTE: -1 INDICATES MISSING DATA)

TIME → DATE ↓	1230 0830 0430	0100 0900 0500	0130 0930 0530	0200 1000 0600	0230 1030 0630	0300 1100 0700	0330 1130 0730	0400 1200 0800	0430 1230 0830	0500 0100 0900	0530 0130 0930	0600 0200 1000	0630 0230 1030	0700 0300 1100	0730 0330 1130	0800 0400 1200	MAX KWD	TOTAL KWHRS
09/25/84	838 1259 1057	839 1250 1034	844 1298 1036	852 1322 1074	828 1298 1077	831 1308 1047	828 1317 1020	830 1318 1008	848 1281 998	888 1265 985	937 1255 919	981 1260 860	1065 1224 851	1136 1184 830	1198 1119 827	1234 1078 802	1322	25218
09/26/84	799 1206 963	794 1278 961	805 1272 962	805 1259 991	792 1272 951	792 1258 831	792 1235 808	789 1230 796	810 1165 761	845 1160 742	894 1145 703	950 1153 669	1032 1104 658	1119 1066 641	1196 1026 632	1264 996 633	1286	23033
09/27/84	620 1004 846	618 1016 829	623 1020 847	624 1031 887	626 1022 864	614 1020 852	613 1028 823	609 1021 812	627 1014 779	647 1018 752	690 1002 701	761 996 680	843 968 662	917 953 653	1001 915 648	1014 876 620	1031	19801
09/28/84	619 1015 859	625 1027 846	625 1024 869	640 1039 879	620 1033 852	622 1024 822	614 1044 795	612 1023 781	636 1028 754	649 1040 734	706 1019 708	763 1007 668	830 988 663	923 963 646	984 936 636	1003 897 627	1044	19860
09/29/84	636 830 722	613 827 730	616 829 728	602 831 712	613 846 706	601 821 703	620 808 687	625 802 679	636 787 680	634 788 680	677 788 670	705 775 650	745 758 648	793 743 620	827 728 625	823 723 611	846	17151
09/30/84	631 742 705	607 744 695	604 794 701	596 800 682	600 792 703	595 808 686	624 814 670	618 775 673	612 771 683	624 763 658	655 763 660	688 756 645	713 770 632	728 748 623	746 720 625	742 709 625	814	16659

TIME	1230 0830 0430	0100 0900 0500	0130 0930 0530	0200 1000 0600	0230 1030 0630	0300 1100 0700	0330 1130 0730	0400 1200 0800	0430 1230 0830	0500 0100 0900	0530 0130 0930	0600 0200 1000	0630 0230 1030	0700 0300 1100	0730 0330 1130	0800 0400 1200

MAXIMUM HALF HOUR NON COINCIDENT DEMANDS IN KILOWATTS

MAX KWD																
838	859	844	852	828	831	828	830	848	888	937	981	1065	1136	1198	1264	
1286	1278	1298	1322	1445	1572	1534	1545	1606	1547	1551	1507	1481	1433	1249		1606
1300	1287	1280	1317	1299	1128	1020	1008	998	985	919	898	891	861	856	841	

TOT KWD																
10074	9998	10013	9945	9880	9843	9866	9862	10184	10434	11157	11941	12826	13651	14674	15058	
15220	15369	15653	15796	16170	16284	16178	16086	16331	16533	16448	16308	16006	15653	14972	14465	628770
14208	14087	14159	14046	13642	13244	12662	12345	12091	11797	11342	10876	10683	10394	10226	10069	

MONTHLY SUMMARY:

MAXIMUM 30 MIN. DEMAND	1606	09/24/84 1230
TOTAL CONSUMPTION RECORDED	628770	KWH
RECORDING START TIME	09/01/84 00:00	(15 MIN. INTERVAL START TIME)
RECORDING STOP TIME	09/30/84 24:00	

USAGE CHARACTERISTICS

MAXIMUM 30 MINUTE DEMAND	1606 KW 09/24/84 1230
AVERAGE MONTHLY DEMAND	873.29 KW
LOAD FACTOR	0.543

Figure 8-2. Example of Metered Demand Profile for a Hospital

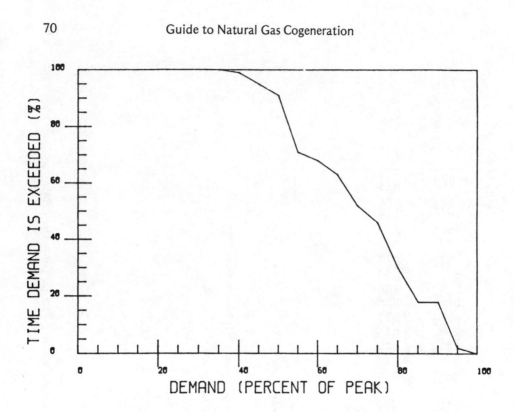

Fig. 8-3. Sample Load Duration Curve

approximately 5 percent of the site's peak demand. For example, the data in Figure 8-2 can be sorted into 50 kW bins or intervals as summarized in Table 8-4. The percentage of time the site demand was in the range defined by each interval or bin is then calculated, and a load duration curve (in this case, for a six-day period) is plotted as Figure 8-4.

Depending on available data, curves for longer time periods, generally for one year, or based on hourly demand readings, may be developed.

Historic utility bills provide a measure of past utility costs; however, in a period when utility costs can increase dramatically, there is no assurance that they are accurate predictors of future costs. The charges on the historic rate may have changed, the fuel adjustment

Table 8-4. Demand Data

Demand Interval (kW)	No. of Readings	Percent of Total Readings	Cumulative Total (%)
551– 600	3	1.0	100.0
601– 650	55	19.1	99.0
651– 700	24	8.3	79.9
701– 750	29	10.1	71.6
751– 800	30	10.4	61.5
801– 850	38	13.2	51.1
851– 900	14	4.9	37.9
901– 950	7	2.4	33.0
951–1000	15	5.2	30.6
1001–1050	33	11.5	25.4
1051–1100	6	2.1	13.9
1101–1150	6	2.1	11.8
1151–1200	5	1.7	9.7
1201–1250	5	1.7	8.0
1251–1300	14	4.9	6.3
1301–1350	4	1.4	1.4
TOTAL	288	100.0	

charges may have changed, or the structure of the rate may have changed. Therefore, the utility or the state utility regulatory body should be contacted to obtain copies of all applicable rates, as well as copies of any pending rate changes.

Estimating the conventional utility costs for new construction is somewhat more troublesome. For more complex structures and industrial processes, the design engineer should be able to provide anticipated energy requirements and projected costs. In some cases, although billing data for specific customers cannot be released, electric utilities may have detailed data on similar types of users.

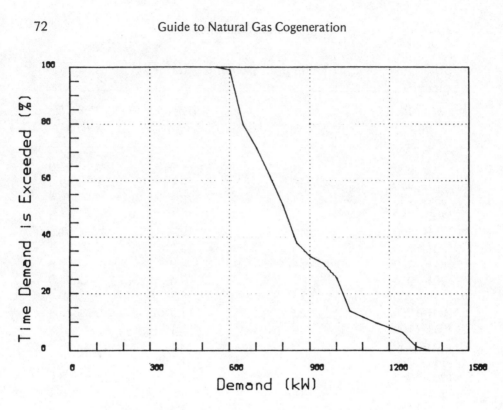

Figure 8-4. Sample Load Duration Curve

FUEL USE DATA

While billing information is a primary source of data on site fuel requirements, particular care is required in using this information. First, the fuel use data should cover the same 12-month period as was used for electrical data. If this is not readily possible, it should be determined whether site (e.g., occupancy, production, etc.) or weather conditions for the two 12-month periods were equivalent, and if not, the data should be adjusted to compensate for differences.

Second, it is necessary to determine how much of the fuel use is displaceable by recovered heat. For example, natural gas used for cooking or in clothes dryers cannot be replaced by heat recovered from an engine, and the site fuel usage must be adjusted to compensate. During the initial site visit, it is important to identify all such nondisplaceable uses.

Several approximations can be used to estimate the amount of displaceable fuel. The simplest of these may be found in a central plant where boiler output is monitored. In this case, it may be assumed that all the fuel used to produce steam is replaceable. A second approach is to inventory all functions, such as cooking and drying, that require fuel rather than heat and to estimate the amount of fuel required by each function or piece of equipment. This total is then subtracted from the billing data total. A third approach, which is appropriate when replaceable fuel use is weather sensitive, is to develop a relationship between fuel use and weather (Figure 8-5, where gas consumption data for heating and cooling are plotted) or, if data are available, between fuel use and heating degree days (see Figure 8-6). The fuel use corresponding to zero degree days is the base usage, which should then be further subdivided into replaceable uses, such as for water heating, and nondisplaceable uses, such as for cooking. Again the estimated nondisplaceable usage should be deducted from billed total fuel usage. This same approach may be applicable to industrial process loads that are not weather sensitive, with an alternate measure, such as tons of output, being used in place of heating degree days.

In those cases where the facility is served by multiple accounts, the actual use for each account may provide some help in determining what fraction of the fuel is replaceable.

Finally, sites using fuel oil present one additional problem. Billing data are based on deliveries and not actual usage. If the end user maintains fuel inventory data, as may be the case for central boiler plants, the calculation of fuel usage is rather straightforward. If, however, no inventory data are available, and a review of the billing information indicates that deliveries may be out of phase with usage, then it may be necessary to allocate actual delivery data to approximate usage. Techniques relating fuel use to weather or output, such as those discussed above, may be appropriate.

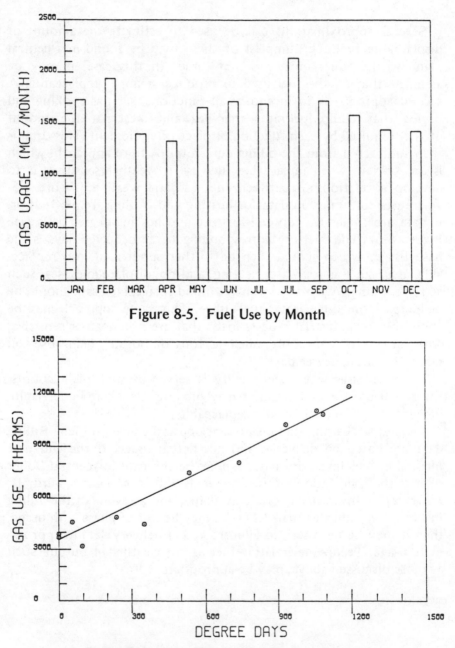

Figure 8-5. Fuel Use by Month

Figure 8-6. Fuel Use vs. Heating Degree Days

Chapter 9

Cogeneration Design and Operating Options

The design of a cogeneration system is an iterative and evolutionary process, with various options identified and then evaluated against site-specific criteria. As one proceeds through the design process, these options become more detailed. Various conceptual design decisions are reviewed in this chapter. More detailed design considerations, such as component sizing and selection, piping layout, etc., are left to the detailed design itself.

The development of a site-specific design is no insignificant task; even the simplest engineering study will cost several thousand dollars. This chapter provides generic data, various "rules of thumb," assumptions and design preferences for use in developing a design as required for a feasibility analysis. Many factors, such as noise suppression and unique structural requirements, are not explicitly considered and, if these factors are thought to be of concern, they should be considered separately. The level of design and analysis described herein is intended for use in determining whether or not an engineer should be retained for a more comprehensive analysis. It is not intended as a substitute for actual design.

In order to conduct either a preliminary economic screening or a more detailed engineering study, it is necessary to:

- Select the type of prime mover—e.g., reciprocating engine, gas turbine or steam turbine;

- Determine the total installed capacity;

- Determine the number and size of the prime movers;

- Determine the utilization of the recoverable heat and the needs for supplemental heating and fuels;

- Determine the required standby capacity and how it will be provided (purchased from the electric utility, load shedding or redundant on-site capacity); and

- Determine whether power will be sold to and/or purchased from the electric utility grid and the electrical system requirements.

The approach taken throughout this section is to provide guidelines that allow the definition of a cogeneration system with the minimal amount of site- and equipment-specific information. These guidelines are intended for use in the walkthrough or screening study and are not intended as a substitute for engineering analyses in the more detailed examination of the project. However, whenever detailed information is available, it should be used to the extent practicable.

DESIGN AND OPERATING OPTIONS

As in any creative project, the design alternatives are limited only by the creativity and ingenuity of the design engineer. However, for purposes of exploring the feasibility of a potential cogeneration system, analysis of a limited number of alternatives will allow one to examine a range of technically possible options. These concepts are described briefly below.

- *Isolated Operation, Electric Load Following*—The facility is independent of the electric utility grid, and the cogenerator is required both to produce all power required on-site and to provide all reserves required for scheduled and unscheduled maintenance. This type of system, which was commonly referred to as a Total Energy System, generally provides the least attractive economic return; and, with the passage of PURPA, the construction of this type of cogeneration facility is extremely rare.

- *Baseloaded, Balanced Electric and Thermal Output*—In general, the size of a balanced cogeneration system is such that both the thermal and electric output are approximately less than or equal to the on-site requirements.

  - An electrically baseloaded system is sized to satisfy that end user electric demand which is always available. The energy end user purchases supplemental power from the utility grid, and, in general, no power is sold to the grid. Supplemental heat is provided by on-site boilers or burners.

  - A thermally baseloaded system is sized so that the required heat is provided from the engine prime mover. This concept can result in the production of more power than is required on-site, and both power sales to the utility and supplemental purchases from the utility may result. In those cases where the avoided cost is higher than the purchased price of power, all power may be sold to the utility.

- *Maximum System*—In implementing PURPA, FERC established criteria that a cogenerator must satisfy in order to obtain QF status and, therefore, eligibility for the benefits identified in PURPA. For oil- and gas-fired systems, the most constraining of these are that at least 5 percent of the annual output of the cogeneration system be thermal energy; and that the net electrical output plus one-half the useful thermal output divided by the oil or gas input (measured in Low Heat Value) and expressed as a percentage be greater than 42.5 percent. Under this option, the cogeneration system prime mover is the largest sized system that will satisfy both these requirements given site specific thermal requirements and equipment characteristics.

Each of these concepts is illustrated in Figure 9-1.

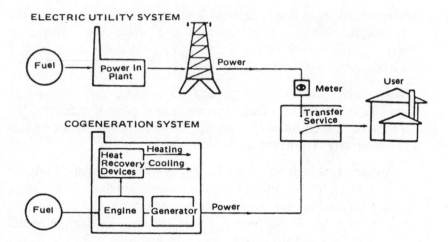

Isolated Cogenerator: User receives all power from cogeneration system *or* electric utility system.

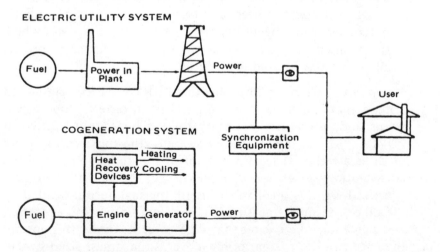

Thermally Base-Loaded Cogeneration: User receives power from cogeneration system *and* cogenerator sells power to the electric utility.

Figure 9-1. Cogeneration Options

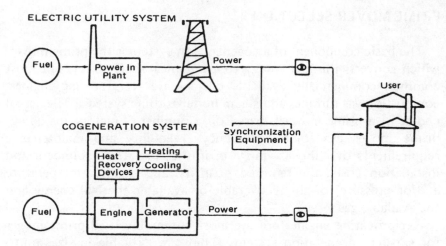

Electrically Base-Loaded Cogeneration: User receives power from cogeneration system and purchases power from the electric utility system.

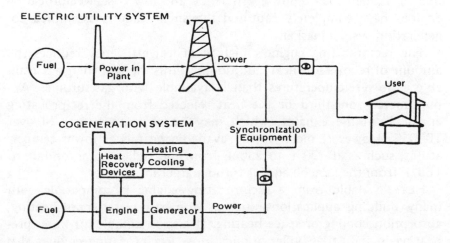

Maximum System: User receives power from the utility. Cogenerator sells output to the utility grid.

Figure 9-1. Continued

PRIME MOVER SELECTION

The basic component of a cogeneration system is the prime mover, which converts the fuel energy to mechanical shaft power. The most common commercially available devices are reciprocating engines, combustion gas turbines and steam boiler-turbine systems. The actual choice of a prime mover is based on a number of factors, including: prime mover availability, efficiency, reliability and maintenance requirements over the size range of interest; relative equipment and installation costs; the required quantity and quality (temperature and/or pressure) of the recoverable or available thermal energy and the available gas service.

Reciprocating engines are the most commonly used prime mover for smaller cogeneration systems. They are available in sizes up to 10 MW, and in larger sizes, they are capable of using a heavy fuel oil as an alternative to interruptible natural gas. In smaller sized applications, the reciprocating engine is usually the prime mover of choice due to its high fuel-to-power efficiency and low cost. Reciprocating engines have completely captured the small, factory-assembled cogeneration system market.

The reciprocating engine's high electrical efficiency reduces the amount of recoverable heat; additionally, this thermal energy is available at lower temperatures than is available with gas turbines. Approximately one-third of the heat rejected from the reciprocating engine is in the exhaust, which may be at temperatures of over 1000°F. However, most of the available heat is at lower temperatures, such as at 220°F to 250°F from the jacket water, and up to 160°F from the lube oil and/or turbocharger intercooler.

Heat available from a reciprocating engine is compatible with many building applications, such as domestic water, pool heating, absorption cooling or space heating. It also is suitable for water preheating in industrial boiler applications. Reciprocating engines also are used in emergency power applications where rapid start-up is required.

Combustion gas turbines have seen increased use in cogeneration applications, and have captured the major share of those newer applications of 3 MW or more. Smaller gas turbines operate at lower mechanical efficiencies than do reciprocating engines and, therefore, produce more heat per unit of power output. All the recoverable heat is available at temperatures of up to 1,200°F in the exhaust, and gas turbines can be used in those facilities where high-pressure steam is required. Gas turbines also are applied where high-temperature air can be used directly, such as in drying applications or as combustion air.

This latter feature is important. The exhaust gases from the gas turbine contain as much as 15 percent to 16 percent oxygen, and this exhaust can be used as combustion air in a supplemental boiler or "duct burner." Supplemental firing can achieve efficiencies of 90 percent to 94 percent as compared to conventional boilers with efficiencies of 75 percent to 84 percent.

Steam turbines are of two types: the condensing type, which operates at below-atmospheric-exhaust pressures and requires vacuum-type condensers and pumps; and the noncondensing type, which operates at exhaust pressures of up to 50 psi. The exhaust steam from a condensing turbine is at very low temperatures, frequently less than 180°F, and, therefore, is of limited usefulness. In contrast, steam from a noncondensing turbine is at a higher quality and more suitable for other uses. Steam for process or heating uses can be extracted from either type of turbine.

Steam cycles sometimes operate at pressures in excess of those allowed for unattended operation by ASME and local codes, and the incremental labor costs for operating engineers should be considered.

Reciprocating engines are available in sizes as small as 5 kW, and in increments of approximately 25 kW for larger sizes. Gas turbines are available in sizes of 350 kW, 500 kW, 800 kW, 1,000 kW and up.

PRIME MOVER SIZING

The three basic design options identified above differ in their approach to sizing, and the resulting installed capacity will vary considerably from one to the other. For an electrically isolated site, the total installed capacity will be derived from a detailed electrical load analysis in which the maximum loading conditions for the plant are defined. For a typical site, with no special electrical loads, the capacity can be estimated based on the historic billing data and an inventory of site motors. For a larger site, above 500 kW, the required capacity will be the sum of the maximum historic peak demand plus the starting requirements of the largest motor. Only the largest motor is considered, and it is assumed that only one large motor will be started at any one time. For smaller sites, where the largest motor starting requirements are significant in comparison to the historic peak, the capacity is 150 percent of the historic peak. Additional capacity for standby service, either on a routine or emergency basis, also must be allowed.

It should be emphasized that the electrically isolated site will, in general, produce the least attractive economic return. While PURPA has, in fact, eliminated the need for this type of design, it does represent an extreme design case and will provide a measure of how well the cogeneration system can perform should the site ever become isolated from the grid.

The balanced, baseloaded system is the second option identified above. Two design criteria are available: either an electrically baseloaded or a thermally baseloaded system. The size of the electrically baseloaded system can be determined using electrical profile data to develop a load-duration curve shown in Figure 9-2. In the electrically baseloaded approach, the installed capacity would be that capacity which is exceeded 90 percent to 95 percent of the time. In Figure 9-2, a system sized for a 95 percent level would have a total installed capacity equal to 38 percent of the site's peak annual demand. Installed capacity would increase only to 39 percent of peak electrical load if the design point were 90 percent. An electrically baseloaded system generally will have no significant power sales to the electric utility, and will generally require supplemental power purchases. In some cases, an electrically baseloaded plant may operate when there is little demand for thermal energy; however, if the electric revenues are less than the system operating cost, it may be desirable to reduce the system's electrical output during these periods.

The second approach to the design of a baseloaded system is to use the site's thermal requirements as a design criterion. An approach similar to that used for the electrically baseloaded system is used to develop a curve such as that shown in Figure 9-3, and the system

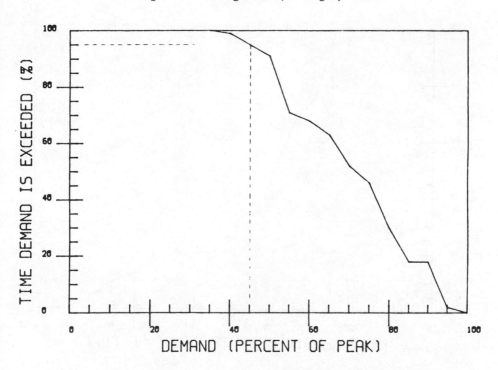

Figure 9-2. Electrical Load Duration Curve—Nursing Home

sized accordingly. Conventional boiler plants with several units generally select the individual boiler size at one-half to two-thirds of the peak capacity; single boiler installations may be 150 percent of the peak demand. While this approach to sizing may be acceptable for boilers, it may result in thermal capacity that is not economically justified when applied to a cogeneration system. Alternative design points, such as one-third or one-half of the peak thermal demand, also should be investigated for cogeneration sizing. In this case, the cogeneration site may sell significant amounts of power to the utility and/or purchase supplemental power.

If available, actual profile data for the site should be used in the design. While generic curves or data from other facilities may be useful approximations, care should be exercised in applying them to any specific site when actual profile data is not available.

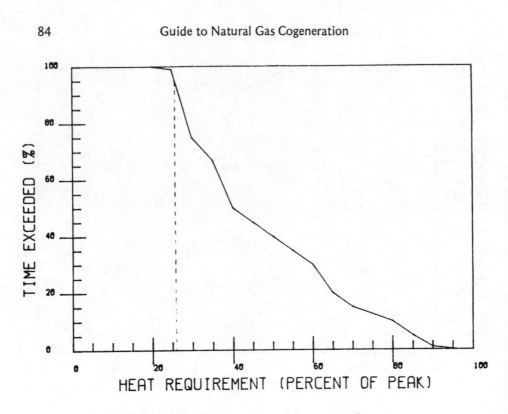

Figure 9-3. Thermal Load Duration Curve

The last design option is the maximum sized system, where the prime mover capacity is determined by the site's thermal requirements, and by the FERC output and efficiency requirements. In general, the 42.5 percent FERC efficiency requirement discussed above will be the primary constraint, and the cogeneration system annual electrical output can be determined using the following equation.

$$\text{Capacity} = \text{Lessor of} \left\{ \begin{array}{c} \dfrac{Q}{2H}\left(\dfrac{e_e}{.425 - e_e}\right) \\ \text{or} \\ Q/He_t \end{array} \right.$$

Where

Q = Site thermal requirement,

H = Hours of operation per year,

e_e = Electrical efficiency, and

e_t = Thermal efficiency.

The total installed capacity can be estimated by dividing the plant's annual electrical output by the system's annual operating hours. In general, gas turbine availability is approximately 8,400 to 8,600 hours per year; reciprocating engines, 7,500 to 8,000 hours. Low-speed engines may have higher availabilities, while high-speed engines (1,200 rpm to 1,800 rpm) will have lower availabilities.

NUMBER OF PRIME MOVERS

The number of prime mvoers or the engine size will be determined by the energy user's reliability and backup requirements, equipment availability and economics. For an electrically isolated cogenerator, the engine size is chosen so that three engines could carry the peak load plus the starting requirements of the largest size motor with sufficient inherent reserve to allow for spikes, such as elevator starting, to occur without affecting other electrical equipment. In addition, one engine should be installed for back-up during the servicing of another prime mover. That is, the actual installed capacity should be at least one-third greater than the calculated nominal capacity. In extremely small installations, the economics may not justify the cost of multiple engines and/or redundant capacity, and, in this case, the site must shed electrical load during an engine outage or cogeneration is not justified for an isolated system.

For the grid-connected design option, the number of prime movers is chosen based on such factors as the site thermal requirements or engine availability. Increasing the size of the engines to decrease their total number may achieve some economies of scale. Two prime movers can be used, with the use of two engines providing some measure of redundancy and an ability to operate at partial output during maintenance procedures. In some cases where prime movers are available in a limited number of sizes, three or more engines may be required.

STANDBY CAPACITY

A cogenerator can choose from several sources of electrical backup or standby capacity: the purchase of capacity from an electric utility; installation of additional on-site capacity, either with or without heat recovery; load shedding; or a combination of any of these. It should be recognized that load shedding is likely to be the least expensive alternative, dependent primarily on the cost of doing without power. If nonessential requirements can be identified and disconnected separately, the need for either redundant capacity or electric utility standby is correspondingly reduced.

In many electric utility service areas, standby capacity can be purchased from the utility at costs that are attractive in comparison to the cost of an outage or of additional installed capacity. Standby rates usually take the form of a fixed monthly charge for the right to purchase standby power and a variable charge based on the actual use of that capacity. The fixed charge usually is based on the amount of capacity the utility is expected to provide and usually is less than or equal to the installed cogeneration capacity.

In some cases, emergency generators may be used as standby for the cogeneration system; however, local codes should be reviewed first. The actual choice of standby or backup will be based on the economics of each available option.

POWER SALES

The value of cogenerated power will be dependent on the disposition of that power. If the end user were to cogenerate all the power that the site would require, and only that amount of power, then the value of that power would be equal to the average cost of purchased power. If, on the other hand, the cogenerator were to produce less than their own requirements, then the value of that power would be the decremental cost of power, which is generally less than the average cost. This is illustrated below with reference to Figure 9-4.

Figure 9-4 shows the cost of purchased power as a function of the electric load factor measured in hours use of demand per month (total kilowatt hours divided by peak kilowatts). The exact shape of

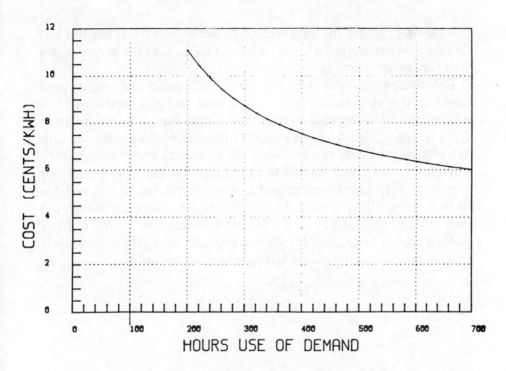

Figure 9-4. Cost of Purchased Power vs. Load Factor

the curve is dependent on both the structure of the electric utility rate and the site's demands. In this case, at a load factor of 350 hours use of demand, the average cost of purchased power would be 8.0¢ per kilowatt hour.

A baseloaded cogeneration system would operate at a high electric load factor, tending to decrease the load factor for purchased power. For illustrative purposes, assume that 50 percent of the end user's electrical energy requirements were cogenerated and the remaining 50 percent were purchased from the utility. Also assume that the load factor would be slightly more than 10.0¢ per kilowatt hour, or 2.0¢ per kilowatt hour more than the average total cost. Thus, if the end user's overall average cost of power were to remain at 8.0¢ per kilowatt hour, the cogenerated power would be worth 2.0¢ per kilowatt less than the average, or 6.0¢ per kilowatt hour. In this illustrative example, where the cogenerated power is used to reduce but not eliminate power purchases, the decremental cost of the power is less than the average cost.

The cogenerator also has the option of selling the cogenerated power to the electric utility at a cost that is equal to the utility's avoided cost. This avoided cost is the cost that the utility would incur if the power delivered by the cogenerator were developed by the utility. The avoided cost can include both a capacity component, which is related to the fixed cost of the capacity that the utility does not require, and an energy component, which is related to the variable costs that the utility avoids. In this illustrative case, even though the average retail cost of power is 8.0¢ per kilowatt hour, if the utility's avoided cost is in excess of 6.0¢ per kilowatt hour, then the project revenue can be maximized by selling power to the utility.

Chapter 10

Factory-Assembled Systems

Cogeneration was developed approximately a century ago and has been applied successfully in industrial plants and in larger commercial, institutional and residential buildings and building complexes. Prior to the 1980s, these systems were designed individually by engineers who had to design both the cogeneration system and its interface with the end users. Each of the various components was specified individually and the system was "built-up" at the site. In the early 1980s, an alternative approach to cogeneration was advanced; complete cogeneration systems were designed, components procured and the system fabricated at a factory. The site engineer has only to select the best equipment for the site's requirements and to design the interface between the cogeneration system and the energy end user.

With the passage of PURPA, various entrepreneurs sought to satisfy the perceived demand for cogeneration systems. One problem that was soon encountered was that, while there were numerous opportunities to apply cogeneration in smaller facilities, the required investments often were prohibitively high. For systems of less than a few hundred kilowatts, the fixed project costs were not justified by the potential reduction in operating costs. The lack of economic viability was true despite the fact that these smaller energy users usually pay more per unit of energy used.

Small factory-assembled or packaged cogeneration systems (PCS), such as shown in Figure 10-1, consisting of a single engine-generator with all ancillary equipment, were promoted as a means to make cogeneration cost effective in smaller applications. The underlying concept was to achieve significant cost reductions by spreading the fixed costs of engineering the cogeneration system over a large number of units and achieving economies through standardized parts and assembly-line production.

The basic concept of a factory-assembled module is to employ assembly-line procedures to produce a standard product. In practice, there is considerable variation in that product, both from vendor to vendor and within a single vendor's product line. Not all modules include the same components, and for such components as engines, generators and controls—which are common to all modules, not all systems include the same type of equipment.

The basic prime mover can be either a reciprocating engine or a combustion turbine. Most reciprocating engine systems are single-fuel systems; however, some can be purchased with either gas and/or oil fuel capability. Turbine-based systems provide a dual-fuel capability.

The basic distinction among generators is whether the module uses a synchronous or an induction generator. Induction generators provide a first cost advantage; however, they generally are not capable of operating in isolation from the electric utility grid. Synchronous generators impose additional control requirements on the module; however, they do provide the capability to operate the cogeneration system during electric utility outages, thus providing a source of backup power and increased value.

Some modules are available with absorption chillers. This capability allows full use of recoverable thermal energy during summer months and can decrease the total system cost through factory installation.

The use of modules provides several advantages to any potential cogenerator. First is the capital cost advantage that modules hold over built-up systems and, particularly, for systems of 100 kW or less. It would be extremely difficult to find an engineering firm that would design and then supervise the construction and installation of a small system at costs competitive with those quoted by vendors.

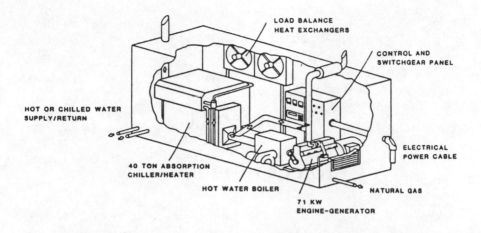

Figure 10-1. Factory-Assembled Cogeneration Module

Second is the ease of installation and the minimum amount of more expensive field work required to complete installation. Modules can be installed in as little as one day, although several days to weeks is a more likely timetable. This short schedule minimizes first cost and reduces the opportunity for field problems. In addition, complete modules can be checked out in the factory, where problems are more readily identified and corrected.

A third advantage is that the cogenerator has a single source that is responsible for the cogeneration system's performance.

Chapter 11

Project Financing

The energy end user has a number of alternatives for financing a viable cogeneration project. Each technique provides various benefits at different costs and risks. The final choice will be dependent on a number of factors, including the availability and cost of capital to the energy end user, and that end user's willingness and ability to accept both the technical and financial risks associated with this type of investment.

FINANCING ALTERNATIVES

Among the more common forms of financing for any capital investment is straightforward end user financing. In this case, the energy end user raises all the required capital; is responsible for the cogeneration system's design, construction and operation; and retains all the financial benefits, including any available tax benefits. The energy end user also accepts all the risks associated with the project. Actual funding may be with 100 percent equity, with funds derived from a capital or operating budget, or with traditional debt financing. More recently, other forms of financing, such as the use of Industrial Development Bonds (IDBs) or leases, also have been used. Small PCS or factory-assembled cogeneration systems, with total capital requirements of a few hundred thousand dollars or less, are sometimes funded out of operating budgets.

Should the energy end user be unable to raise the required capital, unwilling to accept the project risks or unwilling to accept the project rate of return, other forms of financing may be utilized. With each of these techniques, the energy end user gives up some fraction of the project's economic benefits in exchange for third-party responsibility for raising funds, project implementation, system operation or any combination of these activities.

One form of third-party ownership is the installment purchase contract, which may have several of the characteristics of a true lease. The difference, however, is that the end user purchases and owns the cogeneration system, retaining the resulting tax benefits. Among the sources of funding for installment purchases are banks, financial institutions and governmental agencies through the use of Industrial Development Bonds, Municipal Bonds and other tax-exempt instruments.

Another option is the true lease, where the lender purchases and owns the cogeneration system, retaining title to the system and the resulting tax benefits. While the end user may not be directly responsible for the project debt, he generally is required to enter into a leasing arrangement wherein he is ultimately responsible. An advantage of this type of financing is that the end user may be able to obtain more favorable financing costs than would be possible through direct ownership. A second advantage is that the end user may be able to avoid having the project appear on his Balance Sheet. True leases are most commonly used for smaller cogeneration systems.

The Energy Service Contract (ESC) is another form of third-party financing. Under this type of financing, a third party or cogeneration developer agrees to finance, own and operate a cogeneration facility selling power to the end user or to an electric utility and/or selling heat to the energy end user. The energy end user sells off the right to develop a cogeneration project at his facility in exchange for a fee, which may take the form of a discounted rate for heat and/or power (e.g., a 20 percent discount off the cost that the end user would have incurred for heat and/or power), a fixed rate for heat and/or power sales (e.g., electricity to be sold at 6¢ per kilowatt hour and steam at $7.00 per million Btu), or a fixed annual charge. Inflators can be built into the cost structure. The end user usually must commit to take a specified quantity of electricity and/or heat, if available, or if the end user cannot use the energy, to pay a minimum charge. This "take-or-pay" structure is necessary to secure financing.

One frequently employed ESC model is for the third party to sell power to the electric utility rather than to the end user. This arrangement offers several advantages. First, the end user is relieved of the obligation to purchase electricity if available from the cogenerator, thus reducing the end user's risk. Second, PURPA-mandated standby power for cogenerators may not be available to an end user purchasing power from a QF. If this is the case, the end user may incur significant penalties if the cogenerator were to have unplanned outages. Potential cogenerators should consult with appropriate utility personnel to obtain a firm commitment for standby power.

In the past, a number of factors acted together to increase the amount of third-party owned and operated cogeneration, and, with the many forms of business relationships that were developing, the tax status of third-party owners became uncertain. As a response to this uncertainty, in 1983, legislation was passed which established four criteria that had to be met if a third-party project were to be considered an Energy Service Contract for tax purposes. These criteria were:

- The service recipient cannot operate the cogeneration facility;
- The end user cannot incur any significant financial burden if there is nonperformance by the cogeneration system;
- The end user cannot receive any significant financial benefit if the operating costs of the facility are less than those contemplated in the contract; and,
- The energy customer cannot have an option or an obligation to purchase all or part of the project at a price other than the fair market value.

The effect of these requirements is to ensure that any third-party ESC meeting these criteria will not be treated as a lease or sale to the energy end user for tax purposes, and thus will protect the project's financial structure.

QUALIFYING FOR
THIRD-PARTY FINANCING

An end user may elect third-party ownership for any of a number of reasons. The question which then must be answered is, "Does this particular project qualify for third-party financing?" Will the operating cost savings be adequate to provide both the third party's required return on investment and the end user with some financial benefit?

The starting point for any analysis is the realization that the project must be economically viable on its own. While economic viability is a relative term, at a time when the Prime Interest Rate is approximately 10 percent, the minimum internal rate of return that appears acceptable to third-party investors is 20 percent to 25 percent. Therefore, the cogeneration system must produce an internal rate of return of greater than that if the third party is to be able to share some fraction of the savings with the end user.

In order to determine whether the project is financible through a third party, it is necessary to construct a cash flow model of the third party's economics. In general, this projection will be similar to one that would be developed for end user ownership with the following modifications:

- Assume that the third party will be able to fully use any investment tax credits;

- Assume that the third party will structure the project so that deductions are fully usable against income;

- Assume that project financing fees will add from 5 percent to 10 percent to the project capital cost;

- Assume that operating costs will be increased to cover a management fee (5 percent of total operating costs);

- Assume that operating savings are reduced by the end user's share of the savings; and,

- Assume that the third party will be able to secure financing at rates that are 2 percent to 3 percent above the long-term prime rate.

Given these assumptions, the end user can determine whether or not the project can support both the third party and the end user's financial goals. Again, it must be emphasized that third-party relationships and financing will vary considerably and that the above assumptions allow one to conduct a preliminary screening. In addition, it should be recognized that third-party financial requirements will vary considerably, with some developers seeking projects with simple paybacks of two years or less while others will accept paybacks of four to four and one-half years. This latter category includes equipment vendors and utilities that have a second financial interest in the project.

STRUCTURING A THIRD-PARTY ESC

The development of a third-party Energy Service Contract should be undertaken only with professional engineering, legal and financial advice. This section points out some of the economic considerations to be addressed in structuring any cost-sharing contract. In particular, this analysis discusses the cost allocation aspects of such contracts.

Many of the ESCs provide power and heat to the site end user, with charges based on a discounted published utility tariff or posted fuel prices or on a sharing of project "profits." Cogeneration systems generally displace the least expensive blocks of purchased power, and cogenerated power may be worth less than the average cost of power had the site purchased all its power from the utility. The cogenerated power must be valued based on its actual impact on end user costs.

This effect can be illustrated with the following example. Assume that the facility is purchasing power under a demand intensive rate of $24 per kilowatt and energy at 2¢ per kilowatt hour. An all-requirements customer purchasing 200 kW of demand and 80,000 kWh would pay $4,800 in demand charges and $1,600 in energy charges for a total of $6,400, or 8¢ per kilowatt hour. If this same customer installed a 90 kW module producing 50,000 kWh per month, and operated the system so as to reduce billing demands by 90 kW, the site would then purchase 110 kW and 30,000 kWh. Under the same rate schedule, this power would be billed at $2,640 for demand charges and $600 for energy, for a total of $3,240, or 10.8¢ per kilowatt hour. The 50,000 kWh of cogenerated power has reduced the purchased power bill by $3,160, or 6.32¢ per kilowatt hour of cogenerated power.

In a true ESC, the site owner usually would pay a rate for the cogenerated power that is a fraction of the rate he or she would have paid had that power been purchased from the utility. In this example, if a 10 percent discount is assumed, the site owner would pay 90 percent of the amount that the purchased power bill was reduced, or 90 percent of $3,160. The total cost of power to the site owner would be $2,844 as paid to the third-party cogenerator and $3,240 as paid to the electric utility, for a total of $6,084 and a saving of $316.

Chapter 12

Cash Flow Analysis

Cogeneration is a trade-off of an incremental capital investment for longer term revenues. As a result, the economic model must consider how these savings will vary over time, and several factors are important. Among these are relative inflation in the various cost components, financing and tax treatment.

General inflation will have the effect of increasing the operating cost savings over time. If the cost of capital, either an opportunity cost for the owner or the cost of interest, is less than the general level of inflation, the cogeneration system will be an effective hedge against inflation in general. The cogenerator is fixing the capital-related costs of utility service, leaving only the fuel- and maintenance-related costs to vary with inflation.

The relative inflation of the individual cost components is a second concern. If the cogenerator elects to utilize the cogenerated power on site, displacing purchased power, and the cost of purchased power increases faster than the cost of fuel, the cost savings will grow. On the other hand, if fuel costs increase faster than purchased power cost, the operating cost margin will decrease. In this case, a project, with an apparently acceptable simple payback, may not be viable. Cost components include the cost of purchased power, the cost of supplemental power, standby charges, buyback rate, heating fuel costs, power plant costs, staffing, maintenance, insurance and administrative charges. Of these, power and fuel costs are the parameters with the greatest potential for variation over time. Maintenance costs can be controlled through long-term contracts with vendors.

Purchased power and fuel price projections are also the most difficult factors to predict. Indeed, many of the available projections reflect the viewpoint—and sometimes hope—of the organization responsible for their preparation. In preparing a cash flow projection, it is suggested that many alternative sources be reviewed. Gas and electric utilities may have projections that they can provide for planning purposes. In many states, utilities are required to provide projections to the regulatory body as part of rate proceedings. In addition, the state department of energy or regulatory agency may have developed independent utility and fuel cost projections. Finally, some national organizations have prepared national, regional and state level projections which can be used.

In any analysis of projected electric and fuel costs, a two-step procedure is appropriate. First, review available data and projections to identify those assumptions that seem most reasonable for the specific area at that time. There is no right answer! However, unreasonable assumptions are possible. Use these assumptions as the basis for a baseline cash flow. Then vary the assumptions over the range that may be possible, and examine the impact on the overall project economics. If the project is particularly vulnerable to specific combinations of relative inflation, it may be necessary to modify the design, the financing, fuel and power contracts or the third-party relationship to reduce this risk. In some cases, if the risk cannot be reduced to acceptable levels, it may be prudent to drop the project.

Financing is a second area that requires cash flow modeling. While the cogeneration system may have an apparently acceptable payback and internal rate of return, it may not be able to support the debt service during the first few years of the project. Many investment sources are willing to tailor the interest and debt repayment schedule to the cash flow patterns of the project.

Finally, a cash flow analysis is required to accurately model tax impacts. Income taxes will vary with the level of profits that the project generates or with the owner's overall tax status. Property taxes will vary with the value of the project.

In summary, then, cash flow projections for the economic life of the project are essential to the ultimate cogeneration decision. There are a number of microprocessor-based computer programs, either spreadsheets or financial models, that make such calculations simple.

Examples of cash flow analyses are shown in chapters 34 and 35.

Section III

Natural Gas Prime Movers

J. A. ORLANDO, P.E.
GKCO, Incorporated

Chapter 13

Introduction to
Prime Movers

This section of the book examines prime mover technology as it is available today to satisfy the power and thermal requirements of various energy users. It examines applications where the end user can install a prime mover and to thus satisfy those mechanical needs without the inefficiency, and the resulting higher operating costs, of a more conventional utility system.

Each of the prime movers that are commercially available for use in on-site systems including reciprocating internal combustion engines, combustion gas turbines and steam boiler-turbine systems are reviewed in this section.

These prime movers are available in sizes ranging from a few horsepower through to well over 500,000 horsepower. In addition, this Section will review factory-assembled or packaged cogeneration systems, a concept that has brought the benefits of on-site cogeneration systems to even the small commercial and institutional energy user.

The section provides guidance in applying prime mover technology to various end user requirements and in selecting and installing on-site systems. Finally, the Section includes a compendium of equipment performance data and a listing of equipment vendors.

TYPES OF APPLICATIONS

In theory, a prime mover can be used in place of any larger electric motor, whether that motor is used to drive an air or gas compressor, a refrigeration system or to satisfy other needs for mechanical power. In addition, if the prime mover is used to drive a generator, the prime mover output can be used to satisfy smaller requirements for mechanical power by generating electricity for small motors; or it can be used to satisfy nonmechanical needs for energy for lighting, communications as well as other uses of electricity. Finally, in a cogeneration system where the heat that is normally rejected from the prime mover is captured and usefully applied, the prime mover can be used to satisfy requirements for both power and thermal energy.

Chapter 17 lists potential applications for on-site systems, either electrical, refrigeration or direct drive engines.

The primary advantage to the use of prime movers for on-site power is one of economics. Prime movers will be installed in those applications where either the end user's needs for mechanical or electrical power can be satisfied less expensively with power produced by using an on-site engine, or where the needs for reliability and availability in a power supply justify the incremental cost of a prime mover. In many cases, a cogeneration system, wherein heat is recovered and used to reduce fuel purchases, may be economically viable where a simple direct drive system without heat recovery is not.

TERMS AND DEFINITIONS

Depending on the specific needs of the end user at each facility, various characteristics of the prime mover's performance will take on different amounts of importance. Therefore, prior to a detailed discussion of each of the prime movers it is useful to first develop a framework for comparing engines. Among the parameters which should be included are:

- *Mechanical Efficiency:* The simple mechanical efficiency is defined as the engine's mechanical output divided by the fuel input, where both are expressed in consistent units (also referred to as the fuel to power efficiency).

- *Heat Rate:* A measure of generating station thermal efficiency, generally expressed in Btu of fuel input per net kilowatt-hour of output. The heat rate is an alternative means of expressing mechanical efficiency.

- *Recoverable Heat:* The amount of useful heat that can be recovered from the prime mover. The amount of available heat will be dependent on the difference in temperature at which the heat is available from the prime mover and the temperature at which it is required. The amount of usefully recoverable heat will increase as the temperature at which it is required decreases. More low pressure steam at 15 psig can be produced from the exhaust of a turbine than can 250 psig steam.

  Therefore the amount that can be usefully recovered from a prime mover is dependent both on the characteristics of the prime mover and on the needs of the end user.

- *Fuel to Useful Energy Efficiency:* The total amount of mechanical and thermal energy delivered by the prime mover divided by the fuel input, where all parameters are expressed in consistent units.

- *Rating:* The prime mover's power output under specified ambient and operating conditions. Prime movers are generally rated for the following types of duty:

  - *Continuous Operation:* The engine mechanical output which the manufacturer guarantees when the engine is operated with little variation in output for 24 hours a day.

  - *Intermittent Operation:* The mechanical output which the manufacturer guarantees when the engine is operated for a limited number of hours per month or year.

  - *Overload Operation:* The maximum mechanical output over and above the continuous duty rating which the engine is capable of for short periods of time.

  The engine rating is a measure of how much power a prime mover can deliver under specified conditions without excessive maintenance or decreased engine life.

- *Thermal Quality:* This parameter is a measure of the temperature at which the recoverable heat is available, with higher temperature hot water and steam having a higher quality than low temperature hot water and steam.

- *Scheduled Availability:* The fraction of the time that the prime mover is expected to be available for service after deducting the time required for scheduled maintenance procedures.

- *Equipment Life:* The length of time for which the prime mover can be economically maintained.

Other factors which should be considered in selecting a prime mover include the time required from a cold start until the engine can carry a full load; the engine size, weight, and floor space requirements; engine noise and vibration; the capability to change fuels; and, engine emissions. Costs, including equipment, installation, operating and maintenance costs are not the least important of the parameters which must be considered.

Finally, at this point it is useful to review one additional definition, a fuel's heating or calorific value.

When a hydrocarbon fuel such as natural gas is burnt, the products of combustion may include gases such as carbon monoxide (CO), carbon dioxide (CO_2) and water vapor (H_2O). The actual mix of the exhaust products is dependent on the fuel's composition and the combustion conditions.

The amount of water that is present in the exhaust is of particular interest and the total amount of exhaust water vapor is dependent on the amount of hydrogen in the fuel.

That water, created as steam during high temperature combustion, has stored in it the latent heat of vaporization, which is the thermal energy required (approximately 1,000 Btu per pound) to transform water from a liquid to a vapor. Unfortunately, unless the steam is condensed, this stored heat of vaporization is exhausted with the steam and the other products of combustion.

The total quantity of heat that is released during the combustion of a unit of fuel is called the fuel's Higher Heating Value (HHV). The amount of heat that is available from a fuel after the latent heat of vaporization is deducted from the HHV is defined as the fuel's Lower Heating Value (LHV).

Prime mover exhaust temperatures are usually above the temperature at which the water vapor would condense, and therefore the steam, with the heat of vaporization is usually lost to the environment. Therefore the water's heat of vaporization is not available for useful work by the prime mover, and, in analyzing a prime mover's performance, it is necessary to distinguish between a fuel's HHV and LHV. This difference is particularly significant when the fuel is billed on the basis of HHV as is frequently the case for natural gas.

For gaseous fuels the hydrogen content, and the resulting water vapor and most importantly the ratio between the LHV and the HHV, can vary significantly from fuel to fuel. Because it is impossible to generalize or to use a single ratio, most prime mover manufacturers prefer to specify the engine's performance based on a fuel's lower heating value.

Chapter 14

Reciprocating Engines

Internal combustion, reciprocating engines (see Figure 14-1), ranging in size from 10 horsepower through to 60,000 horsepower, have successfully been used for direct drive of compressors, heat pumps, fans and pumps, and, most commonly to drive electric generators. They can use a variety of gaseous fuels, fuel oils and a mixture of gaseous and liquid fuels, and they are relatively efficient in comparison to similarly sized turbines. In addition, they have good heat recovery characteristics and are well-suited for cogeneration applications. In general, reciprocating engines are the only efficient, commercially proven technology that is available in sizes of a few hundred horsepower or less.

DESCRIPTION

Internal combustion, reciprocating engines may be categorized according to several criteria. First, they may be characterized according to their thermodynamic cycle; either Diesel or Otto Cycle. Secondly, they may be characterized as to engine speed; either high, medium or low speed. Thirdly, they can be categorized as to the type of aspiration; either naturally aspirated or supercharged (also referred to as turbocharged). Finally, they can be characterized as to their operating cycle; either two-cycle or four-cycle.

Figure 14-1. Typical Reciprocating Engine

THERMODYNAMIC CYCLE

In an Otto Cycle engine, or as it is sometimes called, a spark ignited engine, the mixture of air and fuel is compressed in the engine cylinder where the fuel/air mix is then ignited by a spark that is usually provided from a spark plug. The burning fuel produces increased pressure and performs work by moving the piston as the cylinder gases expand.

Spark ignited engines are most commonly fired by natural gas; however other fuels including propane, butane, methane recovered from landfills or sewage treatment plants, field gas and gasoline can also be used.

There is a direct relationship between the compression ratio (defined as the ratio of the volume of the uncompressed fuel/air mixture to the minimum volume of the cylinder) and the engine's efficiency, and one way to increase engine efficiency is to increase the compression ratio.

However, the temperature of the fuel/air mixture increases as the mixture is compressed and if the mixture temperature is raised above the ignition point, spontaneous detonation or knocking will occur. Therefore the compression ratio must be low enough so that the temperature does not increase to the point at which the fuel will spontaneously combust. Methane can be compressed to a ratio of almost 15:1, while propane is limited to a ratio of 12:1 and butane to a ratio of 6.4:1. The fuel's maximum allowable compression ratio effectively sets a limit on the engine's maximum efficiency.

In contrast, in a Diesel Cycle engine, the heat produced during the compression of the cylinder gases is used to ignite the fuel. In this cycle the air is compressed until the air temperature is above the fuel's ignition temperature, and at that time, the fuel is injected into the cylinder where it spontaneously ignites and burns. The rate of fuel input is controlled so that the burning gases maintain constant pressure in the cylinder. Engine efficiency is not limited by the need to stay below the compression ratio at which the fuel will self-detonate, and diesel engines are capable of higher compression ratios and therefore higher efficiencies than spark ignited engines.

Diesel engines are generally fueled by liquids ranging from No. 2 diesel oil through to No. 6 residual oil, or by a combination of natural gas and oil.

Natural gas may be used in either an Otto Cycle and spark ignited, or on a Diesel Cycle where a small amount of fuel oil is used as the pilot for gas ignition.

ENGINE SPEED

Both spark ignited and diesel high speed engines (between 900 rpm and 1800 rpm) can achieve mechanical efficiencies approaching 33 percent. As engine speed is decreased, it is possible to achieve higher efficiencies and slower low speed diesel engines (below 300 rpm) are capable of achieving efficiencies of approximately 50 percent. It is this relatively high efficiency that makes reciprocating engines attractive for smaller applications.

When quoted by manufacturers the efficiency is frequently, but not always, based on a fuel's Lower Heating Value (LHV) and care should be exercised in using such data.

ASPIRATION

Engines may be either naturally aspirated or supercharged. Naturally aspirated engines supply air to the piston cylinder at atmospheric pressure and only require that the gas fuel be available at pressures of a few inches of water column.

Supercharging or turbocharging is the delivery of the air/fuel mixture to the cylinder at pressures that are significantly above atmospheric. Supercharged engines require gas supply pressures in the range of 12 to 14 psig.

The primary advantage to be gained by supercharging is an increase in the power output and efficiency of a given cylinder volume by providing more air and therefore oxygen to the cylinder, thus allowing for the combustion of a greater amount of fuel per cycle. Supercharged engines are generally less expensive on a per horsepower or kilowatt basis than are naturally aspirated engines.

Supercharging can be achieved by using a turbocharger which is driven by the engine exhaust gases or by using an air blower which is mechanically driven by the engine. Compressing the air or air/fuel mixture causes its temperature to increase, and in order to achieve maximum compression, an intercooler is required to cool the air/fuel mixture. This cooling can occur either between the turbocharger stages or prior to input to the cylinder. An engine intercooled may require a source of cooling water at 85°F.

OPERATING CYCLE

Another criterion for classifying engines is as either two stroke or four stroke according to the engine's operating cycle. In the four stroke engine as shown by Figure 14-2, four piston strokes or two revolutions of the crankshaft are required for a complete cycle. In the first stroke, air or a mixture of air and gas is taken into the cylinder; with the second stroke then compressing the air or air/fuel mix. The fuel is ignited and the third stroke consists of the expansion of the burning mix; the fourth stroke provides for the exhausting of the combustion products.

In a two stroke engine the first stroke is used to take in and compress the air or air/fuel mixture. The expansion of the ignited mixture and the removal of combustion products occurs in the second stroke. The compression stroke is used to force the combustion products from the engine cylinder.

Four stroke engines provide better part-load performance than do two stroke engines.

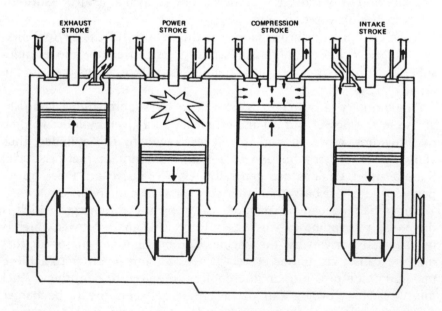

Figure 14-2. Four-Stroke Engine Cycle

HEAT RECOVERY

While specific engine performance will vary, in general, approximately 30 percent of the input energy is converted to mechanical power in a reciprocating engine. The remaining energy is transformed into heat which must be removed from the engine. This heat can be recovered and used to satisfy some thermal requirement as in a cogeneration system; however engine cooling is the primary function.

Figure 14-3 depicts the energy balance for a typical reciprocating engine. The majority of the engine heat is rejected in the engine exhaust and in the engine's jacket coolants. Smaller fractions of heat are rejected by the lube oil cooling system, the turbocharger intercooler or lost by radiation from the hot engine surfaces.

Almost 100 percent of the heat removed from the engine by the jacket cooling water is recoverable, as is the heat from the lube oil and turbocharger intercooler. Only a portion of the exhaust heat can be economically recovered.

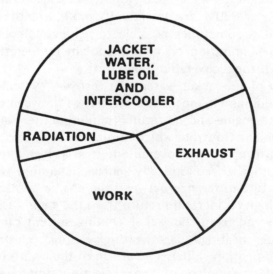

Figure 14-3. Reciprocating Engine Energy Balance

While it is possible to recover all the heat rejected by the lube oil cooler and turbocharger intercooler that heat is available at low temperatures and may not always be useful. In contrast, the engine's exhaust gases are at much higher temperatures and are therefore more useful. Even though all the exhaust heat cannot be recovered, the total quantity of useful exhaust heat may be quite high.

It is important to note that reciprocating engine energy balances will vary from engine to engine and as a function of the engine cooling (hot water or steam), aspiration and loading (Figure 14-4). For example, when operated at 50 percent of the continuous rating, a reciprocating engine's mechanical efficiency will decrease resulting in proportionately more heat in the exhaust, jacket and on engine surfaces. In analyzing specific cogeneration applications it is important to look at the heat balance for specific engines and operating levels because not all the rejected heat is recoverable, nor is that heat which is recoverable all at the same quality or temperature.

The most commonly used source of engine heat is the jacket cooling system which provides cooling for the block, heads and exhaust manifold. Coolant temperatures should be at least 180°F and temperatures of over 220°F are possible. By modifying the engine's seals and gaskets it is sometimes possible to achieve temperatures of up to 250°F. The limitation on engine coolant temperature limits the temperature of the recoverable jacket heat.

The design of the jacket water heat recovery system is a trade-off between a minimum temperature difference between the supply of coolant to the engine and its return (minimum thermal stress on the engine), a coolant flow rate which minimizes engine wear and in general the need to provide recoverable heat at as high a temperature as possible. In order to economically achieve maximum heat recovery a large temperature difference between the engine coolant supply and return and a high temperature return from the engine are desirable.

Heat recovered from the jacket cooling system can be used for domestic water heating, space heating, various industrial processes and for an absorption chiller. The design of the heat recovery system should consider a supplemental heat rejection radiator so that engine heat can be dumped should the site's heat requirement be less than the amount of heat that is available from the engine.

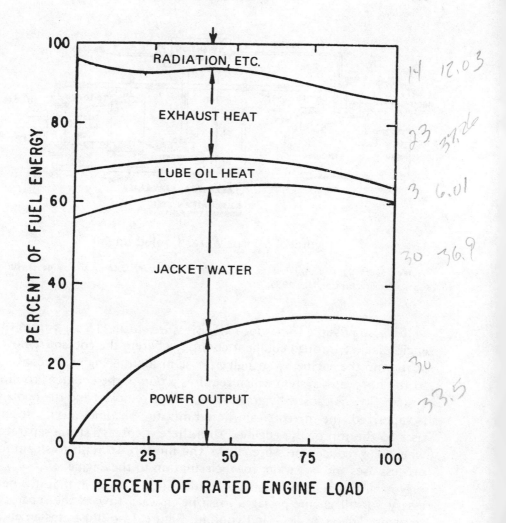

Figure 14-4. Engine Heat Balance vs. Load

In order to minimize the impact of leaks in the hot water distribution system, a heat exchanger should be considered as a means for isolating the engine from the site, rather than have the engine coolant distributed directly to the site. Figure 14-5 is a diagram of a hot water forced circulation cooling system.

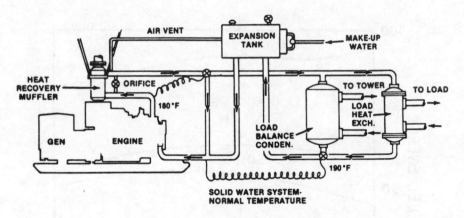

Figure 14-5. Hot Water Cooled Engine

Source: *Spark Ignited Application and Installation Guide*, Caterpillar Engine Division, Caterpillar Tractor Co., May 1986.

It is also possible to produce low pressure steam (15 psig) from the engine jacket through ebullient cooling, wherein the coolant removes heat from the engine by changing phase inside the engine. In contrast to the hot water system which requires a pump, the ebullient cooling system flow is caused by gravity. As heat is removed and the coolant is vaporized, the mixed steam/liquid mixture becomes less dense and rises to the top of the engine. From there it enters a steam separator where the reduced pressure allows the mixture to flash to steam for process uses and hot water for recirculation to the engine.

Not all engines are capable of ebullient operation, in that it is necessary for all engine passages to slope upward toward the separator to assure adequate flow and cooling. Figure 14-6 illustrates an ebullient cooled engine, Figure 14-7 the entire system.

The coolant pressure required in an ebullient system is obtained by locating the steam separator above the engine, and should the system pressure drop quickly, the steam bubbles will expand causing water flow problems within the engine.

Ebullient cooling minimizes thermal stress on the engine because there is little temperature change in the coolant and therefore internal thermal gradients are low. In addition, the ebullient system can

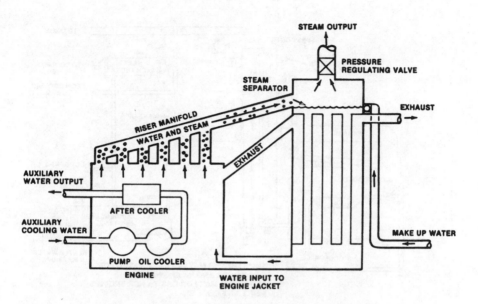

Figure 14-6. Ebullient Cooled Engine

Source: *Spark Ignited Application and Installation Guide,* Caterpillar Engine Division, Caterpillar Tractor Co., May 1986.

provide steam at temperatures of over 240°F and pressures of 15 psig, thus satisfying those end use needs that require low pressure steam.

A second major source of heat from a reciprocating engine are the exhaust gases where temperatures can range from 500°F to over 1,200°F. This heat can be used to produce higher temperatures and steam pressures than are possible from the jacket coolant; however not all the exhaust heat can be recovered.

Temperatures of the exhaust gases as they leave the heat recovery heat exchanger should be high enough to avoid condensation in the stack. In general, temperatures should be above 300°F to 350°F. When lower exhaust gas temperatures are expected a condensate handling system is required, and materials should be selected which are capable of withstanding corrosion from any acids which may develop in the condensate. A general rule of thumb is that approximately 50 percent of the exhaust heat is readily recoverable.

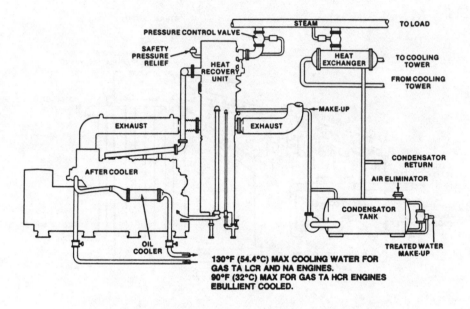

Figure 14-7. Ebullient Cooling Heat Recovery System

Source: *Spark Ignited Application and Installation Guide*, Caterpillar Engine Division, Caterpillar Tractor Co., May 1986.

Heat recovered from the lube oil cooler or from the intercooler is generally very low in temperature and can only be used for a limited number of applications. Lube oil temperatures are maintained in the range of 160°F to 200°F, therefore limiting the temperature of the available heat to somewhat less than these temperatures. Service water heating and industrial processes such as dying and metal baths may be able to use this heat source.

ASSESSMENT

Reciprocating engines have proven to be workhorses for small industrial drive applications. Their overall characteristics are summarized below:

- *Mechanical Efficiency:* Reciprocating engines with efficiencies that typically range from 25 percent to 35 percent are the most efficient commercially available prime mover in smaller sizes.

- *Recoverable Heat:* Full utilization of the recoverable heat requires that heat, at different temperatures, be recovered from several sources including: exhaust, jacket cooling water, lube oil cooler and turbocharger intercooler. As a consequence, while recovery of either jacket water or exhaust heat individually is rather simple, full heat recovery requires more complex piping systems.

- *Overall Energy Efficiency:* With typical heat recovery, reciprocating engine cogeneration systems can achieve overall efficiencies of 80 percent. If the heat that is radiated from the engine is recovered, as is possible in some applications, this added heat recovery can result in system efficiencies of over 90 percent.

- *Thermal Quality:* Approximately one-third of the energy input to the engine is available as low grade heat recovered from the jacket with a maximum temperature approaching 250°F or as steam with a maximum pressure of 15 psig. Only a fraction of the exhaust heat, typically 15 percent of the total input energy, is available at higher temperatures or steam pressures.

- *Rating:* Engines are commercially available in sizes ranging from 10 to 60,000 horsepower. Reciprocating engines are typically capable of operating at an output of 10 percent to 25 percent greater than their continuous duty rating for short periods of time. Manufacturers will derate their engines for use at higher altitudes (approximately 3 percent per thousand feet above sea level) or higher temperatures (approximately 1 percent for every 10°F above the base temperature for which the rating is specified) and when such applications are anticipated, the manufacturer's specification should be examined.

- *Scheduled Availability:* High speed engines burning pipeline quality fuels require approximately forty to eighty hours of planned maintenance per year. In general, after allowing for minor overhauls at intervals of 12,000 to 18,000 hours and major overhauls at intervals of 36,000 hours or more, and for unplanned outages, they can be scheduled for 8,000 hours of operation per year.

Engines burning heavier fuels require more extensive planned maintenance and may be shutdown for overhauls for as much as four weeks per year.

• *Equipment Life:* With proper maintenance, including major overhauls, reciprocating engines are capable of continuous duty operation for periods in excess of 100,000 hours.

In summary the reciprocating engine is capable of a wide variety of stationary power applications, and is particularly attractive in smaller sizes. It provides high efficiencies coupled with long life.

RESEARCH TRENDS

Recognizing the increasing size of the reciprocating engine market, manufacturers have stepped up and broadened their research and development activities. Among the current areas of activity are the use of ceramics to allow higher temperatures and efficiencies in engines and improved turbocharger performance, increased combustion control to improve efficiency and reduce emissions, improved maintainability and the development of efficient, durable engines in very small sizes.

Other areas of commercial development are directed at reducing both equipment cost and operating cost by improving efficiency and power density. Metallurgical improvements directed at reducing engine friction and allowing higher temperature performance, coupled with improved lubricating oils are expected to provide increased efficiency and operating cost reductions.

Another area of current development, the Stirling Engine as shown in Figure 14-8, is an externally fired, reciprocating engine. In this cycle, a low temperature working gas, enclosed in a cylinder is externally heated, causing it to expand as its temperature increases. As the gas expands, it does work by moving the piston. After the high temperature gas is fully expanded, it is cooled and work is used to compress the low temperature gas back to the original volume. Less

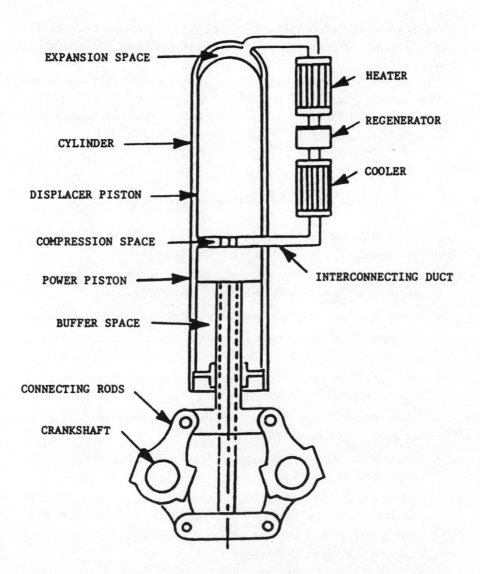

EXPANSION SPACE

HEATER

CYLINDER

REGENERATOR

DISPLACER PISTON

COOLER

COMPRESSION SPACE

POWER PISTON

INTERCONNECTING DUCT

BUFFER SPACE

CONNECTING RODS

CRANKSHAFT

Figure 14-8. Stirling Cycle Engine

Source: *Commercial and Industrial Cogeneration*, Office of Technology Assessment, Congress of the United States, February 1983.

work is required to compress the cool gas than is produced by the expanding hot gas, and the cycle is a net producer of mechanical power.

Stirling engines are expected to be highly efficient in converting fuel to power, reliable, be available in smaller sizes, have long intervals between maintenance and be cost competitive. To date, piston sealing has been a major barrier to commercial availability.

APPENDIX 14-1

RECIPROCATING ENGINE PERFORMANCE DATA

This appendix provides available data on the performance of reciprocating engines. The following parameters are reported.

- *Manufacturer:* Full names and addresses for each manufacturer are provided in a separate appendix (Appendix D).

- *Model Number:* The manufacturer's model number as published.

- *Capacity:* The engine output in kilowatts as rated for continuous duty operation as specified by the manufacturer.

- *Aspiration:* This column indicates whether the engine is naturally aspirated (NA) or turbocharged (TA).

- *Turbo Water Temp:* The temperature of the water supplied to the engine for rating purposes.

- *Compression Ratio:* The compression ratio of the engine for the rated capacity.

- *Rating Conditions:* The conditions under which the engine rating was established, usually 85°F and 29.38 inches of mercury.

- *Engine Speed:* The engine speed in revolutions per minute (rpm) at which the engine was rated.

- *Heat Rate:* The amount of energy required by the engine at rated output expressed in Btu per kilowatt hour of output. Higher Heat Value is used unless noted otherwise.

- *Jacket Water Heat Availability:* The amount of heat rejected in the jacket water at rated capacity expressed in Btu per kilowatt hour of rated output. The temperature at which the heat is available is reported as a comment if that information is available.

- *Intercooler Heat Availability:* For turbocharged engines, the amount of heat rejected in the intercooler at rated capacity expressed in Btu per kilowatt hour of rated output.

- *Lube Oil Heat Availability:* The amount of heat rejected in the lube oil at rated capacity expressed in Btu per kilowatt hour of rated output.

- *Exhaust Heat Availability:* The amount of heat rejected in the exhaust gases at rated capacity expressed in Btu per kilowatt hour of rated output.

- *Exhaust Temperature:* The temperature of the exhaust gases as measured in °F at rated output.

- *Stack Temperature:* The temperature of the exhaust gases entering the stack as measured in °F.

- *Fuel Capabilities:* The fuel for which the engine was rated. Engines are categorized as being rated on gas only (Gas) or on a combination of gas and oil (Dual).

These data were developed from various publications, including manufacturer brochures. The specific source document for each entry is indicated in column 4 of the table.

CODE	INFORMATION SOURCE
0	Manufacturer specs
1	GRI Cogeneration Assessment

If there were no data available, the particular entry was left blank. Finally, it should be noted that while this information is acceptable for use in preliminary analyses it should be used with care. First, the data are subject to change as manufacturers modify equipment; secondly, while manufacturers were contacted and asked to review the published data, no other effort was made to verify the table entries.

The information in Table 15A-1 is listed alphabetically by manufacturer. Table 14A-2 is sorted by engine size.

Table 14A-1. Reciprocating Engine Data By Manufacturer

Row Number	Manufacturer	Model No.	Data Source	Capacity (kW)	Aspiration	Turbo Water Temp (F)	Compression Ratio	Rating Cond	Engine Speed (RPM)
1	Blackstone	KP 3	0	1112	TA	160			600
2	Blackstone	KP 6	0	2223	TA	160			600
3	Blackstone	KP 8	0	2964	TA	160			600
4	Blackstone	KP 9	0	3335	TA	160			600
5	Blackstone	KVP 12	0	4446	TA	160			600
6	Blackstone	KVP 16	0	5928	TA	160			600
7	Blackstone	KVP 18	0	6669	TA	160			600
8	Caterpillar	3306	0	85	NA		8:1	STD	1800
9	Caterpillar	3306	0	100	NA		10.5:1	STD	1800
10	Caterpillar	3306	0	135	TA	90	8:1	STD	1800
11	Caterpillar	G342	0	135	NA		7.5:1	STD	1200
12	Caterpillar	3306	0	150	TA	90	10.5:1	STD	1800
13	Caterpillar	G342	0	150	NA		10.5:1	STD	1200
14	Caterpillar	G342	0	185	TA	90	7.5:1	STD	1200
15	Caterpillar	G342	0	200	TA	90	10.5:1	STD	1200
16	Caterpillar	G379	0	205	NA		7:1	STD	1200
17	Caterpillar	G379	0	230	NA		10:1	STD	1200
18	Caterpillar	G379	0	300	TA	90	7:1	STD	1200
19	Caterpillar	G398	0	315	NA		7:1	STD	1200
20	Caterpillar	G379	0	325	TA	90	10:1	STD	1200
21	Caterpillar	G398	0	350	NA		10:1	STD	1200
22	Caterpillar	G399	8	415	NA		7:1	STD	1200
23	Caterpillar	G398	0	450	TA	90	7:1	STD	1200
24	Caterpillar	G399	0	460	NA		10:1	STD	1200
25	Caterpillar	G398	0	500	TA	90	10:1	STD	1200
26	Caterpillar	G399	0	565	TA	90	7:1	STD	1200
27	Caterpillar	G399	0	650	TA	90	10:1	STD	1200
28	Caterpillar	3516	0	500					1200
29	Caterpillar	3516	0	850	TA	90			1200
30	Caterpillar	3512	0	370					1200
31	Caterpillar	3512	0	610	TA	90			1200
32	Coop-Bes	LSVB-12-SGC	0	3260	TA	90		DEMA	400
33	Coop-Bes	LSVB-12-GDT	0	3600	TA	90		DEMA	400
34	Coop-Bes	LSVB-16-SGC	0	4350	TA	90			400
35	Coop-Bes	LSVB-16-GDT	0	4800	TA	90		DEMA	400
36	Coop-Bes	LSVB-20-SGC	0	5430	TA	90		DEMA	400
37	Coop-Bes	LSVB-20-GDT	0	6000	TA	90		DEMA	400
38	Cummins	G495	0	74			12:1		1800
39	Cummins	G743	0	117			12:1		1800
40	Cummins	G855	0	127			12:1		1800
41	Cummins	GTA743	8	170	TA	85	10:1		1800
42	Cummins	GTA855	0	219	TA	85	10:1		1800
43	Cummins	G1710	0	254			12:1		1800
44	Cummins	GTA1710	0	451	TA	85	10:1		1800

Table 14A-1. Reciprocating Engine Data By Manufacturer

Heat Rate (Btu/kWh)	Jacket Heat (Btu/kWh)	Intercooler Heat (Btu/kWh)	Lube-Oil Heat (Btu/kWh)	Exhaust Heat (Btu/kWh)	Exhaust Temp (F)	Stack Temp (F)	Fuel	Comments	Row Number
10290	770	350	150	1800	730		Dual	Typical	1
10290	770	350	150	1800	730		Dual	Typical	2
10290	770	350	150	1800	730		Dual	Typical	3
10290	770	350	150	1800	730		Dual	Typical	4
10290	770	350	150	1800	730		Dual	Typical	5
10290	770	350	150	1800	730		Dual	Typical	6
10290	770	350	150	1800	730		Dual	Typical	7
12800	4540		720	4370	1115	300	Gas		8
11400	4200		650	3890	1100	300	Gas		9
11860	3660	280	660	4500	1030	300	Gas		10
12210	3920		660	4170	1225	300	Gas		11
11060	3460	340	590	4140	1160	300	Gas		12
11320	3710		590	3530	1180	300	Gas		13
11920	3320	110	610	4350	1120	300	Gas		14
11170	3170	190	600	4230	1130	300	Gas		15
13070	4070		700	4830	1160	300	Gas		16
11390	3920		590	3500	1140	300	Gas		17
12000	3540	240	610	4400	1110	300	Gas		18
12260	4190		650	3920	1200	300	Gas		19
11090	3440	210	590	3780	1100	300	Gas		20
11020	4010		650	4040	1125	300	Gas		21
10030	4250		690	4600	1175	300	Gas		22
11800	3650	240	630	4220	1110	300	Gas		23
11470	4070		620	3630	1080	300	Gas		24
11470	3530	270	610	4080	1060	300	Gas		25
12140	3740	230	640	4480	1100	300	Gas		26
11350	3570	260		3840	1040	300	Gas		27
11390	3560			1340	1040		Gas		28
10880	2580	220		1800	847		Gas		29
11270	3670			1860	1040		Gas		30
10500	2730	620		1750	815		Gas		31
9980	970	590	350		900		Gas		32
9620	930	600	330		850		Dual		33
9980	970	590	350		900		Gas		34
9490	930	600	330		850		Dual		35
9980	970	590	350		900		Gas		36
9380	930	600	330		850		Dual		37
13020	3640				990		Gas	Wet Exh Manifold	38
12430	3650				1130		Gas	Wet Exh Manifold	39
13580	3770				1180		Gas	Wet Exh Manifold	40
13260							Gas	Wet Exh Manifold	41
12050	3740				1150		Gas	Wet Exh Manifold	42
13240	3460						Gas	Wet Exh Manifold	43
11730	3650						Gas	Wet Exh Manifold	44

Table 14A-1. Reciprocating Engine Data By Manufacturer (Continued)

Row Number	Manufacturer	Model No.	Data Source	Capacity (kW)	Aspiration	Turbo Water Temp (F)	Compression Ratio	Rating Cond	Engine Speed (RPM)
45	M.A.N.	E2566E	1	105					1800
46	M.A.N.	E2542E	1	190					1800
47	M.A.N.	E2542TE	1	240					1800
48	M.A.N.	4L20/270G	1	308					900
49	M.A.N.	4L20/27G	1	325					900
50	M.A.N.	5L20/270G	1	385					900
51	M.A.N.	5L20/27G	1	406					900
52	M.A.N.	6L20/270G	1	462					900
53	M.A.N.	6L20/27G	1	487					900
54	M.A.N.	7L20/270G	1	539					900
55	M.A.N.	7L20/27G	1	568					900
56	M.A.N.	8L20/270G	1	616					900
57	M.A.N.	8L20/27G	1	650					900
58	M.A.N.	9L20/270G	1	693					900
59	M.A.N.	9L20/27G	1	732					900
60	M.A.N.	G6V30/45ATCG	1	839					514
61	M.A.N.	G6V30/45BG	1	884					514
62	M.A.N.	G7V30/45ATCG	1	979					514
63	M.A.N.	G7V30/45BG	1	1031					514
64	M.A.N.	G8V30/45ATCG	1	1118					514
65	M.A.N.	G8V30/45BG	1	1178					514
66	M.A.N.	G9V30/45ATCG	1	1258					514
67	M.A.N.	G9V30/45BG	1	1330					514
68	M.A.N.	12V30/45BG	1	1767					514
69	M.A.N.	14V30/45BG	1	2062					514
70	M.A.N.	16V30/45BG	1	2356					514
71	M.A.N.	18V30/45BG	1	2660					514
72	M.A.N.	6L52/55ADG	1	3591					450
73	M.A.N.	6L52/55AG	1	3591					450
74	M.A.N.	7L52/55ADG	1	4190					450
75	M.A.N.	7L52/55AG	1	4190					450
76	M.A.N.	8L52/55ADG	1	4788					450
77	M.A.N.	8L52/55AG	1	4788					450
78	M.A.N.	9L52/55ADG	1	5387					450
79	M.A.N.	9L52/55AG	1	5387					450
80	M.A.N.	10L52/55ADG	1	5985					450
81	M.A.N.	10L52/55AG	1	5985					450
82	M.A.N.	12L52/55ADG	1	7182					450
83	M.A.N.	12L52/55AG	1	7182					450
84	M.A.N.	14L52/55ADG	1	8379					450
85	M.A.N.	14L52/55AG	1	8379					450
86	M.A.N.	16L52/55ADG	1	9576					450
87	M.A.N.	16L52/55AG	1	9576					450
88	M.A.N.	18L52/55ADG	1	10773					450
89	M.A.N.	18L52/55AG	1	10773					450

Table 14A-1. Reciprocating Engine Data By Manufacturer (Continued)

Heat Rate (Btu/kWh)	Jacket Heat (Btu/kWh)	Intercooler Heat (Btu/kWh)	Lube Oil Heat (Btu/kWh)	Exhaust Heat (Btu/kWh)	Exhaust Temp (F)	Stack Temp (F)	Fuel	Comments	Row Number
10560	2370		0	3280			Gas		45
10070	3500		0	1980			Gas		46
9840	2830		0	2050			Gas		47
10440	2180		0	2650			Dual		48
9910	2260		0	2730			Gas		49
10660	2170		0	2640			Dual		50
9910	2260		0	2730			Gas		51
10520	2170		0	2640			Dual		52
9940	2280		0	2730			Gas		53
10440	2170		0	2650			Dual		54
9930	2280		0	2730			Gas		55
10420	2170		0	2640			Dual		56
9840	2270		0	2730			Gas		57
10440	2170		0	2640			Dual		58
9920	2260		0	2730			Gas		59
9830	1710		0	2550			Dual		60
9650	2200		0	2710			Gas		61
10040	1710		0	2550			Dual		62
9650	2200		0	2710			Gas		63
9830	1720		0	2550			Dual		64
9660	2200		0	2710			Gas		65
9960	1710		0	2550			Dual		66
9960	2200		0	2710			Gas		67
9440	2170		0	2710			Dual		68
9560	2170		0	2710			Gas		69
9320	2170		0	2710			Dual		70
9580	2170		0	2710			Gas		71
10580	1710		0	2860			Dual		72
9400	1840		0	2690			Gas		73
10520	1710		0	2860			Dual		74
9350	1840		0	2690			Gas		75
10540	1710		0	2860			Dual		76
9360	1840		0	2690			Gas		77
10520	1710		0	2860			Dual		78
9350	1840		0	2690			Gas		79
10500	1630		0	2860			Dual		80
9320	1750		0	2690			Gas		81
10610	1630		0	2860			Dual		82
9420	1750		0	2690			Gas		83
10580	1630		0	2860			Dual		84
9380	1750		0	2690			Gas		85
10610	1630		0	2860			Dual		86
9420	1750		0	2690			Gas		87
10610	1630		0	2860			Dual		88
9420	1750		0	2690			Gas		89

Table 14A-1. Reciprocating Engine Data By Manufacturer
(Continued)

Row Number	Manufacturer	Model No.	Data Source	Capacity (kW)	Aspiration	Turbo Water Temp (F)	Compression Ratio	Rating Cond	Engine Speed (RPM)
90	Waukesha	220VRG	0	29					1800
91	Waukesha	330VRG	0	45					1800
92	Waukesha	F817	0	75					1200
93	Waukesha	F1197	0	105					1200
94	Waukesha	F1905	0	175					1200
95	Waukesha	H2475	0	225					1200
96	Waukesha	2900G	0	290	NA			STD	1200
97	Waukesha	L3711G	0	350				STD	1200
98	Waukesha	3600G	0	360	NA			STD	1200
99	Waukesha	2900GSI	0	450	TA	85		STD	1200
100	Waukesha	2895GL	0	474	NA			STD	1200
101	Waukesha	5200G	0	500	NA			STD	1200
102	Waukesha	3600GSI	0	550	TA	85		STD	1200
103	Waukesha	3521GL	0	577	NA			STD	1200
104	Waukesha	5900G	0	595	NA			STD	1200
105	Waukesha	7100G	0	725	NA			STD	1200
106	Waukesha	5200GSI	0	800	TA	85		STD	1200
107	Waukesha	5108GL	0	837	NA			STD	1200
108	Waukesha	5900GSI	0	900	TA	85		STD	1200
109	Waukesha	5760GL	0	949	NA			STD	1200
110	Waukesha	9500G	0	975	NA			STD	1200
111	Waukesha	7100GSI	0	1100	TA	85		STD	1200
112	Waukesha	7042GL	0	1154	NA			STD	1200
113	Waukesha	9500GSI	0	1475	TA	85		STD	1200
114	Waukesha	9390GL	0	1539	NA			STD	1200

Table 14A-1. Reciprocating Engine Data By Manufacturer (Continued)

Heat Rate (Btu/kWh)	Jacket Heat (Btu/kWh)	Intercooler Heat (Btu/kWh)	Lube Oil Heat (Btu/kWh)	Exhaust Heat (Btu/kWh)	Exhaust Temp (F)	Stack Temp (F)	Fuel	Comments	Row Number
							Gas	Heat Exch Cooled	90
							Gas	Heat Exch Cooled	91
							Gas	Heat Exch Cooled	92
							Gas	Heat Exch Cooled	93
							Gas	Heat Exch Cooled	94
							Gas	Heat Exch Cooled	95
11880	3590		550	2620	1040	85	Gas	Heat Exch Cooled	96
							Gas	Heat Exch Cooled	97
11720	2830		530	2610	1060	85	Gas	Heat Exch Cooled	98
12340	3330	330	440	2670	1100	85	Gas	Heat Exch Cooled	99
10790							Gas	Heat Exch Cooled	100
11540	3840		540	2540	1020	85	Gas	Heat Exch Cooled	101
12500	3310	340	450	2730	1130	85	Gas	Heat Exch Cooled	102
10710							Gas	Heat Exch Cooled	103
11760	3210		520	2920	1040	85	Gas	Heat Exch Cooled	104
11500	3380		520	2070	1060	85	Gas	Heat Exch Cooled	105
12500	3560	340	440	2520	1070	85	Gas	Heat Exch Cooled	106
10930							Gas	Heat Exch Cooled	107
12340	3320	330	440	2610	1100	85	Gas	Heat Exch Cooled	108
10870							Gas	Heat Exch Cooled	109
12410	3270		400	3170	1180	85	Gas	Heat Exch Cooled	110
12310	3270	330	440	2710	1130	85	Gas	Heat Exch Cooled	111
10750							Gas	Heat Exch Cooled	112
11980	3250	410	440	2840	1220	85	Gas	Heat Exch Cooled	113
11160							Gas	Heat Exch Cooled	114

Table 14A-2. Reciprocating Engine Data By Size

Row Number	Manufacturer	Model No.	Data Source	Capacity (kW)	Aspiration	Turbo Water Temp (F)	Compression Ratio	Rating Cond	Engine Speed (RPM)
1	Waukesha	220VRG	0	29					1800
2	Waukesha	330VRG	0	45					1800
3	Cummins	G495	0	74			12:1		1800
4	Waukesha	F817	0	75					1200
5	Caterpillar	3306	0	85	NA		8:1	STD	1800
6	Caterpillar	3306	0	100	NA		10.5:1	STD	1800
7	M.A.N.	E2566E	1	105					1800
8	Waukesha	F1197	0	105					1200
9	Cummins	G743	0	117			12:1		1800
10	Cummins	G855	0	127			12:1		1800
11	Caterpillar	3306	0	135	TA	90	8:1	STD	1800
12	Caterpillar	G342	0	135	NA		7.5:1	STD	1200
13	Caterpillar	3306	0	150	TA	90	10.5:1	STD	1800
14	Caterpillar	G342	0	150	NA		10.5:1	STD	1200
15	Cummins	GTA743	0	170	TA	85	10:1		1800
16	Waukesha	F1905	0	175					1200
17	Caterpillar	G342	0	185	TA	90	7.5:1	STD	1200
18	M.A.N.	E2542E	1	190					1800
19	Caterpillar	G342	0	200	TA	90	10.5:1	STD	1200
20	Caterpillar	G379	0	205	NA		7:1	STD	1200
21	Cummins	GTA855	0	219	TA	85	10:1		1800
22	Waukesha	H2475	0	225					1200
23	Caterpillar	G379	0	230	NA		10:1	STD	1200
24	M.A.N.	E2542TE	1	240					1800
25	Cummins	G1710	0	254			12:1		1800
26	Waukesha	2900G	0	290	NA			STD	1200
27	Caterpillar	G379	0	300	TA	90	7:1	STD	1200
28	M.A.N.	4L20/270G	1	308					900
29	Caterpillar	G398	0	315	NA		7:1	STD	1200
30	Caterpillar	G379	0	325	TA	90	10:1	STD	1200
31	M.A.N.	4L20/27G	1	325					900
32	Caterpillar	G398	0	350	NA		10:1	STD	1200
33	Waukesha	L3711G	0	350				STD	1200
34	Waukesha	3600G	0	360	NA			STD	1200
35	Caterpillar	3512	0	370					1200
36	M.A.N.	5L20/270G	1	385					900
37	M.A.N.	5L20/27G	1	406					900
38	Caterpillar	G399	0	415	NA		7:1	STD	1200
39	Caterpillar	G398	0	450	TA	90	7:1	STD	1200
40	Waukesha	2900GSI	0	450	TA	85		STD	1200
41	Cummins	GTA1710	0	451	TA	85	10:1		1800
42	Caterpillar	G399	0	460	NA		10:1	STD	1200
43	M.A.N.	6L20/270G	1	462					900
44	Waukesha	2895GL	0	474	NA			STD	1200
45	M.A.N.	6L20/27G	1	487					900

Table 14A-2. Reciprocating Engine Data By Size

Heat Rate (Btu/kWh)	Jacket Heat (Btu/kWh)	Intercooler Heat (Btu/kWh)	Lube Oil Heat (Btu/kWh)	Exhaust Heat (Btu/kWh)	Exhaust Temp (F)	Stack Temp (F)	Fuel	Comments	Row Number
							Gas	Heat Exch Cooled	1
							Gas	Heat Exch Cooled	2
13020	3640				990		Gas	Wet Exh Manifold	3
							Gas	Heat Exch Cooled	4
12800	4540		720	4370	1115	300	Gas		5
11400	4200		650	3890	1100	300	Gas		6
10560	2370		0	3280			Gas		7
							Gas	Heat Exch Cooled	8
12430	3650				1130		Gas	Wet Exh Manifold	9
13580	3770				1180		Gas	Wet Exh Manifold	10
11860	3660	280	660	4500	1030	300	Gas		11
12210	3920		660	4170	1225	300	Gas		12
11060	3460	340	590	4140	1160	300	Gas		13
11320	3710		590	3530	1180	300	Gas		14
13260							Gas	Wet Exh Manifold	15
							Gas	Heat Exch Cooled	16
11920	3320	110	610	4350	1120	300	Gas		17
10070	3500		0	1980			Gas		18
11170	3170	190	600	4230	1130	300	Gas		19
13070	4070		700	4830	1160	300	Gas		20
12050	3740				1150		Gas	Wet Exh Manifold	21
							Gas	Heat Exch Cooled	22
11390	3920		590	3500	1140	300	Gas		23
9840	2030		0	2050			Gas		24
13240	3460						Gas	Wet Exh Manifold	25
11880	3590		550	2620	1040	85	Gas	Heat Exch Cooled	26
12000	3540	240	610	4400	1110	300	Gas		27
10440	2180		0	2650			Dual		28
12260	4190		650	3920	1200	300	Gas		29
11090	3440	210	590	3780	1100	300	Gas		30
9910	2260		0	2730			Gas		31
11020	4010		650	4040	1125	300	Gas		32
							Gas	Heat Exch Cooled	33
11720	2830		530	2610	1060	85	Gas	Heat Exch Cooled	34
11270	3670			1860	1040		Gas		35
10660	2170		0	2640			Dual		36
9910	2260		0	2730			Gas		37
10030	4250		690	4600	1175	300	Gas		38
11800	3650	240	630	4220	1110	300	Gas		39
12340	3330	330	440	2670	1100	85	Gas	Heat Exch Cooled	40
11730	3650						Gas	Wet Exh Manifold	41
11470	4070		620	3630	1080	300	Gas		42
10520	2170		0	2640			Dual		43
10790							Gas	Heat Exch Cooled	44
9940	2280		0	2730			Gas		45

Table 14A-2. Reciprocating Engine Data By Size
(Continued)

Row Number	Manufacturer	Model No.	Data Source	Capacity (kW)	Aspiration	Turbo Water Temp (F)	Compression Ratio	Rating Cond	Engine Speed (RPM)
46	Caterpillar	G398	0	500	TA	90	10:1	STO	1200
47	Caterpillar	3516	0	500					1200
48	Waukesha	5200G	0	500	NA			STO	1200
49	M.A.N.	7L20/270G	1	539					900
50	Waukesha	3600GSI	0	550	TA	85		STO	1200
51	Caterpillar	G399	0	565	TA	90	7:1	STO	1200
52	M.A.N.	7L20/27G	1	568					900
53	Waukesha	3521GL	0	577	NA			STO	1200
54	Waukesha	5900G	0	595	NA			STO	1200
55	Caterpillar	3512	0	610	TA	90			1200
56	M.A.N.	8L20/270G	1	616					900
57	Caterpillar	G399	0	650	TA	90	10:1	STO	1200
58	M.A.N.	8L20/27G	1	650					900
59	M.A.N.	9L20/270G	1	693					900
60	Waukesha	7100G	0	725	NA			STO	1200
61	M.A.N.	9L20/27G	1	732					900
62	Waukesha	5200GSI	0	800	TA	85		STO	1200
63	Waukesha	5108GL	0	837	NA			STO	1200
64	M.A.N.	G6V30/45ATCG	1	839					514
65	Caterpillar	3516	0	850	TA	90			1200
66	M.A.N.	G6V30/45BG	1	884					514
67	Waukesha	5900GSI	0	900	TA	85		STO	1200
68	Waukesha	5760GL	0	949	NA			STO	1200
69	Waukesha	9500G	0	975	NA			STO	1200
70	M.A.N.	G7V30/45ATCG	1	979					514
71	M.A.N.	G7V30/45BG	1	1031					514
72	Waukesha	7100GSI	0	1100	TA	85		STO	1200
73	Blackstone	KP 3	0	1112	TA	160			600
74	M.A.N.	G8V30/45ATCG	1	1118					514
75	Waukesha	7042GL	0	1154	NA			STO	1200
76	M.A.N.	G8V30/45BG	1	1178					514
77	M.A.N.	G9V30/45ATCG	1	1258					514
78	M.A.N.	G9V30/45BG	1	1330					514
79	Waukesha	9500GSI	0	1475	TA	85		STO	1200
80	Waukesha	9390GL	0	1539	NA			STO	1200
81	M.A.N.	12V30/45BG	1	1767					514

Table 14A-2. Reciprocating Engine Data By Size
(Continued)

Heat Rate (Btu/kWh)	Jacket Heat (Btu/kWh)	Intercooler Heat (Btu/kWh)	Lube Oil Heat (Btu/kWh)	Exhaust Heat (Btu/kWh)	Exhaust Temp (F)	Stack Temp (F)	Fuel	Comments	Row Number
11470	3530	270	610	4080	1060	300	Gas		46
11390	3560			1840	1040		Gas		47
11540	3840		540	2540	1020	85	Gas	Heat Exch Cooled	48
10440	2170		0	2650			Dual		49
12500	3310	340	450	2730	1130	85	Gas	Heat Exch Cooled	50
12140	3740	230	640	4480	1100	300	Gas		51
9930	2280		0	2730			Gas		52
10710							Gas	Heat Exch Cooled	53
11760	3210		520	2920	1040	85	Gas	Heat Exch Cooled	54
10500	2730	620		1750	815		Gas		55
10420	2170		0	2640			Dual		56
11350	3570	260		3840	1040	300	Gas		57
9840	2270		0	2730			Gas		58
10440	2170		0	2640			Dual		59
11500	3380		520	2070	1060	85	Gas	Heat Exch Cooled	60
9920	2260		0	2730			Gas		61
12500	3560	340	440	2520	1070	85	Gas	Heat Exch Cooled	62
10930							Gas	Heat Exch Cooled	63
9830	1710		0	2550			Dual		64
10880	2580	220		1800	847		Gas		65
9650	2200		0	2710			Gas		66
12340	3320	330	440	2610	1100	85	Gas	Heat Exch Cooled	67
10870							Gas	Heat Exch Cooled	68
12410	3270		400	3170	1180	85	Gas	Heat Exch Cooled	69
10040	1710		0	2550			Dual		70
9650	2200		0	2710			Gas		71
12310	3270	330	440	2710	1130	85	Gas	Heat Exch Cooled	72
10290	770	350	150	1800	730		Dual	Typical	73
9830	1720		0	2550			Dual		74
10750							Gas	Heat Exch Cooled	75
9660	2200		0	2710			Gas		76
9960	1710		0	2550			Dual		77
9960	2200		0	2710			Gas		78
11980	3250	410	440	2840	1220	85	Gas	Heat Exch Cooled	79
11160							Gas	Heat Exch Cooled	80
9440	2170		0	2710			Dual		81

Table 14A-2. Reciprocating Engine Data By Size
(Continued)

Row Number	Manufacturer	Model No.	Data Source	Capacity (kW)	Aspiration	Turbo Water Temp (F)	Compression Ratio	Rating Cond	Engine Speed (RPM)
82	M.A.N.	14V30/45BG	1	2062					514
83	Blackstone	KP 6	0	2223	TA	160			600
84	M.A.N.	16V30/45BG	1	2356					514
85	M.A.N.	18V30/45BG	1	2660					514
86	Blackstone	KP 8	0	2964	TA	160			600
87	Coop-Bes	LSVB-12-SGC	0	3260	TA	90		DEMA	400
88	Blackstone	KP 9	0	3335	TA	160			600
89	M.A.N.	6L52/55ADG	1	3591					450
90	M.A.N.	6L52/55AG	1	3591					450
91	Coop-Bes	LSVB-12-GDT	0	3600	TA	90		DEMA	400
92	M.A.N.	7L52/55ADG	1	4190					450
93	M.A.N.	7L52/55AG	1	4190					450
94	Coop-Bes	LSVB-16-SGC	0	4350	TA	90		DEMA	400
95	Blackstone	KVP 12	0	4446	TA	160			600
96	M.A.N.	8L52/55ADG	1	4788					450
97	M.A.N.	8L52/55AG	1	4788					450
98	Coop-Bes	LSVB-16-GDT	0	4800	TA	90		DEMA	400
99	M.A.N.	9L52/55ADG	1	5387					450
100	M.A.N.	9L52/55AG	1	5387					450
101	Coop-Bes	LSVB-20-SGC	0	5430	TA	90		DEMA	400
102	Blackstone	KVP 16	0	5928	TA	160			600
103	M.A.N.	10L52/55ADG	1	5985					450
104	M.A.N.	10L52/55AG	1	5985					450
105	Coop-Bes	LSVB-20-GDT	0	6000	TA	90		DEMA	400
106	Blackstone	KVP 18	0	6669	TA	160			600
107	M.A.N.	12L52/55ADG	1	7182					450
108	M.A.N.	12L52/55AG	1	7182					450
109	M.A.N.	14L52/55ADG	1	8379					450
110	M.A.N.	14L52/55AG	1	8379					450
111	M.A.N.	16L52/55ADG	1	9576					450
112	M.A.N.	16L52/55AG	1	9576					450
113	M.A.N.	18L52/55ADG	1	10773					450
114	M.A.N.	18L52/55AG	1	10773					450

Table 14A-2. Reciprocating Engine Data By Size
(Continued)

Heat Rate (Btu/kWh)	Jacket Heat (Btu/kWh)	Intercooler Heat (Btu/kWh)	Lube Oil Heat (Btu/kWh)	Exhaust Heat (Btu/kWh)	Exhaust Temp (F)	Stack Temp (F)	Fuel	Comments	Row Number
9560	2170		0	2710			Gas		82
10290	770	350	150	1800	730		Dual	Typical	83
9320	2170		0	2710			Dual		84
9580	2170		0	2710			Gas		85
10290	770	350	150	1800	730		Dual	Typical	86
9980	970	590	350		900		Gas		87
10290	770	350	150	1800	730		Dual	Typical	88
10580	1710		0	2860			Dual		89
9400	1840		0	2690			Gas		90
9620	930	600	330		850		Dual		91
10520	1710		0	2860			Dual		92
9350	1840		0	2690			Gas		93
9980	970	590	350		900		Gas		94
10290	770	350	150	1800	730		Dual	Typical	95
10540	1710		0	2860			Dual		96
9360	1840		0	2690			Gas		97
9490	930	600	330		850		Dual		98
10520	1710		0	2860			Dual		99
9350	1840		0	2690			Gas		100
9980	970	590	350		900		Gas		101
10290	770	350	150	1800	730		Dual	Typical	102
10500	1630		0	2860			Dual		103
9320	1750		0	2690			Gas		104
9380	930	600	330		850		Dual		105
10290	770	350	150	1800	730		Dual	Typical	106
10610	1630		0	2860			Dual		107
9420	1750		0	2690			Gas		108
10580	1630		0	2860			Dual		109
9380	1750		0	2690			Gas		110
10610	1630		0	2860			Dual		111
9420	1750		0	2690			Gas		112
10610	1630		0	2860			Dual		113
9420	1750		0	2690			Gas		114

Chapter 15

Gas Turbines

While the steam turbine has achieved dominance in electric utility, base loaded, central power plant applications, gas turbines have tended to dominate the base loaded on-site market. This chapter provides a review of commercially available gas turbine technology and the underlying thermodynamic cycles and of the various performance parameters which should be considered in evaluating alternative gas turbines.

Gas turbines (Figure 15-1) are commercially available in sizes ranging from a few hundred horsepower to over a quarter million horsepower. Because of their high availability, low maintenance requirement, multi-fuel capability and the ease of fuel switching, they have become increasingly popular in on-site power applications.

DESCRIPTION

Commercially available turbines may be either aircraft derivative or industrial engines. The aircraft derivative engine has been designed for maximum power output per pound and is generally less expensive than the industrial turbine. Stationary or industrial power turbines are generally larger and designed for longer lives with less rigorous maintenance requirements. As a result, while they may have a higher first cost, industrial turbines may have lower life cycle costs.

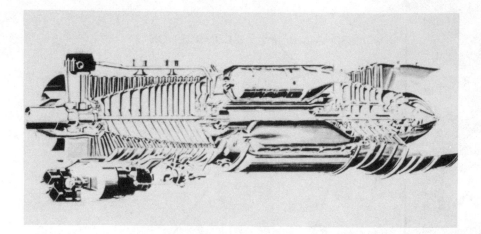

Figure 15-1. Gas Turbines

Source: *Allison Gas Turbines*, Allison Gas Turbine Division, General Motors Corporation, 1984.

Simple gas turbines such as that shown in Figure 15-1 operate through a Brayton Cycle as shown by Figure 15-2. The turbine, as characterized in Figure 15-3 consists of three primary components; the first of which is a compressor which is used to raise the pressure of the working fluid, usually air, to pressures ranging up from four to thirty times atmospheric. The compressed air is then heated to temperatures ranging from 1,500°F to 2,200°F, usually in a combustor, which is the second major component, where fuel is added and ignited. The hot, high pressure gases are then expanded through a power turbine, the third major component, producing power, which is used to drive the working fluid compressor and a generator or other mechanical drive device. Approximately one-third of the power developed by the power turbine can be required by the compressor.

The pressures of the working fluid, usually air, entering the combustion chamber may be ten to fifteen times atmospheric. In order to inject the gas fuel into this chamber, it is necessary to increase the

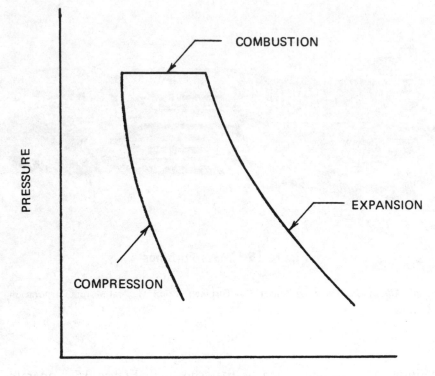

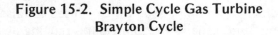

**Figure 15-2. Simple Cycle Gas Turbine
Brayton Cycle**

pressure of the fuel until it is greater than the air supplied to the combustor, and this is usually done with supplemental gas compressors as shown in Figure 15-3. The required gas pressure may exceed 600 psig; however pressures of 200 to 250 psig are more typical. The power required for these compressors, frequently referred to as "parasitic power," must be deducted from the turbine's power output.

The amount of parasitic power required by the compressor will depend on the pressure at which the gas is supplied to the site, the

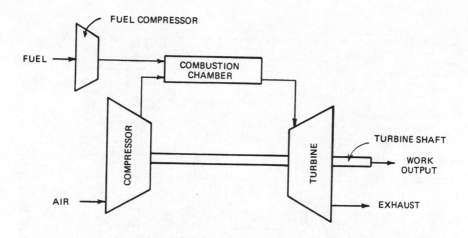

Figure 15-3. Simple Cycle Gas Turbine

pressure required by the turbine and the type of compressor. The compressor may require up to 5 percent of the turbine output; however, the manufacturer should be consulted for more precise information.

Smaller gas turbines are relatively inefficient, operating with mechanical efficiencies (based on the fuel's Higher Heating Value) as low as 15 percent; larger turbines operate at efficiencies approaching 40 percent. Figure 15-4 depicts the approximate relationship between turbine size and efficiency where the efficiency is expressed as a heat rate (Btu per kilowatt hour). Recent development work has resulted in a shift in the curve as small turbines are being made more efficient.

Turbine efficiency increases with an increase in the temperature of the gases to the power turbine. However the turbine's capacity must be decreased for increases in the temperature of air input to the turbine compressor, for increased air pressure to the air compressor caused by air filters and by increases in the turbine's exhaust pressure.

Turbine part load efficiency is very dependent on the design of the specific turbine, and in general, due to the compressor loading, turbine efficiency at very low loads may be poor (Figure 15-5).

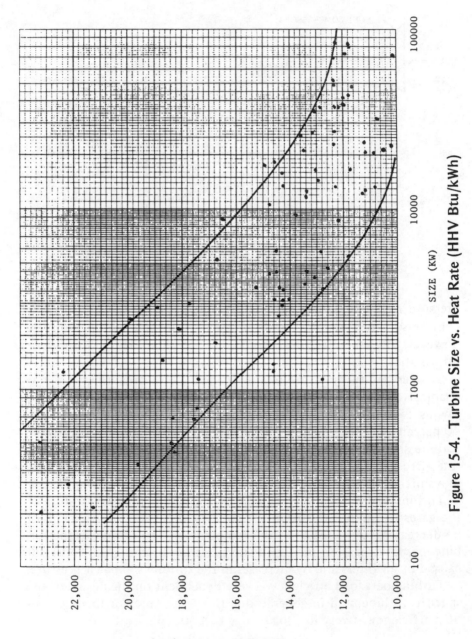

Figure 15-4. Turbine Size vs. Heat Rate (HHV Btu/kWh)

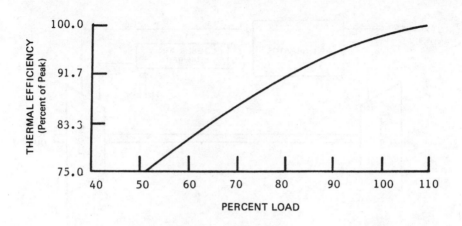

Figure 15-5. Typical Turbine Part Load Efficiency

Efficiency can be increased through the use of regeneration, where heat is recovered from the exhaust and used to preheat compressed gas before it enters the combustor (Figure 15-6). Intercooling, where the air is cooled between compressor stages (Figure 15-7) and reheating, where a second combustor is used to provide heating between stages of the expansion turbine (Figure 15-8) both increase the turbine's power output. Reheating does cause a decrease in mechanical efficiency.

Turbine capacity can also be increased by cooling the compressor inlet air with some manufacturers recommending 55°F inlet air regardless of ambient temperature. In dry climates, evaporative coolers are particularly effective. A typical chiller can reduce the compressor inlet temperature by 60 percent of the difference between the ambient dry bulb and wet bulb temperatures.

Turbines may be either open cycle where the air or working fluid is exhausted from the power turbine to the atmosphere, or closed cycle, as shown in Figure 15-9, where the working fluid is recycled. Closed cycle systems have a higher first cost, however decreased corrosion of turbine blades and improved part load performance provide reduced operating and maintenance costs. In general most applications are based on open cycle turbines.

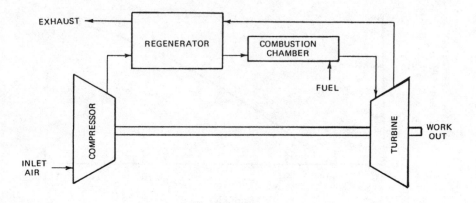

Figure 15-6. Regeneration Cycle

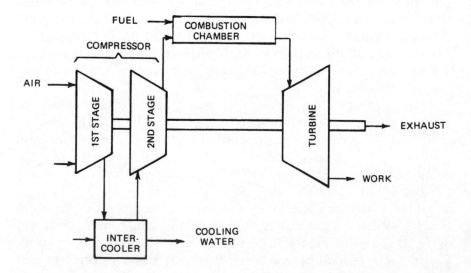

Figure 15-7. Compressor Intercooler Cycle

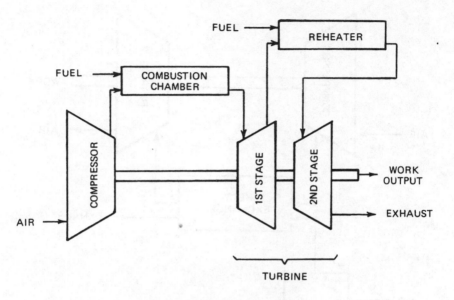

Figure 15-8. Reheat Cycle

HEAT RECOVERY

Almost all the fuel energy that is not converted to shaft power is rejected from the turbine in the exhaust gases which may reach temperatures of over 1,000°F. This exhaust is relatively clean and can be used in direct drying applications. In addition because of its high quality of recoverable thermal energy, when ducted to an unfired boiler a turbine's exhaust can provide steam at pressures of 150 psig to 1,000 psig and they are frequently employed in industrial facilities or central boiler plants. When operated at part load due to the decrease in mechanical efficiency, the amount of exhaust heat per horsepower of output will increase; however that heat will be available at lower exhaust temperatures (Figure 15-10).

Turbines are fired with a large amount of excess air and, as a consequence, the exhaust gas oxygen content is quite high, approaching 16 percent to 18 percent. When heat over and above that which is

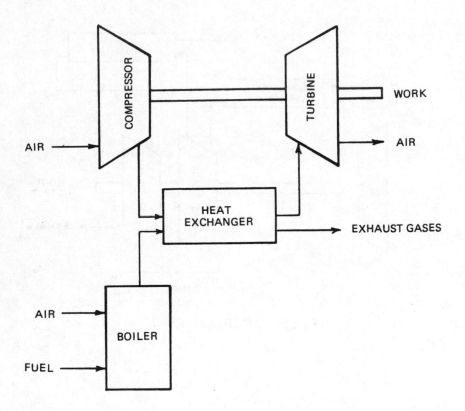

Figure 15-9. Closed Cycle Turbine

directly available in the exhaust is required, this exhaust can support combustion in an "afterburner" or "duct burner" (Figure 15-11). When fired in a duct burner, supplemental fuels can be used at an efficiency of over 90 percent to 94 percent.

Gas turbine, heat recovery boilers are often equipped with a diverter valve so that the exhaust heat can be rejected when there is no thermal load.

Using a combination of duct burners, heat recovery boilers and diverter valves, gas turbines can satisfy a wide range of thermal requirements, while still supplying required power.

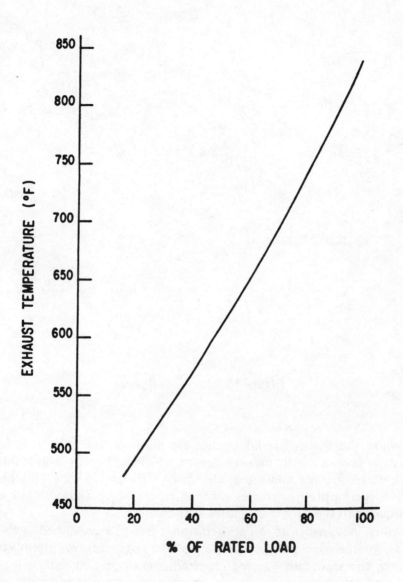

Figure 15-10. Typical Turbine Exhaust Temperatures

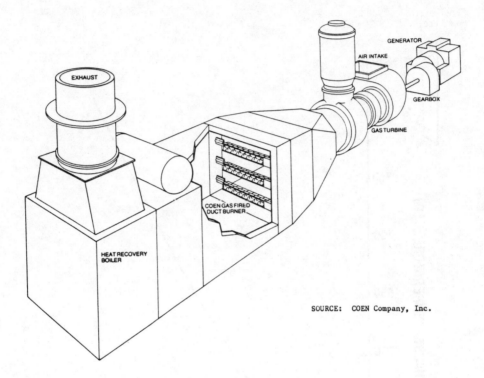

SOURCE: COEN Company, Inc.

Figure 15-11. Duct Burner

When site thermal requirements are minimal, the exhaust can be used to produce high pressure steam which is then used in a steam turbine to provide additional electricity (Figure 15-12). This combined cycle option increases mechanical capacity and efficiency as compared to the simple Brayton Cycle turbine.

When designed into a cogeneration system, a combined cycle allows for the more precise matching of the cogeneration system's output to the seasonally varying requirements of an end user. It should be noted that when operated in an electric output mode only, with no thermal output, a combined cycle is not considered a cogeneration system.

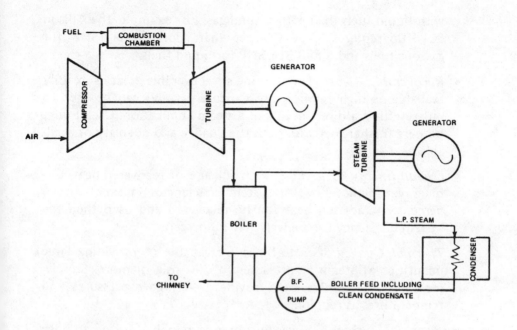

Figure 15-12. Combined Cycle

ASSESSMENT

Gas turbines provide more power per pound than other commercially available prime movers. They are relatively vibration free and quiet and therefore installation costs are lower than for either reciprocating engines or steam turbines. In addition, in typical stationary power applications maintenance shutdowns are scheduled for six month intervals and are relatively short in duration. Even major overhauls can be completed in a short period. Other factors to be considered are reviewed below.

- *Mechanical Efficiency:* Gas turbines with efficiencies ranging from 15 percent to 40 percent are commercially available. Today, only larger size turbines have efficiencies equal to that of a reciprocating engine. Turbines are less efficient when fired with liquid fuels than with natural gas. For example, the Allison 501-KB5 requires 8,430 Btu per Shaft Horsepower (SHP) for gaseous fuels and 8,840 Btu/SHP for liquid fuels.

- *Recoverable Heat:* The turbine's recoverable heat is readily available as high temperature exhaust gas and can be used to satisfy either a direct drying or a steam application. Use of duct burners further increases both the quality and quantity of available heat.

- *Overall Energy Efficiency:* With full use of recovered heat, turbines can achieve overall system efficiencies in excess of 90 percent. If radiated heat can be recovered and used then the system efficiency can be well over 90 percent.

- *Thermal Quality:* Gas turbines are capable of providing large quantities of steam at pressures of 200 psig or more. Supplemental duct burners can provide even higher pressures and temperatures as required.

- *Rating:* A turbine's power output is extremely sensitive to both ambient temperature and altitude. Performance maps such as Figure 15-13 should be consulted to determine turbine site rating. Turbines must also be derated when fired with liquid fuels.

- *Scheduled Availability:* Turbines have extremely high scheduled availabilities. In some installations annual availability has been in excess of 98 percent. Use of liquid fuels will shorten the interval between overhauls with an attendant increase in operating costs.

- *Equipment Life:* With proper maintenance, including major overhauls, turbines are capable of continuous duty operation for periods in excess of 100,000 hours.

Gas turbines are particularly suited for applications requiring high availability and low maintenance or in cogeneration applications where there is a need for higher quantities and qualities of thermal energy.

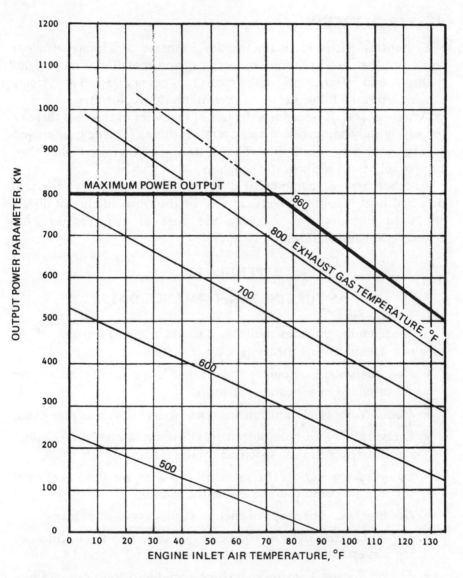

Source: *Solar Gas Turbine Generator Performance,* Solar Turbines Incorporated, 1983.

Figure 15-13. Typical Output Power Available, Gas Fuel

RESEARCH TRENDS

Current activities focus on the development of a more efficient small turbine incorporating recuperators and metallurgy more commonly found in larger engines. Research is being directed at systems ranging from ten kilowatts up to several hundred kilowatts.

A second area of research is the use of ceramics in turbines thereby increasing inlet temperature and overall turbine efficiency. Improved recuperators may result in turbine efficiencies that are higher than comparably sized reciprocating engines.

Another area of commercial development is the design of standardized heat recovery boilers capable of use on a number of different sized turbines thus reducing the cost of heat recovery for cogeneration applications.

APPENDIX 15-1

GAS TURBINE PERFORMANCE DATA

This appendix provides available data on the performance of gas turbines. The following parameters are reported.

- *Manufacturer:* Full names and addresses for each manufacturer are provided in a separate appendix.

- *Model Number:* The manufacturer's model number as published.

- *Capacity:* The turbine output in kilowatts as rated for continuous duty operation as specified by the manufacturer.

- *Compression Ratio:* The ratio of the working fluid as it exits the turbine compressor to ambient pressure.

- *Gas Pressure:* The required fuel pressure expressed in psig.

- *Rating Conditions:* The conditions under which the turbine rating was established, ISO usually consists of 59°F and 14.7 psia.

- *Engine Speed:* The power turbine speed in revolutions per minute (rpm) at which the turbine was rated.

- *Heat Rate:* The amount of energy required by the engine at rated output expressed in Btu per kilowatt hour of rated output. Higher Heating Value is used unless otherwise noted.

- *Exhaust Heat Availability:* The amount of heat rejected in the exhaust gases at rated capacity expressed in Btu per kilowatt hour of rated output.

- *Exhaust Temperature:* The temperature of the exhaust gases as measured in °F at rated output.

- *Exhaust Mass Flow:* The exhaust gases mass flow as measured in pounds per second at rated output.

- *Fuel:* The fuel which was used as the basis for the turbine rating. Turbines were classified as being rated on gas only (Gas) or on a combination of gas and distillate oil (Dual).

These data were developed from various publications, including manufacturer brochures. The specific source document for each entry is indicated in column 1 of the table.

CODE	INFORMATION SOURCE
0	Manufacturer specs
1	GRI Cogeneration Assessment
2	Southern California Gas Company
3	Industrial and Marine Gas Turbines of the World, 1985-1986—Published by Gas Turbine World

The data from Southern California Gas Company were assembled by Mr. Larry W. Carlo of that company.

If there were no data available, the particular entry was left blank. In some cases it may be possible to compute one parameter from data that are available. For example, given the exhaust gas mass flow and temperature, it is possible to estimate the amount of recoverable heat in the exhaust. It may be possible to estimate the required natural gas pressure for gas fueled turbines from the turbine's air compressor ratio.

Finally, it should be noted that while this information is acceptable for use in preliminary analyses it should be used with care.

The information in Table 15A-1 is listed alphabetically by manufacturer. Table 15A-2 is sorted by engine size.

Table 15A-1. Gas Turbine Data by Manufacturer

Row Number	Manufacturer	Model No.	Data Source	Capacity (kW)	Compression Ratio	Req'd Gas Pressure (psig)	Rating Cond	Engine Speed (RPM)
1	Allison	501 KB	0	3316	9.3:1	250	ISO	14200
2	Allison	501 KB5	0	3924	9.3:1	250	ISO	14250
3	Allison	570 KA	0	4877	12.7:1	275	ISO	9000
4	Allison	571 KA	0	5910	12.7:1	275	ISO	13000
5	BBC	L-GT 8/11	3	220000	60.0:1		ISO	3000
6	Centrix Gas Tur	CS 600-2	3	586	5.25:1		ISO	22000
7	Coberra	2348	3	12058			ISO	5200
8	Coberra	2556	3	15554			ISO	4950
9	Coberra	6456	3	24797			ISO	4950
10	Coberra	6462	3	25364			ISO	4800
11	Curtis Wrt	P0020	2	20000	18.0:1	400	ISO	
12	Dresser Clark	DC 990	3	4325	12.0:1		ISO	17800
13	Dresser Clark	DJ-270G	3	21783			ISO	5400
14	Dresser Clark	DJ-125R	3	20580			ISO	5330
15	Dresser Clark	DJ-200R	3	15428			ISO	5400
16	Dresser Clark	DJ-290G	3	25364			ISO	5000
17	Fiat TTG	TG 7	3	8630	6.0:1		ISO	6000
18	Fiat TTG	TG 16	3	18850	7.0:1		ISO	4850
19	Fiat TTG	TG 20	3	44290	11.0:1		ISO	4920
20	Fiat TTG	TG 50	3	105100	12.0:1		ISO	3000
21	Garrett	IM831-800	0	548	10.6:1	210	ISO	41730
22	Gen Electric	G3142	1 & 2	10200				6500
23	Gen Electric	G3142 RECUP	1	9750				6500
24	Gen Electric	MS3002J	3	10450	7.1:1		ISO	6500
25	Gen Electric	MS5002A	3	26280	6.9:1		ISO	4670
26	Gen Electric	MS5001P	3	25890	10.2:1		ISO	5105
27	Gen Electric	MS6001B	3	38100	11.7:1		ISO	5105
28	Gen Electric	MS7001E	3	78700	11.8:1		ISO	3600
29	Gen Electric	MS9001E	3	113150	11.7:1		ISO	3000
30	Gen Electric	LM 500	3	4064	14.5:1		ISO	16000
31	Gen Electric	LM 2500 - 20	2	12860	7.0:1	218		
32	Gen Electric	LM 2500 - GE	3	25130	19.0:1		ISO	9150
33	Gen Electric	LM 5000	3	38858	29.3:1		ISO	3800

Table 15A-1. Gas Turbine Data by Manufacturer

Heat Rate (Btu/kWh)	Exhaust Heat (Btu/kWh)	Exhaust Temp (F)	Exhaust Flow (lb/sec)	Fuel	Comments	Row Number
13280		989	33	Gas	Gen Eff not considered	1
12550		991	35	Gas	Gen Eff not considered	2
12810		1049	42	Gas	Gen Eff not considered	3
11190		999	44	Gas	Gen Eff not considered	4
4560		284	689	Dual		5
15890		977	11	Gas		6
13720		766	168	Gas		7
12880		875	169	Gas		8
10680		887	198	Gas		9
10430		887	198	Gas		10
10890		888	144	Gas		11
12550		895	44	Gas		12
10480		965	149	Gas		13
13180		845	180	Gas		14
12930		861	170	Gas		15
10470		870	199	Gas		16
15680		800	134	Dual		17
14120		768	262	Gas		18
11310		936	353	Gas		19
11960		995	866	Gas		20
17100		1060	8	Gas	Gen Eff not considered	21
13540	8020					22
10520	5040					23
14800		979	115	Gas	Gen Eff Included	24
10830		975	215	Gas	Gen Eff Included	25
13450		916	267	Gas	Gen Eff Included	26
11960		1006	301	Gas	Gen Eff Included	27
11760		995	609	Gas	Gen Eff Included	28
11830		982	876	Gas	Gen Eff Included	29
12010		955	35	Gas		30
11580		788	125	Gas		31
9030		1461	149	Gas		32
8710		1215	276	Gas		33

Table 15A-1. Gas Turbine Data by Manufacturer
(Continued)

Row Number	Manufacturer	Model No.	Data Source	Capacity (kW)	Compression Ratio	Req'd Gas Pressure (psig)	Rating Cond	Engine Speed (RPM)
34	Hispano-Sulza	THM 1304	3	9750	10.0:1		ISO	12000
35	Ingersol Rand	GT-22B	3	4250			ISO	13820
36	Ishikawajima	IM2000	3	12700			ISO	3700
37	Ishikawajima	IM2500	3	21100			ISO	3700
38	Ishikawajima	IM5000	3	33200			ISO	3500
39	Kongsberg	KG 3	3	1540	9.0:1		ISO	35000
40	Kongsberg	KG 5	3	3110	6.5:1		ISO	17400
41	Mitsubishi	MW-101	3	9225	8.0:1		ISO	6000
42	Mitsubishi	MW-191	3	18555	7.0:1		ISO	4912
43	Mitsubishi	MW-151	3	22190	11.0:1		ISO	6548
44	Mitsubishi	MW-251	3	43410	14.0:1		ISO	5460
45	Mitsubishi	MW-501	3	107840	14.0:1		ISO	3600
46	Mitsubishi	MW-701	3	125755	14.0:1		ISO	3000
47	Mitsui Eng	S860	3	13070	12.4:1		ISO	6780
48	Mitsui Eng	S890	3	16810	6.9:1		ISO	5475
49	Norwalk	TG-7	2	480	7.0:1	135	ISO	
50	Onan	560GTU	2	510	11.0:1	200	ISO	
51	Rolls Royce	RB211-24B	1	21587				5000
52	Rolls Royce	RB211-24A	1	20920				5000
53	Rolls Royce	RB211-22	1	19101				4760
54	Rolls Royce	AVON121G	1	14718				4760
55	Rolls Royce	AVON101G	1	13002				5200
56	Rolls Royce	AVON76G	1	11802				5000
57	Rolls Royce	SPEY	1	11658				5900
58	Ruston	TA2500	0 & 3	1865	5.1:1		ISO	11800
59	Ruston	TB5000	0 & 3	3655	6.8:1		ISO	10400
60	Ruston	TORNADO	0 & 3	6230	12.0:1	155	ISO	11085
61	Solar	Saturn	0	800	6.2:1	160	ISO	22124
62	Solar	Centaur	0	3027	9.0:1	200	ISO	14950
63	Solar	Centaur H	0	3599	10.2:1	225	ISO	14750
64	Solar	Mars	0	8562	16.0:1	490	ISO	10780
65	Turbomeca	Astazou IV	3	330	5.6:1		ISO	43500
66	Turbomeca	Turmo III	3	750	4.9:1		ISO	31100
67	Turbomeca	Turmo XII	3	1000	8.2:1		ISO	31240
68	Unit Tech	FT4C-3P	2	27300			ISO	
69	Westinghouse	W101PG	1	8080				6000
70	Westinghouse	W191PG	3	17700	7.5:1		ISO	4912
71	Westinghouse	W251PG	3	41420	14.0:1		ISO	5423
72	Westinghouse	W5010	3	107850	14.1:1		ISO	3600

Table 15A-1. Gas Turbine Data by Manufacturer
(Continued)

Heat Rate (Btu/kWh)	Exhaust Heat (Btu/kWh)	Exhaust Temp (F)	Exhaust Flow (lb/sec)	Fuel	Comments	Row Number
13070		959	101	Dual		34
10480		994	34	Gas		35
11240		773	126	Gas		36
10520		958	147	Gas		37
10080		788	2784	Gas		38
14620		1034	15	Dual		39
17880		898	48	Dual		40
15770		772	151	Gas		41
14390		748	279	Gas		42
12980		959	212	Gas		43
11540		936	360	Gas		44
11130		966	830	Gas		45
11340		950	1017	Gas		46
12360		871	129	Gas		47
14260		957	189	Gas		48
18400		1050	6	Gas		49
17100		923	8	Gas		50
10160	4100					51
10220	4100					52
10310	4010					53
12250	5490					54
12910	5650					55
12720	5330					56
10560	4210					57
17870		940	29	Gas		58
14860		918	46	Gas		59
12120		877	60	Dual		60
17900		813	14	Gas		61
14720		840	39	Gas		62
14540		941	40	Gas		63
12390		870	84	Gas		64
19090		914	6	Dual		65
19750		950	12	Dual		66
16350		842	16	Dual		67
12650		918	314	Gas		68
15340	9550					69
14870		780	270	Gas		70
12030		940	351	Gas		71
11130		966	813	Gas		72

Table 15A-2. Gas Turbine Data by Size

Row Number	Manufacturer	Model No.	Data Source	Capacity (kW)	Compression Ratio	Req'd Gas Pressure (psig)	Rating Cond	Engine Speed (RPM)
1	Turbomeca	Astazou IV	3	330	5.6:1		ISO	43500
2	Norwalk	TG-7	2	480	7.0:1	135	ISO	
3	Onan	560GTU	2	510	11.0:1	200	ISO	
4	Garrett	IM831-800	0	548	10.6:1	210	ISO	41730
5	Centrix Gas Tur	CS 600-2	3	586	5.25:1		ISO	22000
6	Turbomeca	Turmo III	3	750	4.9:1		ISO	31100
7	Solar	Saturn	0	800	6.2:1	160	ISO	22124
8	Turbomeca	Turmo XII	3	1000	8.2:1		ISO	31240
9	Kongsberg	KG 3	3	1540	9.0:1		ISO	35000
10	Ruston	TA2500	0 & 3	1865	5.1:1		ISO	11800
11	Solar	Centaur	0	3027	9.0:1	200	ISO	14950
12	Kongsberg	KG 5	3	3110	6.5:1		ISO	17400
13	Allison	501 KB	0	3316	9.3:1	250	ISO	14200
14	Solar	Centaur H	0	3599	10.2:1	225	ISO	14750
15	Ruston	TB5000	0 & 3	3655	6.8:1		ISO	10400
16	Allison	501 KB5	0	3924	9.3:1	250	ISO	14250
17	Gen Electric	LM 500	3	4064	14.5:1		ISO	16000
18	Ingersol Rand	GT-228	3	4250			ISO	13820
19	Dresser Clark	DC 990	3	4325	12.0:1		ISO	17800
20	Allison	570 KA	0	4877	12.7:1	275	ISO	9000
21	Allison	571 KA	0	5910	12.7:1	275	ISO	13000
22	Ruston	TORNADO	0 & 3	6230	12.0:1	155	ISO	11085
23	Westinghouse	W101PG	1	8080				6000
24	Solar	Mars	0	8562	16.0:1	490	ISO	10780
25	Fiat TTG	TG 7	3	8630	6.0:1		ISO	6000
26	Mitsubishi	MW-101	3	9225	8.0:1		ISO	6000
27	Gen Electric	G3142 RECUP	1	9750				6500
28	Hispano-Sulza	THM 1304	3	9750	10.0:1		ISO	12000
29	Gen Electric	G3142	1 & 2	10200				6500
30	Gen Electric	MS3002J	3	10450	7.1:1		ISO	6500
31	Rolls Royce	SPEY	1	11658				5900
32	Rolls Royce	AVON76G	1	11802				5000
33	Coberra	2348	3	12058			ISO	5200
34	Ishikawajima	IM2000	3	12700			ISO	3700
35	Gen Electric	LM 2500 - 20	2	12860	7.0:1	218		
36	Rolls Royce	AVON101G	1	13002				5200
37	Mitsui Eng	SB60	3	13070	12.4:1		ISO	6780
38	Rolls Royce	AVON121G	1	14718				4760
39	Dresser Clark	DJ-200R	3	15428			ISO	5400
40	Coberra	2556	3	15554			ISO	4950
41	Mitsui Eng	SB90	3	16810	6.9:1		ISO	5475
42	Westinghouse	W191PG	3	17700	7.5:1		ISO	4912
43	Mitsubishi	MW-191	3	18555	7.0:1		ISO	4912
44	Fiat TTG	TG 16	3	18850	7.0:1		ISO	4850
45	Rolls Royce	RB211-22	1	19101				4760

Table 15A-2. Gas Turbine Data by Size

Heat Rate (Btu/kWh)	Exhaust Heat (Btu/kWh)	Exhaust Temp (F)	Exhaust Flow (lb/sec)	Fuel	Comments	Row Number
19090		914	6	Dual		1
18400		1050	6	Gas		2
17100		923	8	Gas		3
17100		1060	8	Gas	Gen Eff not considered	4
15890		977	11	Gas		5
19750		950	12	Dual		6
17900		813	14	Gas		7
16350		842	16	Dual		8
14620		1034	15	Dual		9
17870		940	29	Gas		10
14720		840	39	Gas		11
17880		898	48	Dual		12
13280		989	33	Gas	Gen Eff not considered	13
14540		941	40	Gas		14
14860		918	46	Gas		15
12550		991	35	Gas	Gen Eff not considered	16
12010		955	35	Gas		17
10480		994	34	Gas		18
12550		895	44	Gas		19
12810		1049	42	Gas	Gen Eff not considered	20
11190		999	44	Gas	Gen Eff not considered	21
12120		877	60	Dual		22
15340	9550					23
12390		870	84	Gas		24
15680		800	134	Dual		25
15770		772	151	Gas		26
10520	5040					27
13070		959	101	Dual		28
13540	8020					29
14800		979	115	Gas	Gen Eff Included	30
10560	4210					31
12720	5330					32
13720		766	168	Gas		33
11240		773	126	Gas		34
11580		788	125	Gas		35
12910	5650					36
12360		871	129	Gas		37
12250	5490					38
12930		861	170	Gas		39
12880		875	169	Gas		40
14260		957	189	Gas		41
14870		780	270	Gas		42
14390		748	279	Gas		43
14120		768	262	Gas		44
10310	4010					45

Table 15A-2. Gas Turbine Data by Size
(Continued)

Row Number	Manufacturer	Model No.	Data Source	Capacity (kW)	Compression Ratio	Req'd Gas Pressure (psig)	Rating Cond	Engine Speed (RPM)
46	Curtis Wrt	P0020	2	20008	18.0:1	400	ISO	
47	Dresser Clark	DJ-125R	3	20580			ISO	5330
48	Rolls Royce	RB211-24A	1	20920				5000
49	Ishikawajima	IM2500	3	21100			ISO	3700
50	Rolls Royce	RB211-248	1	21587				5000
51	Dresser Clark	DJ-2706	3	21783			ISO	5400
52	Mitsubishi	MW-151	3	22190	11.0:1		ISO	6548
53	Coberra	6456	3	24797			ISO	4950
54	Gen Electric	LM 2500 - GE	3	25130	19.0:1		ISO	9150
55	Coberra	6462	3	25364			ISO	4800
56	Dresser Clark	DJ-2906	3	25364			ISO	5000
57	Gen Electric	MS5001P	3	25890	10.2:1		ISO	5105
58	Gen Electric	MS5002A	3	26280	6.9:1		ISO	4670
59	Unit Tech	FT4C-3P	2	27300			ISO	
60	Ishikawajima	IM5000	3	33200			ISO	3500
61	Gen Electric	MS6001B	3	38100	11.7:1		ISO	5105
62	Gen Electric	LM 5000	3	38850	29.3:1		ISO	3800
63	Westinghouse	W251PG	3	41420	14.0:1		ISO	5423
64	Mitsubishi	MW-251	3	43410	14.0:1		ISO	5460
65	Fiat TTG	TG 20	3	44290	11.0:1		ISO	4920
66	Gen Electric	MS7001E	3	78700	11.8:1		ISO	3600
67	Fiat TTG	TG 50	3	105100	12.0:1		ISO	3000
68	Mitsubishi	MW-501	3	107840	14.0:1		ISO	3600
69	Westinghouse	W501D	3	107850	14.1:1		ISO	3600
70	Gen Electric	MS9001E	3	113150	11.7:1		ISO	3000
71	Mitsubishi	MW-701	3	125755	14.0:1		ISO	3000
72	BBC	L-GT 8/11	3	220000	60.0:1		ISO	3000

Table 15A-2. Gas Turbine Data by Size
(Continued)

Heat Rate (Btu/kWh)	Exhaust Heat (Btu/kWh)	Exhaust Temp (F)	Exhaust Flow (lb/sec)	Fuel	Comments	Row Number
10890		888	144	Gas		46
13180		845	180	Gas		47
10220	4100					48
10520		958	147	Gas		49
10160	4100					50
10480		965	149	Gas		51
12980		959	212	Gas		52
10680		887	198	Gas		53
9030		1461	149	Gas		54
10430		887	198	Gas		55
10470		870	199	Gas		56
13450		916	267	Gas	Gen Eff Included	57
10830		975	215	Gas	Gen Eff Included	58
12650		918	314	Gas		59
10080		788	2784	Gas		60
11960		1006	301	Gas	Gen Eff Included	61
8710		1215	276	Gas		62
12030		940	351	Gas		63
11540		936	360	Gas		64
11310		936	353	Gas		65
11760		995	609	Gas	Gen Eff Included	66
11960		995	866	Gas		67
11130		966	830	Gas		68
11130		966	813	Gas		69
11830		982	876	Gas	Gen Eff Included	70
11340		950	1017	Gas		71
4560		284	689	Dual		72

Chapter 16

Steam Turbines

Steam turbine systems have achieved market dominance in larger on-site applications. While the steam turbine is generally thought of in the context of large, central station, electric utility power plants, they have, nevertheless, been very successfully used in on-site applications. This is particularly so in the industrial sector and for systems of twenty megawatts or more. This chapter provides a short summary of steam turbine technology.

DESCRIPTION

Boiler/steam turbine systems. operating through a Rankine Cycle are commonly found in larger, electric utility power plants. As shown in Figure 16-1 this system usually consists of a boiler, where fuel is used to heat a working fluid, usually water, producing high temperature and high pressure superheated steam.

As in the gas turbine, the working fluid is expanded through the power turbine producing mechanical power; however, unlike the gas turbine, in a Rankine Cycle, the steam that is exhausted from the turbine is usually condensed and recycled through feedwater pump to the boiler. By condensing the exhaust steam and increasing the

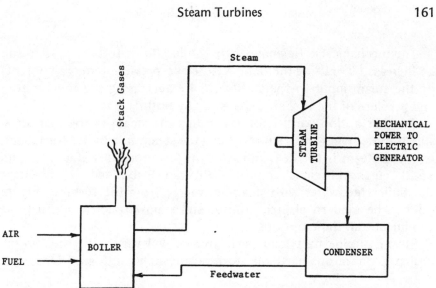

Figure 16-1. Boiler Steam Turbine System

difference between the steam turbine's inlet and outlet temperatures, it is possible to increase the steam cycle efficiency over that of a noncondensing turbine.

The system of Figure 16-1 is classified as a condensing turbine system, wherein the steam is exhausted at pressures below atmospheric. Other types of turbines include:

- *Back Pressure Turbines:* The steam is only partially expanded and exhausted from the turbine at pressures above atmospheric.

- *Extraction Turbines:* The steam is extracted from the turbine at some point prior to the turbine exhaust.

- *Induction Turbines:* The turbine has inlets for steam at two different pressures.

Various combinations of the above concepts are also possible. Because of the range of available turbine concepts, when used in a cogeneration system, steam turbines allow a great deal of flexibility in satisfying an end user's steam requirements.

The mechanical efficiency of the boiler/steam turbine system can be increased by raising the temperature and pressure (energy content) of the steam input to the turbine or by decreasing the temperature and pressure of the turbine exhaust or by both.

Another option consists of expanding the steam in the high pressure turbine stages and then returning the steam to a boiler for reheating. The steam is then expanded in the lower pressure stages (Figure 16-2). Further efficiency improvements can be obtained by preheating the boiler feedwater with steam extracted from the turbine (Figure 16-3). The modern electric utility, steam powerplant (Figure 16-4) can include all these options.

Steam turbine part load performance will vary with the turbine design. Data for one particular turbine are shown in Figure 16-5.

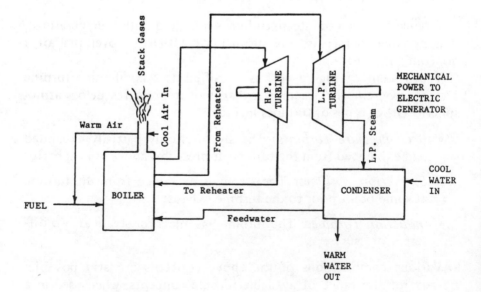

Figure 16-2. Reheat Cycle

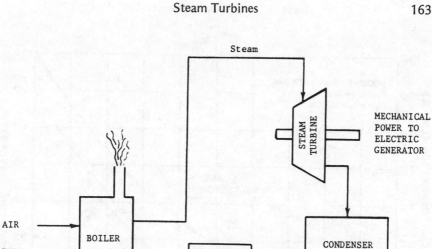

Figure 16-3. Preheat Cycle

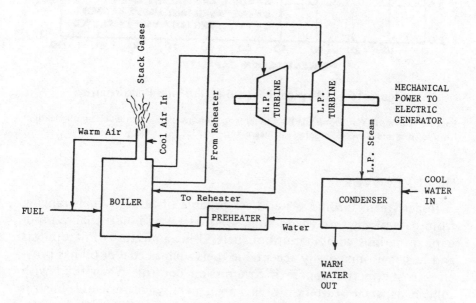

Figure 16-4. Modern Steam Power Plant

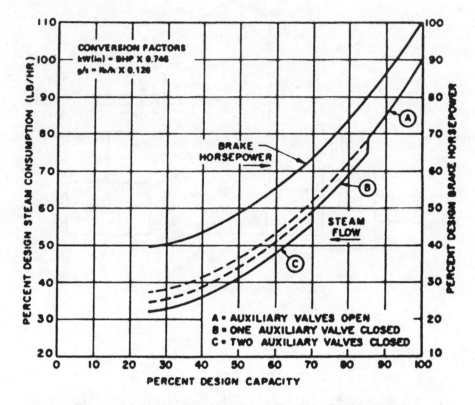

Figure 16-5. Typical Part-Load Turbine Performance

Source: *ASHRAE Handbook*, 1983 Equipment Volume, American Society of Heating, Refrigerating and Air-Conditioning Engineers, Inc., 1983.

HEAT RECOVERY

Boiler/steam turbine systems are capable of burning fuels ranging from natural gas through to coal, solid wastes and industrial byprod-ucts, or various combinations of fuels. Using a mixture of natural gas and less environmentally acceptable fuels such as coal or oil has prov-en a cost-effective approach to emission control. When fired with solid fuels, steam turbine systems are generally cost-effective in sizes of ten megawatts or more, although units as small as ten kilowatts are available.

A noncondensing turbine can readily provide low pressure steam (atmospheric pressure) for process requirements. Where higher pressures are required, the turbine can be designed to allow for extraction at the required pressure. These turbines are usually referred to as back pressure turbines. The overall effect of higher exhaust steam pressures is a reduction in the turbine's mechanical efficiency.

Noncondensing turbines have frequently been used in cogeneration applications where steam is readily and relatively inexpensively available at pressures of several hundred pounds per square inch gage or more and is only required at low pressures in the range of 15 psig to 50 psig. Because of their relatively modest incremental costs, these systems have proved to be cost effective in very small sizes.

Turbines can be designed with one or more extraction points thus providing steam at one or more different pressures. Steam extraction for process use will decrease the turbine's mechanical efficiency, and the steam that is not required for process is fully expanded in the lower pressure stages of the turbine to maximize electrical efficiency. Extraction turbines can satisfy a broad range of thermal and mechanical energy requirements and are frequently used in cogeneration systems.

ASSESSMENT

Steam turbines have been used in both central plant and decentralized power applications. They have also been used in refrigeration applications where they drive mechanical compressors. Overall, steam turbines can be characterized as follows:

- *Mechanical Efficiency:* In large sizes of 10 MW or more steam turbines achieve efficiencies of 30 percent or more.

- *Recoverable Heat:* Steam turbines provide the capability to extract steam at almost any required temperature or pressure. Heat extraction is accompanied by a decrease in mechanical efficiency and power output.

- *Overall Energy Efficiency:* Steam turbine cogeneration systems are capable of achieving overall system efficiencies of up to 75 percent.

- *Thermal Quality:* Steam turbines provide the greatest capability for matching the thermal output with the end user needs. Extraction and noncondensing turbines allow recovery of almost any desired steam pressure, although this heat recovery is accompanied by a loss in mechanical efficiency and output.

  Heat availability from a modern condensing turbine with below atmospheric exhaust pressures is limited to a maximum of 180°F to 190°F.

- *Scheduled Availability:* Steam turbines are capable of availabilities similar to the gas turbine. In addition the turbine's maintenance requirements are not affected by the fuel being used in the boiler although other components of the system may be. Major overhauls do, however, require more time than with the gas turbine.

- *Equipment Life:* With proper maintenance, steam turbines are capable of economic lives of 100,000 hours or more.

RESEARCH TRENDS

One area of commercial development is the use of the Kalina Cycle wherein two working fluids such as water and ammonia are used, thus increasing overall efficiency. A second area of research is metallurgy and water quality as a means of minimizing corrosion and extending turbine blade life.

Increased activity is being directed at minimization of blade corrosion, early detection of fatigued turbine blades and the rebuilding of aged steam turbines.

Chapter 17

Packaged Cogeneration Systems

Cogeneration was developed approximately a century ago and has been successfully applied in industrial plants and in larger commercial, institutional and residential buildings and building complexes. Prior to the 1980s, these systems were individually designed by engineers who were responsible for both the cogeneration system and its interface with the end users. Each of the various components such as the engine-generator, the heat recovery boilers and heat exchangers, radiators, controls and meters were individually specified and the system was "built-up" at the site.

During the 1970s and into the early 1980s an alternative approach to cogeneration was developed. Under this concept, complete cogeneration systems are designed, components procured and the system fabricated at a factory. The packaged cogeneration system (PCS) or module is then delivered to the site as an integrated unit. The site engineer is responsible for the design of the interface between the cogeneration system and the energy end user. This chapter examines PCS or as they are also called "factory-assembled" systems.

THE MODULE CONCEPT

The origins of the PCS concept can be traced to the early 1970s and efforts to develop more cost-effective utility options for residential and commercial buildings. Those early efforts did not meet with significant success, and it was only after the passage of PURPA with its incentives, that entrepreneurs sought to satisfy the perceived residential and small commercial demand for cogeneration systems.

One immediate problem was that while there were numerous opportunities to apply cogeneration in smaller facilities the required investments were often prohibitively high. With total installed costs approaching $2,000 per kilowatt for systems of less than a few hundred kilowatts there was little economic market for cogeneration systems. The fixed engineering, legal, financing and installation costs were not justified by the potential reduction in operating costs. The lack of economic viability was true despite the fact that these smaller energy users usually paid more per unit of energy used than did residential or larger industrial customers.

One response to this problem was the PCS, a small cogeneration system consisting of a single engine-generator with all ancillary equipment, assembled and shipped from the factory as a unit. The underlying concept was to achieve significant cost reductions by spreading the fixed costs of engineering the basic cogeneration system over a large number of units and achieving economies through standardized parts and assembly line production. Larger PCS installations could be designed with multiple modules. The developmental goal was to bring the PCS installed cost down to the point where it was competitive with larger, engineered, built-up systems.

In general, this goal was achieved through a combination of cost reduction techniques that included:

- Incorporating state-of-the-art technological improvements, particularly with regard to switchgear and on-site microprocessor-based controls
- Limiting the number of available options
- Providing a specialized resource pool that could address more complex problems such as financing or the connection of the cogeneration system with the electric utility grid
- Utilizing simpler equipment such as induction generators
- Eliminating the need and the cost for a power or HVAC engineer to design the entire cogeneration system

The use of modules provides several advantages to any potential cogenerator. First and foremost is the capital cost advantage. Second is the ease of installation and the minimum amount of expensive field work required to complete installation. The short installation schedule minimizes first cost and reduces the opportunity for field problems. In addition, complete modules can be checked out in the factory where problems are more readily identified and corrected.

A third advantage is the ability to select from modules that have electric interconnects which have been approved by the electric utility, minimizing the potential cogenerator's site specific efforts required to intertie with the grid.

A fourth advantage is that the potential cogenerator has a single source which is responsible for the cogeneration system's performance. Incidental to this single source responsibility is the ability for a potential cogenerator to check out an entire cogeneration system and vendor. In the past, it has only been possible for a potential owner to verify the capabilities of an engineer, a particular piece of equipment or a contractor. With modules, it is now possible to check out the performance of a PCS by consultation with other cogenerators who have purchased and installed the specific module being considered.

A second order effect of PCS development was to increase the size of the market by eliminating the need for local engineers to design the entire cogeneration system. By purchasing a complete system, the engineer or architect did not have to have in-house cogeneration design expertise or have to incur the cost of a cogeneration engineering subcontractor.

COMPONENTS

The basic PCS concept is to employ assembly line procedures to produce a standard product as illustrated by Figure 17-1. In practice there is considerable variation in that product. Not all modules include the same components and for those components such as engines, generators and controls, which are common to all modules, not all systems include the same type of equipment. The full range of equipment options and the various alternatives are identified below.

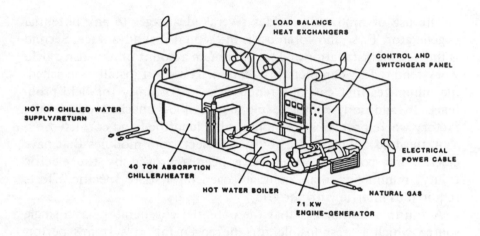

Figure 17-1. Factory-Assembled Cogeneration Module

The basic prime mover can be either a reciprocating engine or a gas turbine. In general, most of the prime movers offered in PCS have been previously used in commercial cogeneration and stationary power applications, although some PCS prime movers have not.

The basic distinction among generators is whether the module uses a synchronous or an induction generator. Induction generators provide a first cost advantage, however they are generally not capable of operating in isolation from an active source of power such as the electric utility grid. Unless they are otherwise equipped for self-excitation, they cannot be used to provide emergency or standby power. Synchronous generators impose additional control requirements on the PCS, however they are generally more efficient than induction generators and they do provide the capability to operate the cogeneration system during electric utility grid outages.

Reciprocating engine cogeneration systems may utilize heat recovered from the lube oil cooler, the jacket, the exhaust and, if turbocharged, from the intercooler. Some PCS are limited to heat recovery from the jacket water and the exhaust, while other systems capture all available heat.

Recovered heat may be provided in the form of hot water, steam or both. In general, when steam is delivered, it is developed by heat recovered from the high temperature exhaust, although some systems provide 15 psig steam through ebullient cooling of the engine block.

Most systems have a built-in capability for heat rejection either as a standard component of the PCS or as an option. A built-in heat rejection radiator allows the module to continue operating during periods of low site thermal requirements, or as a backup generator when there are no thermal requirements. This capability can be quite useful in minimizing utility demand charges. During periods of low thermal load, without a heat rejection radiator, delivery of the heat into storage or engine shutdown is required.

Some PCS are available with absorption chillers as a factory installed option. This capability allows full use of recoverable thermal energy during summer months and can decrease the total system cost through factory installation of the chiller.

Most PCS provide the capability for electric and/or thermal load following. If the module does not have a load tracking capability, then during periods of low electric load, electricity must be sold to the utility or to another user or, the module must be shut down.

Microprocessors can serve the dual function of controlling the prime mover output, deciding whether to use purchased power or cogenerated power and also to provide protection from equipment failures. Microprocessors can also monitor and track engine performance and shut the prime mover down when specified thresholds are exceeded.

RESEARCH TRENDS

The PCS industry appears to be well-suited to take advantage of several technology improvements including advances in prime movers, controls, heat recovery and absorption chillers. Recent advances in the use of microprocessor-based controls and utility interconnections have the potential for significant impact on both capital and operating costs, as do foreseeable reciprocating engine advances.

Turbochargers are another area of development. Although the technology was commercially available in the early 1960s, it was only recently that turbocharging has become commonly applied. Improved turbo efficiency, reliability and increased intercooler performance will contribute both to increased power densities and to higher efficiencies.

To date most small cogeneration applications have been based on reciprocating engines. The limited availability of small turbines and their mechanical efficiency are the two primary reasons for this trend. Small turbine cogeneration applications have been limited to those facilities where the end user can utilize all the recoverable heat that is available, and requires that heat as high temperature hot air or as steam at pressures in excess of 15 psig. There is increased research into the development of more efficient, small gas turbines with Allison, Garrett, Kawasaki and Solar all participating.

Current developmental work indicates that a larger number of gas turbines will be available, and that the fuel to power conversion efficiency will be significantly higher than is currently available. Some manufacturers are projecting mechanical efficiencies of over 40 percent.

Research on the use of microprocessor-based controls is more directed at the application of a commercially available product than at the development of the technooogy itself. That is, the hardware for microprocessor-based automated control systems is available for numerous applications covering a broad range of activities.

The single predominant equipment trend appears to be the development of expanded product lines by the module vendors. As the market is developing, the individual vendors are offering additional module sizes and increasing the number of options.

Chapter 18

Prime Mover Applications

Gas-fueled prime movers have proven successful in three types of on-site applications; electrical power generation; refrigeration systems and direct drive systems such as for fans, air compressors, pumps, etc. In this chapter each of these application areas is reviewed.

ELECTRIC POWER GENERATION

The use of engine generator sets to produce power on-site, either with or without heat recovery, is a well-established concept. Three power generation concepts are discussed below: cogeneration, stand-by and peak shaving.

Cogeneration Systems

Cogeneration is simply power generation with heat recovery. The removal of many institutional barriers to cogeneration with the passage of the Public Utilities Regulatory Act (PURPA) of 1978 brought about a dramatic increase in the number of such applications. Today the amount of cogenerated electricity is steadily increasing (Figure 18-1).

As discussed in previous chapters, the application of heat recovery equipment on gas engine generator sets allows the use of energy that might otherwise be rejected to the environment. These prime movers

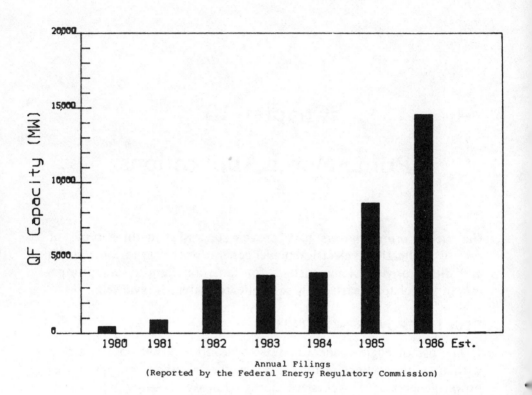

Figure 18-1. Cogeneration Capacity

are capable of providing hot water over a wide range of temperatures, or steam, or both, as may be required by an end user.

Cogeneration systems allow the use of this recovered heat to displace fuel that would otherwise be used in a boiler, and they are well-suited for commercial, institutional and industrial facilities with steady, nonseasonal thermal requirements. Cogeneration has been successfully applied in many different types of buildings including:

- Hospitals
- Nursing homes
- Hotels and motels

- Schools, colleges and universities
- Multifamily dwellings
- Shopping centers
- Health clubs
- Municipal buildings
- Public safety buildings
- Computer centers
- Refrigerated warehouses
- Fast food restaurants
- Office buildings

In general, most buildings or building complexes with central heating/-cooling plants may be adaptable to cogeneration.

In addition, industrial facilities that are characterized as having long hours of operation or occupancy (4,000 or more hours per year); high thermal load factors; or constant process requirements are potential cogeneration candidates. Finally, facilities with highly sensitive or critical process requirements may also be well-suited for cogeneration.

Industrial applications of cogeneration include:

- Chemical plants
- Food processing plants
- Paper mills
- Breweries
- Laundries
- Plastic extruders
- Textile mills
- Foundries
- Metal finishers
- Dairies
- Wastewater treatment plants

- Oil refineries
- Meat packing plants
- Ceramic and brick manufacturers
- Bottling plants
- Automotive assembly plants

As indicated above PURPA provided the impetus that resulted in a significant growth in cogeneration applications. One major reason for this is that PURPA, through rules developed by the Federal Energy Regulatory Commission (FERC), provided cogenerators with access to the electric utility grid. This ability to interconnect with the grid and to purchase standby and supplemental power at nondiscriminatory rates or to sell power to the electric utility at the utility's avoided cost allowed those cogeneration systems that met FERC efficiency requirements to be designed so as to provide the maximum economic return to the cogenerator. Current practice is to generally design cogeneration systems with high capacity factors.

Standby Power Systems

Cogeneration is but one form of on-site power production. Many commercial and institutional facilities install engine generator sets either as a backup for critical or emergency electrical loads or as a higher quality alternative to power purchased from the utility grid. This latter requirement is frequently encountered in computer centers, where power conditioning equipment is required to eliminate voltage variations, transients and harmonics that may be found in grid supplied power.

Emergency generators required to protect sensitive loads provide another opportunity for on-site operation. In this case the engine generator may be coupled both to the critical load and to a less sensitive, continuous load. During normal operation the engine generator's output is used to drive a chiller, compressor, or base electrical requirements. However, during an emergency period, this base load is dropped and the generator output is transferred to the critical load. This concept, known as Continuous Duty/Standby (CD/S) operation, was applied during the late 1960s, and allows the use of the engine generator set both to provide needed standby and to decrease purchased power costs. A CD/S system can also be used for peak shaving.

Peak Shaving Systems

Another potential application of engine generators is as a means for electric peak shaving. With many utilities moving toward highly differentiated time of day rates or to rate structures that are characterized by high demand charges (demand charges in excess of $20 per kilowatt are becoming more common), end users can significantly reduce the cost of purchased power by reducing their peak demands.

In most cases, demand reduction can be readily accomplished by shedding or cycling noncritical loads, by reducing operations, or by installing less energy intensive equipment. Engine-generator sets may be used to reduce the peak demand imposed on the utility grid without reducing process or occupant energy requirements.

One final note about on-site power applications: when the cost of the engine-generator set may be justified based on the need for a more reliable power source, higher quality power, or to protect critical loads, the potential for heat recovery and cogeneration should be tested. In this case, for the incremental cost of the heat recovery system and more sophisticated controls and switchgear, the on-site powerplant may be converted into a significant cost-cutting measure. In cases where the emergency generator is required by a local code, the code should be checked to determine whether the engine can be operated continuously while still providing the required standby.

REFRIGERATION

Engines have been successfully applied in refrigeration systems. In a mechanical compression system, an engine is used instead of a motor in order to drive the compressor. Reciprocating engines and turbines have been used in both centrifugal and reciprocating compressors.

Heat recovered from an engine or turbine can be used as the energy source for an absorption chiller. In this case, the prime mover may provide mechanical drive for a mechanical chiller while the heat is used in an absorption chiller.

Development of double effect chillers with a Coefficient of Performance of approximately 1.0 and combined Chiller/Heaters capable of being fired directly from a turbine exhaust, have made absorption chilling more cost effective.

In some cases, the refrigeration plant may consist of a mix of mechanical compressors, either electric or engine driven, and absorption chillers. This combined plant allows maximum flexibility and provides a balanced mechanical and thermal load.

The subject of refrigeration is covered in detail in the American Gas Association *Natural Gas Cooling Manual.*

DIRECT DRIVE

Prime movers may be used for direct drive of air compressors, pumps, fans, and for many applications that would otherwise require a large horsepower motor. These systems enjoy a cost advantage over electric power applications in that they do not require a generator or the expensive switchgear, controls, and electric utility interconnection that may be required for an engine-generator set.

Direct drive devices can be backed up with motor drives and utility supplied power thus assuring higher reliability than is possible with either an on-site engine or utility power alone.

The major disadvantage of such systems is their lack of flexibility as compared to an engine-generator set.

Section IV

Natural Gas for
Efficient Electric Generation

Chapter 19

Natural Gas Use for Efficient Electric Generation: Introduction

Natural gas cogeneration is a highly-efficient way of generating electricity, and much of the technology can be used directly by utilities. Conversely, technologies being developed by the electric utility industry can be valuable to cogeneration, particularly in the larger and more efficient system ranges.

Utilities, although a major consumer of natural gas, have not (until recently) valued the premium characteristics of gas, including its environmental, operational, and efficiency attributes. But this is changing, and there is also increased acceptance by electric utilities of cogenerated electric power production where economically attractive.

This section addresses the use of natural gas by utilities as a primary fuel to generate electricity. Many of the advantages natural gas has for utilities are equally valuable to cogenerators.

In 1985, 3 Tcf of natural gas was consumed by electric utilities, representing 19 percent of all gas delivered to consumers. The bulk of this gas was consumed as a boiler fuel in both baseload and intermediate load facilities, with the remainder consumed by relatively low efficiency peaking turbine units. In the traditional electric utility market, gas competes head-to-head with residual fuel oil, often in dual-fuel capable boilers. In addition, given that "least cost dispatch" is the prevalent operating rule for electric utilities, natural gas plants also compete against coal generated electricity.

Over the last fifteen years, the electric utility industry has added new capacity through the construction of large nuclear and coal-fired units. The nuclear option is no longer available and the coal option is expensive and environmentally difficult. While its ability to construct these types of new plants is thus being foreclosed, the electric industry is faced with the prospect of tight capacity margins in the early 1990s—with some regions likely to experience capacity shortfalls even with aggressive construction and cogeneration programs.

Technological advances and environmental sensitivities have resulted in the emergence of expanded use of gas by the utility sector. That is, gas is no longer merely a source of "Btu's" indistinguishable from coal or oil. Rather, gas offers additional value to utilities in the forms of greater efficiency (less fuel purchased), environmental enhancement (reducing the need to purchase pollution control equipment), and operational superiority (increasing efficiency and reducing operating costs). For example, the development and commercialization of combined-cycle gas turbine systems has improved the thermal efficiency of these gas plants to around 45 percent. Combined-cycle plants are often No. 2 oil capable, but are not usually No. 6 oil capable (No. 6 fuel oil would also have negative operational characteristics). In addition, the higher thermal efficiency, coupled with short lead times and low capital costs, makes these plants very competitive with coal plants (efficiencies around 32 percent) from a least cost electricity dispatch perspective.

On the environmental front, the select use of gas with coal (or other polluting fuels) offers utilities a more cost-effective approach to reduce emissions than mechanical devices such as scrubbers. This quality of gas is becoming more important as an increasing number of states adopt acid rain legislation, and as the adoption of federal legislation becomes more likely. Further, the co-firing of natural gas with coal, one form of select use, has indirectly led to a greater awareness of another premium characteristic of gas—its operational superiority. Mixing gas with coal not only reduces the pollution attributable to coal-only firing, but it also reduces the operational problems associated with using coal such as slagging.

This section addresses three applications for increased gas use in electric generation: (1) new combined-cycle plants; (2) repowering existing plants into combined-cycle plants; (3) retrofitting existing coal or oil plants to use gas in a "select use" mode.

(1) *New Combined-cycle*—four announced plants, highly efficient use of gas, facilities not easily resid capable. Problem: need of FUA exemption.

(2) *Combined-cycle Repowering*—upgrading of existing units to more efficient combined-cycle type operation. Quick and inexpensive, FUA exemption may not be needed. Actively being marketed by General Electric.

(3) *Select use*—uses environmental and operational attributes of gas to penetrate plants currently served exclusively by coal or oil. Broad applications nationally, two plants recently converted to select use in Ohio and Colorado. Select use applications sensitive to national and local environmental regulations.

In aggregate, these three new electric generation applications represent the largest potential growth market for the gas industry. Combined-cycle and select use alone can add between 600 Bcf and 1.3 Tcf of demand by 1990 and 2 Tcf by the year 2000.

There are compelling reasons why this is important to all segments of the natural gas industry. This new gas demand is ideal from a load balancing perspective in that it is either year-round or summer peaking. This new gas demand is urgently needed to alleviate the serious take-or-pay problem and to provide incentives to develop the nation's largest economically feasible energy resource base, which will, in turn, benefit cogenerators.

There are equally compelling reasons why this new gas market is vital to the electric industry. It offers the best option to meet growing electric demand in a timely and economic manner most acceptable to the environmental and utility regulatory framework. There are two steps that need to be pursued immediately and simultaneously. One is the recognition in all environmental regulations and/or legislation of the vital contribution which natural gas can make. Second, there should be an organized and accelerated marketing effort.

All of the foregoing will be addressed in the following chapters.

Chapter 20

Analysis of New Natural Gas Technologies for Electric Generation

MICHAEL I. GERMAN
American Gas Association

NEW COMBINED-CYCLE POWERPLANTS

Natural gas-fired combined-cycle electric plants are becoming the primary technological choice of electric utilities for new baseload generation. For the last 15 years, nuclear and coal plants were the overwhelming choice of electric companies adding new generation capacity.

The economics and technical characteristics of gas plants compare very favorably with coal plants. On average, the cost of producing electricity by a combined-cycle gas plant will be about 80 percent of the cost of producing electricity by a new coal plant. In addition, a new gas plant will have one-third the capital cost, one-third or less the planning and construction lead time, and between one-half and one-one-thousandth of the environmental impact. The only real advantage that coal currently enjoys is that electric utilities can contract for coal supplies with fixed price escalators that run the life of the electric plant. This single coal advantage—which may be addressable through new gas contracting practices—must be weighed against the economic, technical, efficiency, and environmental advantages of gas-fired combined-cycle plants.

The above cited A.G.A. projections on the economic advantages of gas plants over coal plants are premised on baseload operations—an assumption unfavorable to natural gas. Because new coal plants have capital costs about three times higher than new combined-cycle gas plants, the cost of electricity from an intermediate load (e.g., 30 percent capacity) combined-cycle plant would be only 60 percent of the cost of electricity from an intermediate load coal unit.

The prices paid by electric utilities for coal and natural gas in our economic analysis were based on the A.G.A.-TERA Base Case 1985-I, and they do not reflect the decline in oil and natural gas prices in 1986. Gas prices employed in the analysis increase from $3.94 per MMBtu in 1990 to $9.68 in 2020, while coal prices rise over the same period from $2.08 per MMBtu to $4.50 ($1985). The average price of natural gas to electric utilities is currently $2.17/MMBtu. Thus, the favorable economics of combined-cycle gas plants illustrated above are based on extremely conservative marketplace assumptions.

The gas-powered combined-cycle system has negligible sulfur dioxide and particulate emissions. In contrast, the coal system emits over 4.1 thousand tons of SO_2 per year (with an 85 percent effective scrubber system) and 206 tons of particulate matter per year (with a 99 percent effective electrostatic precipitator). Emissions of NO_x by the combined-cycle system are 57 percent of the coal unit's annual NO_x emissions—1,352 tons per year versus 2,385 tons per year.

In terms of noncombustible solid wastes, the coal-based system would generate 168,000 tons of sludge per year and 650,000 tons of ash. Natural gas combustion results in no sludge or ash production. In addition, the combined-cycle option would require only half the water of the coal system—15 million gallons per year versus 30 million gallons.

Gas-fired combined-cycle power generation offers a clean, economical, efficient, and reliable source of electricity. Combined-cycle units can be constructed quickly, and additional units may be added as needed over time in a modular fashion. These features of rapid and modular construction protect the electric utility and the electric ratepayer from the economic burden of over-building, as well as the risk of inadequate capacity.

NATURAL GAS REPOWERING

A significant portion of U.S. electric utility generating capacity is approaching the end of its design life, roughly 30 to 35 years. By 1990, approximately 25 percent of all fossil fuel generating capacity will be at least 30 years old. As generating equipment ages, it presents plant operators with a variety of problems, including reduced efficiency and reliability. The efficiency of a 30-year-old unit is about 6 to 8 percent less than when the unit was placed in service—i.e., it requires 6 to 8 percent more fuel (costing millions of dollars) for the same output of electricity. There is a similar degradation in unit reliability over time as indicated by the forced outage rate. A 10-year-old coal-fired plant has a forced outage rate of about 5 percent, but this increases to nearly 15 percent by age 35.

The primary attractions of combined-cycle powerplants, as previously noted, are their low capital and operating costs, short construction lead times, clean operation and modular design. Combined-cycle *repowering* refers to the integration of new and used equipment at an existing site, with the final equipment configuration resembling a new gas-fired combined-cycle unit. The type of repowering employed will vary from site to site. For example: (1) *peaking turbine repowering* refers to the addition of a steam turbine and heat recovery unit to an existing gas turbine, with the efficiency improvement allowing the unit to convert from peaking to baseload operation; (2) *heat recovery repowering* is the replacement of an old coal boiler with a gas turbine and heat recovery unit, leaving the existing steam turbine in place; and (3) *boiler repowering,* in which the exhaust from a new gas turbine is fed into an existing coal boiler, replacing existing forced draft fans and air heaters.

The cost of producing electricity in a repowered unit is only about 60 percent of the cost of producing electricity in a new coal-fired powerplant—3.7¢/kWh to 3.9¢/kWh versus 6.2¢/kWh. Although the delivered price of natural gas is 30 to 40 percent greater than the price of coal, this disadvantage is more than offset by the lower capital charges and the lower operation and maintenance expenses applicable to repowering. The capital charge for a repowered unit is very low—0.6¢/kWh to 1.0¢/kWh versus 3.3¢/kWh for a new coal unit—since the repowered unit uses some existing equipment and does not require pollution control equipment. Similarly, the operation and maintenance expense of repowered units is only about one-third that of new coal units which require coal storage and handling, as well as the maintenance of pollution control equipment and sludge and ash disposal. Finally, the fuel cost advantage enjoyed by coal units is lessened because of the 10 to 30 percent higher efficiency of repowered gas units.

SELECT GAS USE

Select gas use refers to the use of gas for environmental compliance. There are many forms which select use may take, from the co-firing of gas and coal in utility boilers, to "bubbling," to the seasonal use of gas for episodic pollution control or the reduction of sulfur and nitrogen oxides (associated with acid deposition). Exhibit 20-1 presents a number of select use examples.

Co-firing gas and coal in the same boiler offers a variety of environmental and operational benefits. Not only does the use of gas reduce emissions of sulfur dioxide, particulate matter, nitrogen oxides, sludge, ash and many other pollutants, but it can also enhance boiler performance and result in lower maintenance costs and fewer plant breakdowns. Co-firing also provides cleaner and quicker start-ups and insurance against disruptions in coal supplies. The advantages of co-firing coal units with gas are being explored by Consolidated Natural Gas at the 570 megawatt Cheswick powerplant north of Pittsburgh. Consolidated hopes to show that by using 5 to 10 percent gas in the previously all-coal boiler, plant performance can be increased and emissions reduced. Preliminary test results have been very encouraging, and gas sales of 800 Bcf in two years are possible.

Exhibit 20-1

Example of Select Gas Use Applications

Company	State	Gas Used in Conjunction With
Atlantic City Electric	NJ	Coal
Bellefield Boilers	PA	Coal
Burlington Electric	VT	Coal
Public Service Electric & Gas	NJ	Coal
Springfield Utilities	MO	Coal
U.S. Steel	PA	Coal
Andre's Greenhouse	PA	Oil
Arbogast-Bastian	PA	Oil
Boston Edison	MA	Oil
Narragansett Electric	RI	Oil
Scott Paper	PA	Oil
J. H. Thompson	PA	Oil
3M	AL	Oil
Uniroyal	CT	Oil
British Petroleum Oil	PA	Refinery Gas
Gulf Oil	PA	Refinery Gas
Sun Oil	PA	Refinery Gas
U.S. Steel	PA	Various Fuels

Sources: U.S. Environmental Protection Agency, Office of Policy, Planning and Evaluation, "Status Report on Emission Trading Activity," November, 1983; and A.G.A. internal research.

The select use of gas at existing facilities offers a low cost alternative to reduce emissions of sulfur dioxide and nitrogen oxides, the primary precursors of "acid rain." While A.G.A. has never taken a stand on the need for acid rain legislation, we continue to persuade policy makers that *if* legislation is passed—and many feel this may happen in the next session of Congress—plant operators should be allowed flexibility to choose their least costly compliance strategy. Such an approach would favor select use.

In analyzing the cost-effectiveness of select use versus other SO_2 reduction strategies, A.G.A. found that substituting gas for high sulfur coal could reduce SO_2 emissions at a cost ranging from less than $100 per ton removed to more than $600 per ton on a regional basis —competitive with other control options and less expensive than scrubbing in most cases. Scrubbing can cost over $3,000 per ton of SO_2 removed (see Exhibit 20-2).

Substituting gas for high sulfur oil would be an even less costly approach (as contrasted with mechanical pollution removal) due to the lower fuel cost differential, if any, between gas and high sulfur oil. Again, the gas price declines of 1986 are not reflected in this analysis. One innovative means of reducing SO_2 emissions would be the substitution of gas for coal or oil on a seasonal rather than a year round basis. Controlling SO_2 emissions at specific periods of the year may be more effective than year-round controls, as indicated by the M.I.T. Energy Laboratory findings at their Brookhaven monitoring station—70 percent of annual acid deposition falls from May through October and 54 percent from June through September (see Exhibit 20-3). This coincides with the fact that pipeline availability is greatest in summer months, with a national capacity utilization rate of approximately 78 percent, and regional rates ranging from 51 percent to 93 percent, concurrent with the peak deposition period (see Exhibit 20-4). (Although there is significant variation from system to system within individual regions.)

CONVERSION FROM COAL AND OIL TO NATURAL GAS

There have been a number of electric powerplant conversions from fuel oil to natural gas. A recent study by M.I.T. for A.G.A. of the operating characteristics of certain of these plants using natural gas shows very minor efficiency losses and no derating. In addition, such conversions to natural gas involve a reduction in parasitic energy losses (e.g., preheating and pumping oil). There have also been a number of conversions from coal to gas—mostly in the 1950s and

Exhibit 20-2

Estimated Cost of Reducing Sulfur Dioxide Emissions by Replacing High Sulfur Coal[1] with Natural Gas in Powerplants—By Region

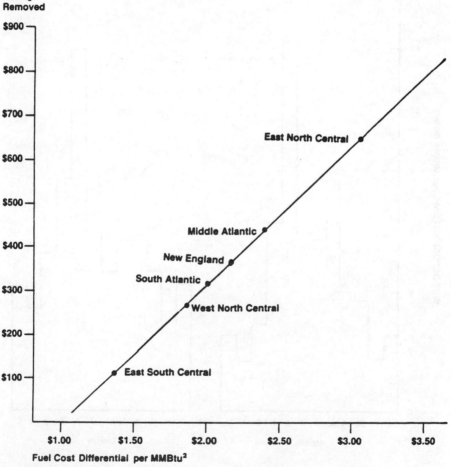

[1] Based on high sulfur (3.5%) coal; chart includes only those regions where high sulfur coal represents a significant portion (10%–30%) of total coal consumption.

[2] Cost of gas minus cost of high sulfur coal—based on cost of fuel delivered to powerplants in June of 1984.

Exhibit 20-3

Monthly Acidic Deposition as Measured at Brookhaven, New York

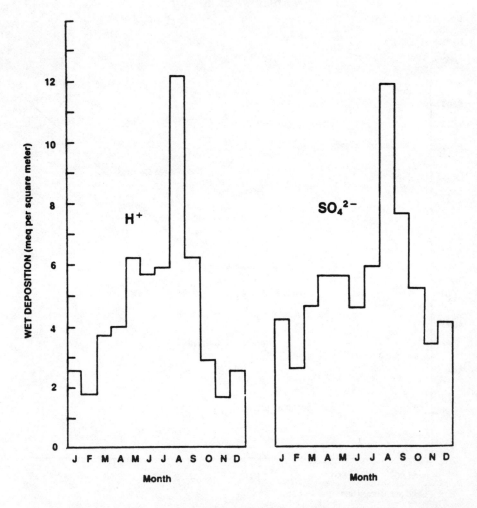

Note: Acid content of rain measured by hydrogen ion (H⁻) concentration. SO₂ is oxidized to form SO₄²⁻ ions deposited in raindrops.

Source: James Fay, Director of Environmental Programs, MIT Energy Laboratory, *Controlling Acid Rain*, (Cambridge, MA) p. 29.

Exhibit 20-4

Comparison of Daily Sendout by Gas Utilities in Peak Month versus Summer Months—by Census Region

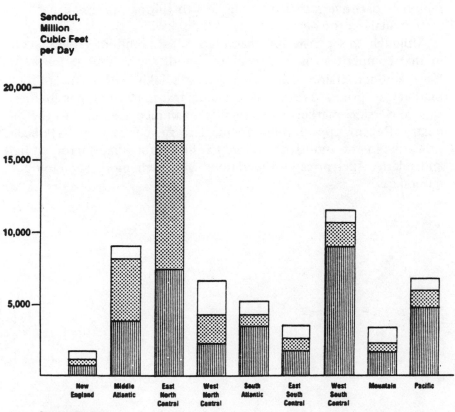

Sendout,
Million
Cubic Feet
per Day

Census Region

Top of bar represents average daily sales in the peak month from the two-year
period 1981-1982.

  Excess summer pipeline capacity.

  Peak month storage withdrawals from underground, LNG, SNG and propane
facilities.

  Average daily sales plus storage injections in the summer months (April-
September) of 1982.

1960s. Again, only minor efficiency losses are encountered in converting coal boilers to gas (3 to 5 percent), and there is little or no derating. Offsetting benefits such as a reduction in the energy required to handle coal and run pollution control equipment may well negate any efficiency losses resulting from the conversion. The capital cost required to convert coal or oil boilers to full natural gas capability are minimal—in the range of $5 to $30 per kW.

Although this conversion market is not a premium value market in that competition is with coal and resid, the volume potential is huge, and conversions can be made very quickly. Given current spot market gas prices, as well as evolving transportation possibilities, this conversion market offers significant immediate relief to the industry. Recent opportunities for coal to gas conversions have been primarily in the industrial sector. In this sector, load factor, siting, and relative fuel price considerations are much more favorable for natural gas.

Chapter 21

Economic and Environmental Advantages of Natural Gas-Fueled Combined-Cycle Electric Power Generation

PAUL L. WILKINSON
Manager, Policy Analysis
American Gas Association

The period from the early 1960s through the mid-1970s was one in which a number of tenets were generally accepted by utility planners, energy forecasters, and policy makers: the demand for electricity was expected to increase at a steady and predictable rate; natural gas and oil were not considered viable options to fuel new electric utility plants due to price and availability expectations; and large, central station coal and nuclear units were assumed to be the most cost-effective means of generating electricity for the future. All of these basic tenets have been proven wrong.

The rapidly escalating capital costs of nuclear and coal-fired units, as well as various associated environmental and regulatory barriers, have diminished the attractiveness of these options. As a result of these and other factors, no new nuclear units have been ordered since

1978, and over 100 units have been cancelled. Similarly, the number of new coal-fired units projected to come on-line over the next decade has fallen steadily from 198 units projected in 1981 to only 107 units projected in 1984. In addition, the demand for electricity has become less predictable due to conservation, cogeneration, increasing fluctuations in the business cycle, and other factors. This uncertainty in electricity requirements, along with double digit interest rates, makes large, capital intensive coal and nuclear plants financially unattractive options.

Finally, the outlook for natural gas (from both supply and price perspectives) has improved dramatically since the late 1970s, and significant advances have been made in electrical generation technologies which use gas. For example, gas-fired combined-cycle turbine units have become even more efficient and reliable. These units can be installed quickly (12 to 24 months versus 8 to 15 years for coal and nuclear units), and they may be brought on-line in small incremental blocks. This modular construction feature offers greater flexibility to utility planners, and the relatively low capital cost of combined-cycle units reduces financial uncertainties.

The purpose of this chapter is to compare two electric power generating options, a new coal-fired powerplant and a new natural gas-fired combined-cycle unit, from both economic and environmental perspectives. The comparison is based on a 240 megawatt baseload system.

EXECUTIVE SUMMARY

New natural gas-fired combined-cycle units offer an efficient and economical alternative to electricity produced by new nuclear or coal-fired units. In addition, combined-cycle power generation is attractive from an environmental perspective, with minimal impacts in terms of air pollutant emissions, solid waste generation and water consumption.

Economic Results

The 240 megawatt (MW) combined-cycle unit modeled in this chapter produces electricity at a cost which is only 54 to 92 percent of the cost of electricity produced by a new, comparably sized coal unit. Combined-cycle gas units enjoy economic advantages because capital charges (return on investment, capital recovery, taxes, etc.) and non-fuel operation and maintenance expenses are about one-third those of coal units, more than offsetting coal's fuel cost advantage. (See Exhibit 21-1.)

- Under the Base Case assumptions (65 percent capacity factor, 10 percent real after tax discount rate) the levelized cost of producing electricity via combined-cycle gas operation is 5.4¢ per kilowatt hour (¢/kWh), only 79 percent of the cost of producing electricity in a new coal-fired unit—6.8¢/kWh.

  - Based on installed capital costs of $408 million for the coal unit versus $120 million for the combined-cycle system (240 MW unit), the capital recovery portions of their total levelized costs are 3.4¢/kWh and 1.1¢/kWh for the coal and combined-cycle gas units, respectively.

  - Operation and maintenance of the combined-cycle system entails a cost of 0.3¢/kWh, 38 percent of the 0.8¢/kWh charge for O&M of the coal unit and ancillary pollution control equipment.

  - The levelized fuel component of the combined-cycle system is 4.0¢/kWh versus 2.6¢/kWh for the coal unit. The impact of the higher cost of gas per MMBtu relative to coal is dampened since the more efficient combined-cycle unit requires less energy input for equal electricity output—10.8 TBtu per year versus 13.7 TBtu per year for the coal unit.

- The Base Case pricing assumptions reflect national average gas and coal prices for the electric utility sector—the cost of gas per MMBtu is roughly twice the cost of coal. Sensitivity analyses were performed under high fuel price differential assumptions (cost of gas increased by 15 percent, cost of coal decreased by 15 percent) and low fuel price differential assumptions (cost of gas decreased by 15 percent, cost of coal increased by 15 percent). The sensitivity analyses approximate the regional variations in regional coal and gas prices relative to the national

Exhibit 21-1

Summary Comparison of the Levelized Cost of Producing Electricity in a 240-MW Powerplant: Coal-Fired Boiler vs. Natural Gas-Fired Combined-Cycle (1985 ¢/kWh)

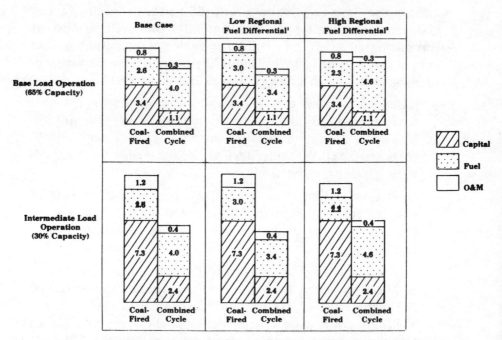

Note: Does not include the cost of electricity transmission and distribution which may account for 30% of the delivered cost to consumers from new facilities.

[1] Base Case cost of gas times 85%, cost of coal times 115%.
[2] Base Case cost of gas times 115%, cost of coal times 85%.

average—the price of coal is currently about one-third the price of gas in the highest differential region, and three-fourths the price of gas in the lowest differential region.

— Under the low fuel price differential assumptions, the levelized total cost of the combined-cycle gas operation drops to only 67 percent of coal operation—4.8¢/kWh versus 7.2¢/kWh.

- The high fuel price differential assumptions reduce the economic advantage of the combined-cycle gas system to 92 percent of the coal system's total levelized cost—6.0¢/kWh versus 6.5¢/kWh.

• Operating in an intermediate load mode (30 percent capacity versus 65 percent baseload operation in the Base Case) enhances the attractiveness of combined-cycle operation as the capital component of the total levelized cost (where combined cycle has a 3:1 advantage over coal) becomes relatively more important, and the fuel component (where coal has a 1.5:1 advantage) becomes relatively less important.

- The total levelized cost of the combined-cycle gas system under Base Case pricing assumptions is 6.8¢/kWh, 61 percent of the total cost of coal operation—11.1¢/kWh.

- The total levelized cost of the combined-cycle gas option— assuming intermediate load and a low fuel price differential— drops to only 54 percent of the coal option total cost, and it increases to 69 percent assuming a high fuel price differential.

Environmental Results

The annual emissions of air pollutants of primary concern—sulfur dioxide (SO_2), nitrogen oxides (NO_x) and total suspended particulates (TSP)—were calculated for the 240 MW coal-fired (2 percent sulfur, 10 percent ash coal) and combined-cycle units assuming each would at least meet applicable federal emission standards. Each unit's production of sludge and ash, and its annual water consumption were also calculated. Source: These projections have been made using the Wharton Economic Forecasting Associates Long-Term Model. The assumptions and simulations were prepared by the American Gas Association and do not represent the views of Wharton Econometric Forecasting Associates.

• Even when equipped with a flue gas desulfurization system for SO_2 control and an electrostatic precipitator for TSP control, the coal unit emits over 4 thousand tons of SO_2 and 200 tons of TSP per year. Gas-fired combined-cycle units emit negligible amounts of SO_2 and TSP—less than 0.3 percent of the coal unit's combined tonnage.

- Emissions of NO_X from the combined-cycle plant are 57 percent of the coal unit's emissions. These emissions from the combined-cycle unit could be further reduced, but unit efficiency would decline.

- The coal plant would produce 60 thousand tons of ash per year and 168 thousand tons of sludge (a semi-solid scrubber by-product). Combined-cycle units produce neither of these non-combustible solid wastes.

- Treated water consumption by the combined-cycle plant would be 15 million gallons per year, one-half of the consumption of the coal plant.

Site specific factors such as relative fuel costs, interest rates (individual company's cost of debt), state and local pollution regulations, taxes, etc., must be considered when comparing such major projects. This analysis in no way precludes the need for detailed engineering, economic and environmental analyses for specific projects.

BACKGROUND

A combined-cycle gas turbine powerplant combines the operation of a gas turbine, a steam turbine and a heat recovery steam generator. Appendix 21-1 provides an illustrative diagram of a "typical" combined-cycle system. Natural gas (or fuel oil) is fed into a gas turbine which drives a generator. Waste heat from the turbine is captured by the heat recovery steam generator which provides steam for the steam turbine. The steam turbine, in turn, also drives a generator. In the system modeled in this analysis (which is a common configuration), two-thirds of the electric power is provided by the gas turbine and one-third by the steam turbine.

One of the primary attributes of combined-cycle gas turbine power generation is its inherent energy efficiency. Whereas two-thirds of the energy input to conventional fossil fuel boiler systems is normally lost to the environment (30 to 34 percent efficiency), the waste heat from the gas turbine in a combined-cycle unit is captured and utilized. Thus, combined-cycle system efficiency is in the 45 percent range. This greater system efficiency results in a fuel requirement for combined-cycle systems which is only about 80 percent of the fuel requirement of a conventional boiler system.

Another major advantage of combined-cycle systems over coal-fired or nuclear powerplants is their relatively low capital cost—approximately one-third the cost per kilowatt of coal-based capacity, and about one-fifth the cost per kilowatt of nuclear capacity. The cost of combined-cycle systems, most of which are largely prepackaged, has risen far less rapidly than field-erected coal and nuclear units. In addition, combined-cycle systems can be brought on-line quickly and in stages. For example, two-thirds of the electric capacity of a 100 MW combined-cycle system (the gas turbine component) can be on-line in 12 to 18 months, with the balance operating in an additional 6 months.[1] In contrast, new coal units require approximately 8 years for design and construction.[2] It is impossible to estimate the lead time required for a new nuclear plant in light of current regulatory, environmental and other problems.

Combined-cycle power generation does not represent a new technology, although innovations and improvements continue to enhance this method of electricity production. There are currently over one hundred combined-cycle gas systems in operation worldwide with a combined capacity of approximately 17,000 MW, (7,000 MW of which is located in the U.S.). The systems range in size from less than 10 MW to several hundred MW. A 2,000 MW plant currently under construction in Japan will be the world's largest combined-cycle operation, with 14 systems at one site. The systems on-line today operate in a variety of modes: baseload (above 4,000 hours per year), intermediate load (1,000 to 4,000 hours per year), and peaking (under 1,000 hours per year). According to the Electric Power Research Institute (EPRI), roughly 30 percent of the combined-cycle capacity in the U.S. is peaking capacity, 20 percent is intermediate load, and 50 percent is baseload.[3]

---

[1] See Footnotes at the end of the chapter.

A movement toward combined-cycle units began in the early 1960s, and there were some questions as to unit reliability and downtime. However, product quality has shown steady improvement, and the reliability index of combined-cycle units today is at least as high as that of steam systems. The world's largest combined-cycle system manufacturer, General Electric, has published data which indicates that two 200 MW combined-cycle units would have an "equivalent reliability" of 95.4 percent. In contrast, the reliability profile of a 400 MW steam plant (based on National Electric Reliability Council data for the period 1972 through 1981) is 86.8 percent. ("Equivalent availability" takes account of the fact that a combined-cycle unit is rarely shut down entirely—work can be performed on portions of multi-turbine systems while other parts of the system continue to operate. Large coal units, conversely, are forced to shut down entirely.) The reliability claims of the manufacturer are substantiated by EPRI, whose *Technical Assessment Guide* predicts a 75.7 percent operating availability for a new large coal unit versus 93.0 percent for a combined-cycle system.[4]

In the mid-to-late 1970s electric utilities in the U.S. began to rely almost exclusively on large coal- and nuclear-powered central generating stations for new capacity. This movement was due in part to the Arab oil embargo of 1973, as well as to restrictions by some gas utilities on the sale of gas for electric generation (due to gas supply uncertainty), and to constraints imposed by the Powerplant and Industrial Fuel Use Act of 1978 (FUA). Gas supply uncertainty of the mid-1970s was the result of over two decades of federal price controls on natural gas which kept the price of gas at artificially low levels and discouraged drilling. The phased decontrol of natural gas prices established by the Natural Gas Policy Act (NGPA) has resulted in an orderly transition to a free market, and both the near-term and long-term outlooks for gas supply and price are very favorable.

The FUA prohibited electric utilities from constructing new gas-fired combined-cycle units with input capacities in excess of 100 MMBtu per hour. One electric utility, Kissimmee Municipal Electric System in Florida, is known to have obtained an FUA exemption, and their 51 MW gas-fired combined-cycle unit is currently under construction. However, it is not known how many other utilities were dissuaded by the regulatory uncertainty of FUA from constructing such units. Exhibit 21-2, which lists the domestic combined-cycle units constructed by Westinghouse and General Electric from 1968 through January 1983, presents an interesting perspective on this operation. These two manufacturers have built 21 combined-cycle units in the U.S. with a combined capacity of over 5,000 MW—*all* commenced construction prior to the enactment of FUA with the exception of the Cool Water Project completed last year, which is exempt from FUA because it operates on gas derived from coal.

METHODOLOGY AND ASSUMPTIONS

This section describes the methodology and assumptions employed in the comparison of a new 240 megawatt (MW) coal-fired electric powerplant versus a comparably sized gas-fired combined-cycle unit, both from an economic and environmental perspective. The alternatives were compared on a plant life (30 years) basis, for the period 1990 through 2020 (although it is unlikely that a new coal unit could be brought on-line in only five years). The 240 MW size was chosen because it reflects a movement by utilities to bring new capacity on-line more quickly and in smaller sizes than was popular in the 1960s and 1970s. Nuclear power was not considered a feasible option in this time frame or size range. Operating characteristics of the two systems are shown in Exhibit 21-3.

Economics

A levelized cost approach was used to compare the two options. Levelization is a mathematical technique which converts the varying annual charges over the life of a project into a single constant annual charge. The total levelized cost of producing electricity via coal-firing and combined-cycle operation in 1985 dollars was calculated, as were each of the three major cost components—capital, fuel, and operation and maintenance.

Each of the 240 megawatt units was assumed to operate at a 65 percent capacity factor ("baseload"), and their annual electrical output would thus be 4.7 trillion Btu. Two intermediate load cases were examined for sensitivity analysis, one at 30 percent capacity

Exhibit 21-2

Partial Listing[1] of U.S. Electric Utility Combined-Cycle Units to Come On-Line Between January, 1968 and January, 1983

Utility	Plant Rating (MW)	Date of Commercial Operation	Number of Total System Operating Hours
Wolverine Electric	21	1968	102,735
City of Clarksdale	21	1972	73,022
City of Hutchinson	11	1972	64,435
Southwestern Public Service	30	1973	N.A.
Central Iowa Power	31	1973	N.A.
St. Joseph L&P	63	1973	N.A.
Public Service of Oklahoma	250	1973	N.A.
Duquesne P&L	330	1974	16,124
Houston L&P	600	1974	109,265
Salt River Project	288	1974	70,131
Ohio Edison	225	1974	16,842
Central Iowa Power	33	1975	N.A.
El Paso Electric	250	1975	N.A.
Arizona Public Service	249	1976	20,913
Florida P&L	500	1976	N.A.
Iowa Illinois G&E	105	1977	8,832
Jersey Central	340	1977	8,542
Puerto Rico EPA	640	1977	32,489
Western Farmers	300	1977	96,179
Portland G&E	550	1977	2,432
Massachusetts Municipal	340	1981	3,676
Total MW	5,177		

N.A.: Not available.

Note: All units commenced construction prior to enactment of Fuel Use Act in 1978.

[1] Combined-cycle units installed by two major U.S. manufacturers—Westinghouse and General Electric—represents about three-fourths of total U.S. combined-cycle capacity.

Exhibit 21-3

Characteristics of a 240 Megawatt Electric Generation Station—
Coal-Fired versus Gas-Fired Combined-Cycle

	Coal-Fired Steam Plant	Gas-Fired Combined Cycle
Operating Characteristic		
Rated Capacity (MW)	240	240
Capacity Utilization (%)	65	65
Unit Efficiency (%)	34	43
Heat Rate (Btu/kWh)	$10,000^2$	$8,000^1$
Annual Energy Input (TBtu)	13.744	10.819
Annual Energy Output (TBtu)	4.669	4.669
Economic Parameters		
Installed Capital Cost ($1985/kW)	$\$1,700^3$	$\$500^2$
Operation and Maintenance—Total (mills/kWh)	8.5	2.7
— Fixed ($/kW/yr)	$\$19.00^2$	$\$6.50^2$
— Variable (mills/kWh)	5.2^2	2.6^2
Other Factors		
Lead Time (years)		
— 160 MW	n.a.^{4}	1^5
— 240 MW (Full Load)	9^2	2^5
Expected System Life (years)		
— Pollution Control	15	n.a.
— All Other	30^2	30^2

---

[1] General Electric, *STAG Combined Cycle Product Line and Performance Characteristics*, GER 3401, (Schenectady, NY: 1984) pp 6-7. Based on STAG 207E with two pressure heat recovery steam generator.

[2] Electric Power Research Institute, *Technical Assessment Guide*, (Palo Alto, CA: May 1982) Appendix B-55 and B-79.

[3] Electric Power Research Institute, *Ibid.*, Appendix B-55. Total capital cost scaled up by 35% to account for the higher cost per kW of constructing a single unit 240 MW facility versus a dual unit 1000 MW facility, as per discussion with EPRI.

[4] n.a.—not applicable.

[5] Gas turbine portion of system operational in 12 months, with remainder on-line in 24 months according to: General Electric, *General Electric 60 Hz STAG Combined Cycle Power Plants*, GEA–11387, (Schenectady, NY: 1984) p. 60831.

utilization and one at 50 percent. The combined-cycle unit, which includes a two pressure heat recovery steam generator, is 43 percent efficient, and the input to the gas turbine would therefore be 10.8 trillion Btu per year. Approximately two-thirds of the unit's total output is derived from the gas turbine, and the remainder is provided by the steam turbine. The less-efficient coal unit operates at 34 percent efficiency, and it would therefore require a greater energy input —13.7 trillion Btu. Efficiency assumptions were based on General Electric's *STAG Combined Cycle Product Line and Performance Characteristics*,[5] and the *Technical Assessment Guide* of the Electric Power Research Institute (EPRI).[6]

The installed capital cost of the combined-cycle unit is $500 per kilowatt (kW), versus $1,700 per kW for the coal unit. Operation and maintenance charges for the combined-cycle system are 2.7 mills per kWh, based on a fixed component of $6.50 per kW per year, and a variable component of 2.6 mills per kWh. The fixed component of the coal system is $19.00 per kW per year, and the variable component is 5.2 mills per kWh, for a total operation and maintenance charge of 8.5 mills per kWh. No real escalation was assumed for the O&M of either system. All capital and operation and maintenance estimates were taken from the EPRI, *Technical Assessment Guide,* with costs converted from 1981 to 1985 dollars based on the implicit GNP inflator. Two additional adjustments were made to the EPRI data. First, the installed cost per kW of the coal unit was scaled up by 27 percent due to the smaller size of the unit examined in this analysis relative to the EPRI base size—240 MW versus 1000 MW— and it was scaled up by an additional 8 percent to reflect the cost penalty for building a single unit versus two-unit staged construction. The scale-up factors were provided by EPRI via phone communications in March of 1985 with an author of the *Technical Assessment Guide.* (The EPRI estimate for a 1000 MW unit is approximately $1,250/kW, indicative of the economies of scale not available at the 240 MW size. Even at the dated and conservative $1,250/kW cost level, only in one of the six cases shown in Exhibit 21-1 would the coal option be less costly than gas-fired combined-cycle.) The second adjustment to the EPRI data was a reduction in the capital cost of the combined-cycle system by $25 per kW based on more recent equipment costs contained in General Electric's *STAG Combined Cycle Product Line and Performance Characteristics.*[7]

The prices paid by electric utilities for coal and natural gas were based on the A.G.A.-TERA Base Case 1985-I, Scenario DM8450Z.[8] The A.G.A.-TERA Model is a detailed energy forecasting model developed and maintained by the American Gas Association. Fuel price projections were not available beyond 2000, and therefore the growth rate in fuel prices projected for the period 1995 through 2000 was assumed to hold through 2020—2.7 percent per year (real) for utility coal and 3.4 percent per year for gas delivered to powerplants. Gas prices employed in the analysis increase from $3.94 per MMBtu in 1990 to $9.68 in 2020, while coal prices rise over the same period from $2.08 per MMBtu to $4.50 ($1985). Two alternative fuel pricing scenarios were examined for a sensitivity analysis —a high fuel price differential scenario in which the price of coal was reduced by 15 percent and the cost of gas was increased by 15 percent, and a low fuel price differential scenario in which the cost of coal was increased by 15 percent and the cost of gas was reduced by 15 percent.

In recognition of the "time value of money," expenses in future years are less "costly" than today's expenses. A 10 percent discount factor (real, after tax) was used in this analysis to obtain the present value of future costs. The discount factor should roughly equate to the long-term cost of capital of electric utilities. Discount factors of 7 percent and 13 percent were used in the sensitivity analysis.

Other financial assumptions inherent in the analysis include: a 50:50 debt:equity ratio, a 12 percent cost of debt, 15 percent cost of equity, 15 year depreciation of equipment (except for pollution control equipment which is granted a 5 year accelerated depreciation), and a 10 percent investment tax credit (both systems). In addition, all equipment was assumed to have a 30 year life, with the exception of pollution control equipment required for the coal system which has only a 15 year life.

Environmental

The environmental impacts of the alternative generating systems operating at a 65 percent capacity factor were estimated for three primary air pollutants—sulfur dioxide (SO_2), particulate matter (TSP) and nitrogen oxides (NO_X). It was assumed that currently applicable federal emission standards for these air pollutants would be met. In addition, water consumption and the production of sludge and ash were also calculated. The coal system was assumed to operate on a commonly available coal type—2 percent sulfur, 10 percent ash—with a flue gas desulfurization (scrubber) system for SO_2 control, and an electrostatic precipitator for TSP control.

A new coal-fired electric utility powerplant is subject to the new source performance standards of 1978 (NSPS) as set out in 40 *CFR* Subpart Da.[9] The NSPS for SO_2 mandates a sliding scale for SO_2 control: facilities emitting less than 0.60 pounds of SO_2 per MMBtu must have reduced potential (uncontrolled) emissions by at least 70 percent, while facilities emitting from 0.60 to 1.20 pounds per MMBtu must have reduced potential emissions by 90 percent. The scrubber employed in this analysis was assumed to be 85 percent effective, reducing uncontrolled emissions from 3.8 pounds of SO_2 per MMBtu combusted to 0.6 pounds per MMBtu. The particulate NSPS is 0.03 pounds per MMBtu (99 percent reduction), and the NO_X NSPS requires a 65 percent reduction in total potential emissions with a cap of 0.60 pounds per MMBtu. The controlled coal plant emissions: SO_2—0.60 pounds per MMBtu; TSP—0.03 pounds per MMBtu; and NO_X—0.35 pounds per MMBtu, were multiplied by the total annual coal input (13.7 trillion Btu) to obtain annual emissions. Uncontrolled air pollutant emissions were obtained from the U.S. Environmental Protection Agency's (EPA's) *Compilation of Air Pollutant Emission Factors.*[10]

The only air emissions from the combined-cycle unit originate in the gas turbine portion of the system, since there is no supplemental firing of the steam turbine. SO_2 and NO_X emissions from stationary electric utility gas turbines in excess of 100 MMBtu per hour and built after 1982 are subject to NSPS, but there is no standard for particulate emissions. The NO_X standard is approximately 75 parts per million of NO_X, which equates to 0.25 to 0.30 pounds of NO_X per MMBtu. SO_2 is limited to 0.015 percent of the exhaust gas total volume, but this standard is far in excess of actual SO_2 emissions from a gas turbine. Thus, uncontrolled SO_2 emissions from a gas boiler (0.0006 pounds per MMBtu according to U.S. EPA, *Compilation of Air Pollutant Emission Factors*)[11] were used to estimate the negligible SO_2 emissions from a turbine. Uncontrolled TSP emissions from a gas turbine operating on distillate oil are 0.004 pounds per MMBtu according to: Radian Corp, *Emissions from Stationary Gas Turbines*,[12] and emissions would be lower when operating on gas. It was assumed that the reduction when operating on gas would be 61 percent based on uncontrolled TSP emission factors for gas turbines contained in EPA's *Compilation of Air Pollutant Emission Factors* (0.004 pounds per MMBtu times 39 percent = 0.0016 pounds per MMBtu when operating on gas).

Coal-fired electric powerplants produce a number of solid wastes including ash and sludge. Fly ash is collected by electrostatic precipitators or baghouses, while bottom ash is formed by fuel combustion. Sludge is a semi-solid waste which results primarily from scrubbing for SO_2. These two noncombustible solid wastes must be disposed of in landfills or ponds, and according to the U.S. EPA, *Energy/Environment Fact Book*,[13] a 1000 MW powerplant produces 700,000 tons of sludge per year and 250,000 tons of ash. These volumes were scaled down by 76 percent to estimate the wastes produced by a 240 megawatt unit.

One other environmental impact was considered in this analysis—the consumption of treated water. Treated water consumption for the coal and combined-cycle units was obtained from General Electric's *60 Hz STAG Combined Cycle Power Plants*.[14] Data cited in the report were for a 450 MW unit, and were scaled down to a 240 MW unit assuming a linear relationship between plant size and water usage.

RESULTS OF ANALYSIS

The following sections present the economic and environmental results of this analysis.

Economic Results

The total levelized cost of the combined-cycle system, as indicated in Appendix 21-2, is $73.3 million per year, or 5.4¢/kWh (all costs are in constant 1985 dollars) only 79 percent of the coal system's annual cost—6.8¢/kWh. Nearly half of the coal system's total cost is attributable to the annual capital charge of 3.4¢/kWh, versus only 20 percent for the combined-cycle unit (1.1¢/kWh). The annual capital charge of the combined-cycle system is based on an installed cost of $120 million, less than one-third the installed cost of the coal unit —$408 million. Approximately $72 million (17.6 percent) of the coal unit's total cost is attributable to required pollution control equipment which would have to be replaced after 15 years.

The annual operation and maintenance charge is 0.8¢/kWh for the coal unit, nearly three times the combined-cycle unit's O&M level— 0.3¢/kWh. Operation and maintenance expenditures were held constant in real terms, and any real escalation would have a more adverse impact on the coal unit than on the gas unit.

The levelized fuel cost is 4.0¢/kWh for the combined-cycle system versus 2.6¢/kWh for the coal system. This difference of 1.4¢/kWh is more than offset by the combined-cycle system's 2.3¢/kWh capital advantage and 0.5¢/kWh operation and maintenance advantage. It should be noted that the differential in coal and natural gas prices is moderated by the fact that the combined-cycle system uses 21 percent less fuel than does the coal system (10.8 TBtu per year versus 13.7 TBtu per year).

A sensitivity analysis was performed for three of the primary variables in the analysis—fuel prices, capacity utilization rate, and the discount factor. Results of this analysis are presented in Appendix 21-2.

The Base Case of the analysis was performed using national average coal and natural gas prices, with a delivered coal price just over half the price per MMBtu of natural gas. This fuel price differential may be higher or lower, however, depending on plant location. For example, the cost of coal is about three-fourths the cost of gas in the South Atlantic region, and about one-third the cost of gas in the Mountain region. In order to reflect this variation, a higher fuel price differential scenario was established by decreasing the cost of coal by 15 percent and increasing the cost of gas by 15 percent, and a lower fuel price differential scenario was formulated similarly—by increasing the cost of coal and decreasing the cost of gas by 15 percent, respectively. The lower fuel price differential scenario increases the levelized cost advantage of combined-cycle operation over coal to 2.4¢/kWh (combined-cycle total cost is 66 percent of coal), while increasing the differential reduces the cost advantage to 0.5¢/kW (combined-cycle total cost is 92 percent of coal).

A 65 percent capacity factor (baseload operation) is assumed in the Base Case of this analysis. Reducing the facility to intermediate load—30 percent and 50 percent capacity utilization—has a significant positive impact on the relative cost of the combined-cycle system. This is due to the fact that the capital component of the total levelized cost (where combined-cycle has a 3:1 advantage over coal) is unchanged regardless of the level of operation, whereas the fuel cost component (where coal has 1.5:1 advantage over combined-cycle) becomes less significant as the level of operation is reduced. The Base Case advantage of 1.4¢/kWh for the combined-cycle option is increased to 1.7¢/kWh at a 50 percent capacity factor, and 2.0¢/kWh at a 30 percent capacity factor. That is, at the 30 percent capacity level the annual cost of the combined-cycle unit is only 60 percent of the coal system's total cost—3.1¢/kWh versus 5.1¢/kWh.

The results of the analysis are fairly insensitive to changes in the discount factor. The Base Case factor of 10 percent was reduced to 7 percent, and increased to 13 percent for the sensitivity cases. The higher discount factor reduces all costs of both systems slightly, while the lower discount factor has the opposite effect. There is little change in the relative positions of the two options. Income tax features, including accelerated depreciation and a 10 percent investment tax credit, tend to reduce the sensitivity of the high capital cost coal system to fluctuations in the discount factor.

Environmental Results

Exhibit 21-4 presents the environmental results of this analysis. As indicated, the gas-powered combined-cycle system has negligible

Exhibit 21-4

Environmental Comparison of Natural Gas-Fired Combined-Cycle Generating Station and Coal-Fired Station
(Annual Impacts, 240 Megawatts)

	Coal-Fired Steam Plant	Gas-Fired Combined Cycle
Air Pollutant Emissions (tons/year)		
Sulfur Dioxide	4,123[1]	3[2]
Particulate Matter	206[1]	9[3]
Nitrogen Oxides	2,385[1]	1,353[4]
Other Pollutant Emissions (thousands of tons/year)[5]		
Sludge	168	none
Ash	60	none
Treated Water Consumption (millions of gallons/year)[6]	30	15

[1] Based on new source performance standards for coal-fired electric powerplants (2% sulfur coal assumed): SO_2—0.6 pounds per MMBtu, TSP—0.03 pounds per MMBtu, and NO_x—65% reduction. See: 40 *CFR* Subpart Da, Part 60.40a through 60.43a.

[2] Data on sulfur dioxide emissions from gas turbines not available, assumed to be equal to SO_2 emissions from gas boiler—.0006 pounds per MMBtu. See: U.S. EPA, *Compilation of Air Pollutant Emission Factors,* (Research Triangle Park, NC: August 1982) p. 1.4-3.

[3] Gas turbines operating on distillate oil have particulate emissions of 0.004 pounds per MMBtu according to: Radian Corp. for the U.S. EPA, *Emissions from Stationary Gas Turbines, New Source Performance Standards,* (Research Triangle Park, NC: November 1984) p. 4-78. Emissions when operating on gas would be approximately 39% of this amount based on: U.S. EPA, *Compilation of Air Pollutant Emission Factors,* (Research Triangle Park, NC: August 1977) p. 3.3.1-2.

[4] Based on new source performance standards for electric utility stationary gas turbines: NO_x—75 parts per million. See 40 *CFR* Subpart GG, Part 60.330 through 60.333.

[5] U.S., Environmental Protection Agency, Research and Development-Energy Minerals and Industry, *Energy/Environment Fact Book,* (Washington, DC: March 1978) p. 24.

[6] General Electric, GEA-11387, *60 Hz STAG Combined Cycle Power Plants,* (Schenectady, NY) p. 568316A.

sulfur dioxide and particulate emissions. In contrast, the coal system emits over 4.1 thousand tons of SO_2 per year (with an 85 percent effective scrubber system) and 206 tons of particulate matter per year (with a 99 percent effective electrostatic precipitator). Emissions of NO_X by the combined-cycle system are 57 percent of the coal unit's annual NO_X emissions—1,352 tons per year versus 2,385 tons per year.

In terms of noncombustible solid wastes, the coal-based system would generate 168,000 tons of sludge per year and 650,000 tons of ash. Natural gas combustion results in no sludge or ash production. In addition, the combined-cycle option would require only half the water of the coal system—15 million gallons per year versus 30 million gallons.

CONCLUSIONS AND RECOMMENDATIONS

Gas-fired combined-cycle power generation offers a clean, economical, efficient, and reliable source of electricity. Combined-cycle units can be constructed quickly with shop-assembled units, and additional units may be added as needed over time in a modular fashion. These features of rapid and modular construction protect the utility and the ratepayer from the economic burden of overbuilding, as well as from the unacceptable risk of inadequate capacity.

FOOTNOTES

[1] General Electric, GEA-11387, *General Electric 60 Hz STAG Combined Cycle Power Plants* (Schenectady, NY: 1983), p. 60831.

[2] Burns and Rose, Inc., *Preliminary Assessment of the Modular Block Plant Concept,* (Arlington, VA: April, 1984) p. 3.

[3] Duncan and Dolbec, Electric Power Research Institute, *Operation and Technology of Combined Cycle Power Plants in the USA,* (presented in Chicago, IL: April 26, 1982) p. 2.

[4] Electric Power Research Institute, *Technical Assessment Guide,* (Palo Alto, CA: May, 1982) pp. B-55, B-79.

[5] General Electric, GER 3401, *STAG Combined Cycle Product Line and Performance Characteristics*, (Schenectady, NY: 1984) pp. 6-7.

[6] Electric Power Research Institute, *op. cit.*, pp. B-55, B-79.

[7] General Electric, GEA-11387, *op. cit.*, p. 568322.

[8] American Gas Association, *A.G.A.-TERA Base Case 1985-I*, DM8450Z, (Arlington, VA: January, 1985).

[9] *Code of Federal Register*, 40 *CFR* Subpart D, Part 60.42 through 60.44.

[10] U.S. Environmental Protection Agency, *Compilation of Air Pollutant Emission Factors*, (Research Triangle Park, NC: August, 1982) supplement 13.

[11] U.S. Environmental Protection Agency, *ibid.*, p. 3.3.1-2.

[12] Radian Corp., *Emission from Stationary Gas Turbines*, (Research Triangle Park, NC: November, 1984) p. 4-78.

[13] U.S. Environmental Protection Agency, *Energy/Environment Fact Book*, (Washington, DC: March, 1978) p. 24.

[14] General Electric, GEA-11387, *op. cit.*, p. 568316A.

Appendix 21-1

Illustrative Diagram of a 240-MW Natural Gas-Fired Combined-Cycle Electric Powerplant

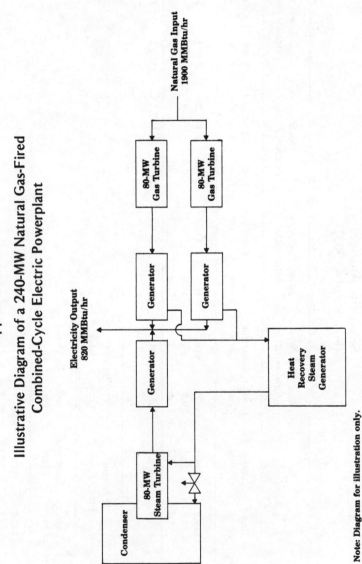

Note: Diagram for illustration only.

Appendix 21-2
Sensitivity Analyses
(Millions of $1985)

Levelized Cost Component	Base Case		Higher Fuel Price Differential		Lower Fuel Price Differential		Higher Discount Factor		Lower Discount Factor	
	Coal-Fired	Combined Cycle	Coal-Fired	Combined Cycle	Coal-Fired	Combined Cycle	Coal-Fired	Combined Cycle	Coal-Fired	Combined Cycle
Capital	$45.9	$15.1	$45.9	$15.1	$45.9	$15.1	$44.6	$15.0	$47.1	$15.2
Fuel	36.2	54.5	30.8	62.7	41.6	46.3	34.8	51.4	37.4	57.5
O&M	11.6	3.7	11.6	3.7	11.6	3.7	11.6	3.7	11.6	3.7
Total	$93.7	$73.3	$88.3	$81.5	$99.1	$65.1	$91.0	$70.1	$96.1	$76.4

Levelized Cost Component	30 Percent Capacity Factor		50 Percent Capacity Factor	
	Coal-Fired	Combined Cycle	Coal-Fired	Combined Cycle
Capital	$45.9	$15.1	$45.9	$15.1
Fuel	16.7	25.2	27.8	41.9
O&M	7.8	2.6	10.0	3.3
Total	$70.4	$42.9	$83.7	$60.3

Scenario	Annual Capacity Factor	Discount Factor[1]	Assumptions Fuel Prices ($1985/MMBtu) Natural Gas	Coal
Base Case	.65	.10	$3.94 in 1990, $9.68 in 2020	$2.08 in 1990, $4.50 in 2020
Higher Differential	.65	.10	Base Case X 1.15	Base Case X .85
Lower Differential	.65	.10	Base Case X .85	Base Case X 1.15
Higher Discount	.65	.13	Base Case	Base Case
Lower Discount	.65	.07	Base Case	Base Case
30% Capacity	.30	.10	Base Case	Base Case
50% Capacity	.50	.10	Base Case	Base Case

[1] Real after tax.

Chapter 22

Repowering With Natural Gas

PAUL L. WILKINSON
Manager, Policy Analysis
American Gas Association

This chapter examines the concept of combined-cycle .repowering with natural gas as one possible solution to the impending dilemma facing electric utilities—tight capacity margins in the 1990s and the inordinate expense of traditional powerplants. Combined-cycle repowering refers to the production of electricity through the integration of new and used equipment at an existing site, with the final equipment configuration resembling a new gas-fired combined-cycle unit (i.e., gas turbine, waste heat recovery unit and steam turbine/-generator). Through the utilization of improved waste heat recovery and gas-fired equipment, repowering provides both additional capacity and increased generating efficiency.

Three modes of repowering are considered: (1) *peaking turbine repowering* refers to the addition of a steam turbine and heat recovery unit to an existing gas turbine, with the efficiency improvement allowing the unit to convert from peaking to baseload operation; (2) *heat recovery repowering* is the replacement of an old coal boiler with a gas turbine and heat recovery unit, leaving the existing steam turbine in place; and (3) *boiler repowering*, in which the exhaust from a new gas turbine is fed into an existing coal boiler, replacing existing forced-draft fans and air heaters. These three options are compared with the option of adding new coal-fired boilers on the basis of economics, energy efficiency and environmental impacts.

EXECUTIVE SUMMARY OF ANALYSIS RESULTS

Repowering electric utility powerplants with natural gas can provide highly efficient generating capacity in about one-fourth the time required to construct a new coal-fired powerplant, and electricity can be generated in repowered units at a cost which is only about 60 percent of the production cost of new coal units. In addition, repowering offers major environmental benefits, reducing pollutant emissions by thousands of tons per year.

Economic Results

The three natural gas repowering options presented in this analysis produce electricity at a cost which is approximately 60 percent of the cost of producing electricity in a comparably sized new coal-fired powerplant. As indicated in Exhibit 22-1, the cost of producing electricity via repowering ranges from 3.7 cents per kilowatt hour (¢/kWh) to 3.9¢/kWh, versus 6.2¢/kWh for a new coal-fired powerplant. (Results are stated on a levelized cost basis in constant 1986 dollars.) Although the fuel cost component of the repowering options is 30 to 40 percent greater than those of the coal-based units, this disadvantage is more than offset by the lower capital charges and operation and maintenance expenses applicable to repowering.

- The capital charge (return on investment, capital recovery, taxes, etc.) of the repowering units ranges from 0.6¢/kWh to 1.0¢/kWh, only 18 to 30 percent of the capital charge applicable to a new coal-fired unit—3.3¢/kWh. This significant advantage is due to a number of factors:

    - The repowered units use some existing equipment, whereas the coal unit is new from the ground up.

Exhibit 22-1

Summary Economic Comparison of Natural Gas Repowering
Versus a New Coal-Fired Electric Powerplant
(1986 ¢/kWh)

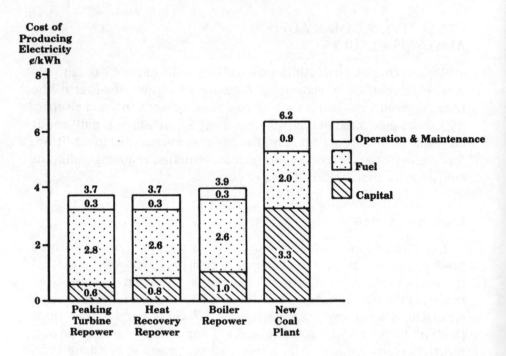

Note: **Based on levelized annualized costs**

— New coal units must include control equipment to limit the
emission of sulfur dioxide, particulate matter and other pollutants. These flue gas desulfurization units (scrubbers) and
electrostatic precipitators or baghouses (not needed in repowered gas units) increase the cost of coal units by 15 to
25 percent.

• The fuel cost component for repowering systems ranges from
2.6¢/kWh to 2.8¢/kWh, versus 2.0¢/kWh for a coal unit. Although the delivered price of natural gas is 65 to 70 percent

greater than the delivered price of coal over the period, the fuel cost component of the repowered units is only 34 to 40 percent greater than the coal unit's fuel cost.

— The cost advantage of coal relative to gas is reduced due to the higher efficiency of the repowered units—38 to 44 percent versus 34 percent for the coal units.

• The non-fuel operation and maintenance expense of repowering is only one-third the expense for a new coal unit—0.3¢/kWh versus 0.9¢/kWh. Coal units are larger and require greater service. In addition, there are significant expenses associated with the storage and handling of coal, as well as the maintenance of pollution control equipment and disposal of sludge and ash.

Environmental Results

The annual emissions of air pollutants of primary concern—sulfur dioxide (SO_2), nitrogen oxides (NO_X), and total suspended particulates (TSP), as well as the annual production of sludge and ash, were estimated for each of the power production options under the assumption that applicable federal emission standards would be met. (See Exhibit 22-2.)

• Peaking turbine repowering and heat recovery repowering emit virtually no sulfur dioxide or particulate matter, while boiler repowering results in net reductions in SO_2 and TSP emissions of 1,060 and 50 tons per year, respectively. (This net reduction is due to the fact that the coal portion of the repowered unit, which continues to operate, consumes about 14 percent less coal than when operating only on coal.) The coal options emit 1,300 to 3,400 tons per year of SO_2, and 65 to 170 tons per year of TSP.

• The net increase in nitrogen oxide (NO_X) emissions attributable to repowering is 110 to 1,290 tons per year, versus 750 to 1,980 tons per year for the comparably sized coal options.

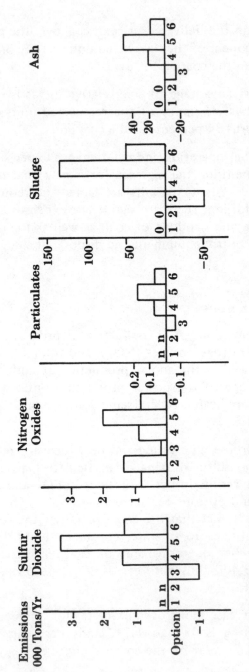

Exhibit 22-2

Summary Environmental Comparison of Natural Gas Repowering
Versus a New Coal-Fired Electric Powerplant
(Emissions, Thousand Tons/Year)

Electric Power Production Options
1. Peaking Turbine Repower......... 80 MW
2. Heat Recovery Repower........... 198 MW
3. Boiler Repower.................... 75 MW
4. New Coal Plant................... 80 MW
5. New Coal Plant................... 198 MW
6. New Coal Plant................... 75 MW

Notes: n=Negligible, less than 10 tons per
year. Emissions shown are on an incremental basis; boiler repowering
results in a net reduction in emissions
due to reduced coal consumption in
repowered coal boiler.

- The new coal units modeled in this analysis would produce from 52,500 to 138,000 tons per year of sludge (a scrubber by-product), and 18,750 to 49,300 tons per year of ash. There would be no net increase in sludge or ash production attributable to repowering, and the boiler repowering option would result in net decreases in the production of sludge and ash of 53,300 and 19,000 tons per year, respectively.

It should be noted that repowering is extremely site specific. The engineering possibilities and economic impacts of repowering will vary greatly from site to site. For example, recoverable equipment may have a useful remaining life at one site but not at another. Also, the equipment at a particular site may or may not be optimally sized to combine with new equipment for repowering. Therefore this study is in no way intended to preclude the need for detailed site specific economic and engineering analyses. Rather, it is intended to present a preliminary overview within the broad range of capacity expansion possibilities open to electric utilities.

BACKGROUND

Because of the heavy financial burden associated with the construction of large new coal-fired or nuclear powerplants, and because there is continued uncertainty with respect to the level of electricity demand in the 1990s and beyond, utilities have become reluctant to commit to large new baseload powerplants. Instead, they are increasingly pursuing strategies of extending the life of existing units and/or adding capacity in smaller, less capital intensive increments. One such strategy is combined-cycle repowering with natural gas.

The basic components of a gas-fired combined-cycle powerplant are a gas turbine/generator, a waste heat recovery unit, and a steam turbine/generator. The primary attractions of combined-cycle powerplants are their low capital and operating costs, short construction lead times, clean operation and modular design (capacity can be added as required). Combined-cycle *repowering* refers to the integration of new and used equipment at an existing site, with the final equipment configuration resembling a new gas-fired combined-cycle unit. The type of repowering employed will vary from site to site.

SCENARIO DESCRIPTIONS

Peaking Turbine Repowering—Exhibit 22-3. The first hypothetical scenario is that of an electric utility which needs to add 80 megawatts (MW) of capacity, and which already has a number of underused gas peaking turbines on-line. There is currently over 46 gigawatts of gas turbine generating capacity in the U.S., about 7 percent of total installed capacity. However, most of these units are operated as peak shavers, and many are used less than 100 hours per year —sometimes started up only for testing purposes.[3] This scenario examines the option of adding a heat recovery steam generator (HRSG), steam turbine/generator and an air-cooled condenser to an existing gas turbine. Total capacity is increased from 53 MW to 80 MW. However, because unit efficiency increases from 27 percent for peaking units to 42 percent for the repowered system, the repowered system can economically be run in a baseload mode. Thus, capacity is effectively increased by a full 80 MW. The option is to install a new 80 MW coal-fired powerplant.

Heat Recovery Repowering—Exhibit 22-4. The second scenario is that of a utility wishing to retire and replace a relatively small (60 MW) old coal-fired boiler, while adding 138 MW of new capacity. The boiler is replaced with a new gas turbine/generator (138 MW) and HRSG, leaving the existing steam turbine/generator (60 MW) in service. The option is to add a new 198 MW coal-fired unit.

Boiler Repowering—Exhibit 22-5. The third scenario is that of a utility with a relatively new and large (452 MW) coal-fired boiler, which has two goals—to increase capacity by 75 MW while improving overall system efficiency. A gas turbine/generator (75 MW) is added in the repowering option, and the exhaust from the turbine/-generator provides combustion air to the existing coal boiler. The gas turbine replaces the function of the existing forced-draft fan and air heater. An economizer is also added to provide additional feedwater heating, replacing the air heater. Boiler repowering has been used extensively in Europe, but only on a limited basis in the U.S. Interest in this mode of repowering is increasing in the U.S., however, in part because the gas turbine and boiler complement each other in the reduction of NO_X. NO_X reduction is achieved by reducing the flame temperature in the boiler, and the burners in the boiler may also be operated in a rich/lean mode, allowing a zone for NO_X decomposition. The option in this scenario is to add a 75 MW coal-fired unit.

METHODOLOGY AND ASSUMPTIONS

This section describes the methodology and assumptions employed in the comparison of natural gas repowering versus the addition of new coal-fired capacity for each of the three scenarios discussed above. Comparisons are made on the basis of both economics and environmental impacts.

Economics

A levelized cost approach was used to compare the various options over the 15-year period 1986–2000 ("overnight" construction was assumed). Levelization is a mathematical technique which converts the varying annual charges over the life of a project into a single constant annual charge. The total levelized cost of producing electricity via coal-firing and natural gas repowering in 1986 dollars was calculated, as were each of the three major cost components—capital, fuel, and operation and maintenance.

In recognition of the "time value of money," expenses in future years are less "costly" than today's expenses. A 7 percent discount factor (real, after tax) was used in this analysis to obtain the present value of future costs. This rate is equivalent to a 12.3 percent nominal pre-tax rate. The discount factor should roughly equate to the long-term cost of capital of electric utilities.

Other financial assumptions inherent in the analysis include: a 50:50 debt:equity ratio, a 10 percent cost of debt, 14.5 percent cost of equity, and a 20-year depreciation of equipment (consistent with the proposed new tax law). In addition, all equipment was assumed to have a 30-year life.

The installed capital cost for the coal option in each of the three scenarios is $1,325 per kW. The capital cost estimate is based on the EPRI *Technical Assessment Guide*[4] as adjusted for inflation—based

Exhibit 22-3

Peaking Turbine Repowering Characteristics versus a
New Coal-Fired Electric Powerplant

	Peaking Turbine Repower	New Coal-Fired Powerplant
Operating Characteristics		
Capacity "Addition" (MW)	80	80
Capacity Utilization (%)	65	65
Unit Efficiency (%)	42	34
Heat Rate (Btu/kWh, HHV)	8,200	10,000
Annual Energy Input (BBtu)	3,715	5,590
Annual Energy Output (BBtu)	1,560	1,560
Economic Parameters		
Installed Capital Cost ($1986/kW)	245	1,325
O&M — Total (mills/kWh)	2.9	9.1
— Fixed ($/kW/yr)	6.7	19.4
— Variable (mills/kWh)	1.7	5.7

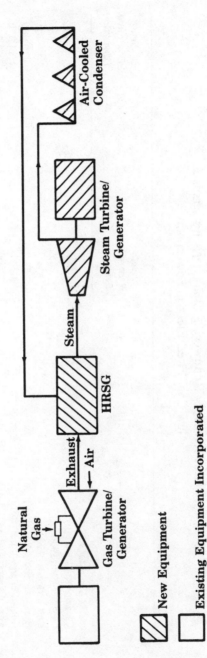

Schematic—Peaking Turbine Repowering

Exhibit 22-4

Heat Recovery Repowering Characteristics versus a New Coal-Fired Electric Powerplant

	Heat Recovery Repower	New Coal-Fired Powerplant
Operating Characteristics		
Capacity Addition (MW)	198	198
Capacity Utilization (%)	65	65
Unit Efficiency (%)	44	34
Heat Rate (Btu/kWh, HHV)	7,775	10,000
Annual Energy Input (BBtu)	8,740	11,310
Annual Energy Output (BBtu)	3,845	3,845
Economic Parameters		
Installed Capital Cost ($1986/kW)	310	1,325
O&M – Total (mills/kWh)	2.9	9.1
– Fixed ($/kW/yr)	6.7	19.4
– Variable (mills/kWh)	1.7	5.7

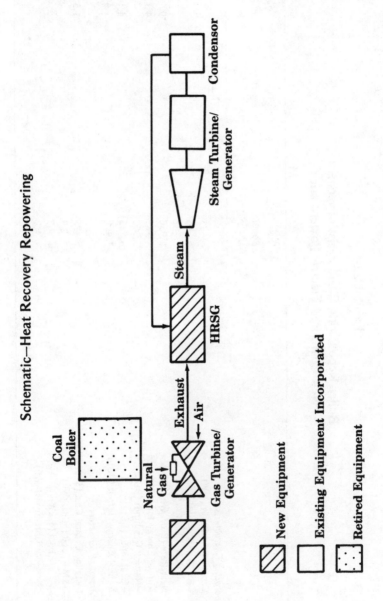

Schematic—Heat Recovery Repowering

Exhibit 22-5

Boiler Repowering Characteristics versus a New Coal-Fired Electric Powerplant

	Boiler Repower	New Coal-Fired Powerplant
Operating Characteristics		
Capacity Addition (MW)	75	75
Capacity Utilization (%)	65	65
Unit Efficiency (%)[1]	38	35
Heat Rate (Btu/kWh, HHV)[1]	8,950	9,775
Annual Energy Input (BBtu)[2]	1,785	4,170
—Natural Gas Input (BBtu)[2]	5,320	—
—Coal Input (BBtu)[2]	(3,535)	4,170
Annual Energy Output (BBtu)[2]	1,460	1,460
Economic Parameters		
Installed Capital Cost ($1986/kW)	425	1,325
O&M—Total (mills/kWh)	2.9	9.1
—Fixed ($/kW/yr)	6.7	19.4
—Variable (mills/kWh)	1.7	5.7

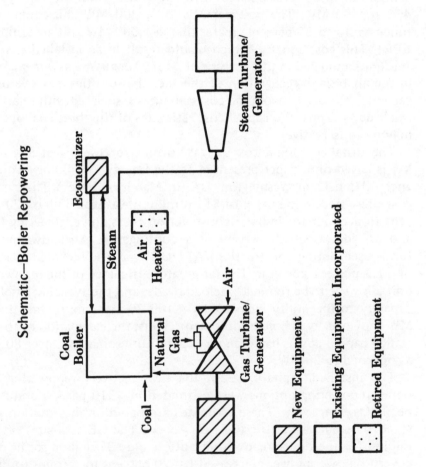

Schematic—Boiler Repowering

on growth in the implicit GNP inflator from 1980 through 1986. Building coal units of the sizes indicated in this analysis, 75 to 198 MW, would be significantly more costly per MW added than would be the 1,000 MW unit analyzed by EPRI. There are significant dis-economies of scale associated with the construction of units below 400 to 500 MW. The cost per MW of a 200 MW unit could be as much as 30 to 35 percent greater than a 1,000 MW unit according to EPRI. This cost penalty, which would result in an installed cost for the units modeled in this analysis of $1,750 per kW, was *not* included in the analysis in recognition of the fact that utilities may overbuild capacity in order to avoid the cost penalties associated with relatively small units. Thus, the capital cost estimates of the three coal options may be conservative.

The installed capital cost of gas turbine repowering of $245 per kW is based on a paper presented at the Energy Technology Conference XIII held in Washington, DC, in March of 1985.[5] The system cost presented in the paper of $11.8 million was scaled up by 20 percent to account for indirect construction costs and engineering fees, and an additional 10 percent as a contingency. An allowance for funds used during construction (AFUDC) was also included at a rate of 12.3 percent per year. The far greater efficiency of the repowered unit allows it to be run in a baseload mode, and the available baseload capacity to the utility increases by a full 80 MW, not by just the 26 MW provided by the *new* equipment. Thus, the calculation of the in-stalled capital cost is based on an expenditure spread out over 80 MW as opposed to 26 MW.

The installed capital costs of the heat recovery repowering and boiler repowering options were estimated at $310 per kW and $425 per kW, respectively. These estimates are based on information pro-vided by General Electric (GE) to A.G.A. The GE estimates of $19 million for the boiler repowering option and $37 million for heat re-covery repowering were increased by 20 percent to account for indi-rect construction costs, engineering and design, and by an additional 10 percent as a contingency factor. AFUDC was also added in at a rate of 12.3 percent per year.

Operation and maintenance charges for each of the three gas re-powering options is 2.9 mills per kWh, comprised of a fixed component of $6.7 per kW per year, and a variable component of 1.7 mills per kWh. The fixed and variable components of the coal units are $19.4 per kW per year and 5.7 mills per kWh, respectively, for a total O&M charge of 9.1 mills per kWh. All O&M estimates were based on the EPRI *Technical Assessment Guide*,[6] with costs converted from 1981 to 1986 dollars on the basis of growth in the implicit GNP inflator. No real escalation (increase above inflation) was assumed for the O&M cost component of any of the six options analyzed.

The prices paid by electric utilities for coal and natural gas were based on the A.G.A.-TERA Model Reference Case 1986-I, as set out in *Historical and Projected Natural Gas Prices: 1986 Update*[7] published by the American Gas Association. The A.G.A.-TERA Model is a detailed energy forecasting model developed and maintained by the American Gas Association. Gas prices employed in the analysis increased from $3.05 per MMBtu in 1986 to $3.89 in 2000, while coal prices rise over the same period from $1.85 per MMBtu to $2.28 ($1986). Due to the depressed state of the energy markets, current prices of both natural gas and coal are temporarily less than those set out in the TERA Reference Case (powerplants, on average, paid $1.62 per MMBtu for coal and $2.62 per MMBtu for gas over the first five months of 1986).

Environmental

The environmental impacts of the alternative generating systems operating at a 65 percent capacity factor were estimated for three primary air pollutants—sulfur dioxide (SO_2), particulate matter (TSP) and nitrogen oxides (NO_x). It was assumed that currently applicable federal emission standards for these air pollutants would be met. In addition, the production of sludge and ash were calculated. The coal system was assumed to operate on a commonly available coal type—2 percent sulfur, 10 percent ash—with a flue gas desulfurization (scrubber) system for SO_2 control, and an electrostatic precipitator for TSP control. All environmental results are presented on an incremental basis—i.e., the absolute increase or decrease in pollution resulting from putting the new unit on line. For example, since the heat recovery repowering option reduces coal consumption, and the increased consumptions of natural gas produces no SO_2, there would be a *reduction* in SO_2 emissions with this option.

A new coal-fired electric utility powerplant is subject to the new source performance standards of 1978 (NSPS) as set out in 40 *CFR* Subpart Da. The NSPS for SO_2 mandates a sliding scale for SO_2 control: facilities emitting less than 0.60 pounds of SO_2 per MMBtu must have reduced potential (uncontrolled) emissions by at least 70 percent, while facilities emitting from 0.60 to 1.20 pounds per MMBtu must have reduced potential emissions by 90 percent. The scrubber employed in this analysis was assumed to be 85 percent effective, reducing uncontrolled emissions from 3.8 pounds of SO_2 per MMBtu combusted to 0.6 pounds per MMBtu. The particulate NSPS is 0.03 pounds per MMBtu (99 percent reduction), and the NO_X NSPS requires a 65 percent reduction in total potential emissions with a cap of 0.60 pounds per MMBtu. The controlled coal plant emissions: SO_2—0.60 pounds per MMBtu; TSP—0.03 pounds per MMBtu; and NO_X—0.35 pounds per MMBtu, were multiplied by the total annual coal input to obtain annual emissions. Uncontrolled air pollutant emissions were obtained from the U.S. Environmental Protection Agency's (EPA's) *Compilation of Air Pollutant Emission Factors.*[8]

The only air emissions from combined cycle units originate in the gas turbine portion of their systems, since there is no supplemental firing of the steam turbine. SO_2 and NO_X emissions from stationary electric utility gas turbines in excess of 100 MMBtu per hour and built after 1982 are subject to NSPS, but there is no standard for particulate emissions from any turbine unit, nor are there any SO_2 or NO_X standards for units constructed prior to 1982. The NO_X standard for the heat recovery and boiler repowering options is approximately 75 parts per million of NO_X, which equates to 0.25 to 0.30 pounds of NO_X per MMBtu. NO_X emission from the peaking turbine repowering unit, the gas turbine of which is assumed to be pre-1982 and not subject to NSPS, are based on an uncontrolled emission rate of 0.41 pounds per MMBtu. SO_2 is limited to 0.015 percent of the exhaust gas total volume, but this standard is far in excess of actual SO_2 emissions from a gas turbine. Thus, uncontrolled SO_2 emissions from a gas boiler (0.0006 pounds per MMBtu according to U.S. EPA, *Compilation of Air Pollutant Emission Factors,*[9] were used to estimate the negligible SO_2 emissions from a turbine in each of the repowering scenarios. Uncontrolled TSP emissions from a gas turbine operating on distillate oil are 0.004 pounds per MMBtu according to Radian Corp., *Emissions from Stationary Gas Turbines,*[10] and emissions would be lower when operating on gas. It was assumed that the reduction when operating on gas would be 61 percent based on uncontrolled TSP emission factors for gas turbines contained in EPA's *Compilation of Air Pollutant Emission Factors* (0.0004 pounds per MMBtu times 39 percent = 0.0016 pounds per MMBtu for turbines when operating on gas).

Coal-fired electric powerplants produce a number of solid wastes including ash and sludge. Fly ash is collected by electrostatic precipitators or baghouses, while bottom ash is formed by fuel combustion. Sludge is a semi-solid waste which results primarily from scrubbing for SO_2. These two noncombustible solid wastes must be disposed of in landfills or ponds, and according to the U.S. EPA, *Energy/Environment Fact Book,*[11] a 1000 MW powerplant produces 700,000 tons of sludge per year and 250,000 tons of ash. These volumes were scaled down proportionately to estimate the wastes produced by the 75, 80 and 198 megawatt units considered in this analysis.

ANALYSIS RESULTS

The economic and environmental results of this analysis are presented in the two sections below, and they are summarized in Exhibits 22-6 through 22-8.

Economic Results

In the three scenarios examined, natural gas repowering was found to have a levelized annualized cost of only 60 to 65 percent of the comparable new coal plant option. Peaking turbine repowering has a total annual cost of $16.7 million versus $27.9 million for a new coal unit. Similarly, heat recovery repowering has a total annual cost of $41.7 million relative to $69.1 million for a new coal unit, and the comparison for boiler repowering versus coal is $16.8 million versus

Exhibit 22-6

Comparison of Economic and Environmental Results—
Peaking Turbine Repowering versus a
New Coal-Fired Powerplant

	Peaking Turbine Repower	New Coal-Fired Powerplant
Levelized Annual Cost (*$1986 MM*)		
Capital	$ 2.8	$14.7
Fuel	12.6	9.0
O&M	1.3	4.2
Total	$16.7	$27.9
Pollutant Emissions (*Tons/year*)		
Sulfur Dioxide	1	1,375
Nitrogen Oxides	760	800
Particulates	3	70
Sludge	—	56,000
Ash	—	20,000

$25.9 million. In each scenario the advantages of repowering in terms of lower capital and operating costs more than offsets the fuel cost advantage of the straight coal units.

The most striking part of the comparison is in the capital cost component where repowering is only 19 to 32 percent as costly as constructing a new coal unit. This advantage is due to a number of factors: repowering salvages useful existing equipment whereas the coal unit is new from the ground up; repowered units do not require expensive pollution control equipment as do new coal units (e.g., scrubbers and precipitators); coal equipment, as a rule, is larger and more costly than gas equipment; and, the new tax law does not treat capital intensive projects as kindly as existing law (e.g., repeal of the investment tax credit). The annualized capital cost of peaking turbine repowering is only $2.8 million versus $14.7 million for a new

Exhibit 22-7

Comparison of Economic and Environmental Results—
Heat Recovery Repowering versus a
New Coal-Fired Powerplant

	Heat Recovery Repower	New Coal-Fired Powerplant
Levelized Annual Cost ($1986 MM)		
Capital	$ 8.7	$36.6
Fuel	29.7	22.2
O&M	3.3	10.3
Total	$41.7	$69.1
Pollutant Emissions (Tons/year)		
Sulfur Dioxide	2	3,395
Nitrogen Oxides	1,290	1,980
Particulates	9	170
Sludge	—	138,075
Ash	—	49,320

coal unit. Heat recovery and boiler repowering have capital costs of $8.7 million and $4.4 million, respectively, versus their new coal counterparts of $36.6 million and $13.8 million.

The annualized fuel cost is greater for each of the repowering options than for the comparable new coal unit. The fuel cost associated with repowering ranges from 34 percent greater in the case of heat recovery repowering to 40 percent greater in the case of peaking turbine repowering. The price of gas delivered to powerplants over the 1985 to 2000 period is 65 to 70 percent greater than the cost of coal. However, the full advantage of this cost differential is not realized by the coal options because of their relatively lower efficiencies.

In each scenario the non-fuel operation and maintenance expense of the repowering options is roughly one-third the annualized O&M of the new coal unit. Coal units are more costly to maintain, in part

Exhibit 22-8

Comparison of Economic and Environmental Results—
Boiler Repowering versus a New Coal-Fired Powerplant

	Boiler Repower	New Coal-Fired Powerplant
Levelized Annual Cost		
($1986 MM)		
Capital	$ 4.4	$13.8
Fuel	11.2	8.2
O&M	1.2	3.9
Total	$16.8	$25.9
Pollutant Emissions		
(Tons/year)		
Sulfur Dioxide	(1,060)	1,290
Nitrogen Oxides	110	750
Particulates	(50)	65
Sludge	(53,310)	52,500
Ash	(19,040)	18,750

because they simply are dirtier and require constant upkeep to ensure smooth operation. There are also significant expenses associated with the storage and handling of coal, as well as the maintenance of pollution control equipment and disposal of sludge and ash.

Environmental Results

A comparison of the relative environmental impacts of gas repowering versus the addition of new coal-fired capacity is also presented in Exhibits 22-6 through 22-8. As indicated by the exhibits, the three repowering options all provide increased generating capacity with negligible increases (and in the case of boiler repowering, a net decrease) in the production of SO_2, TSP, sludge and ash. There is an increase in NO_x emissions attributable to repowering, but in all cases it is less than the increase that would result from adding new coal capacity.

Peaking turbine repowering would produce 1 ton per year of SO_2, 3 tons of TSP, and no sludge or ash. In contrast, the new 80 MW coal unit would produce 1,375 tons of SO_2, 70 tons of TSP, 56,000 tons of sludge and 20,000 tons of ash. NO_x emissions from peaking turbine repowering are 95 percent of coal unit NO_x emissions, 760 tons per year versus 800 tons per year. The gas turbines in this scenario are not subject to NSPS, as the units are assumed to be pre-1982, and they are thus higher in NO_x emissions relative to coal than in the other scenarios.

The other environmental results are similar, with the exception of NO_x emissions, for heat recovery repowering. The annual emissions of the respective gas repowering and coal alternatives are: SO_2, 2 tons versus 3,395; TSP, 9 tons versus 170; sludge, none versus 138,075 tons; and ash, none versus 49,320 tons. NO_x emissions of the repowering option are only 65 percent of the coal unit NO_x emissions—1,290 tons versus 1,980—due to both the higher efficiency and tighter standards of the gas-based unit.

Boiler repowering results in pollutant emissions which are not only less than those of the coal alternative, but emissions from the coal portion of the repowered unit which continues to operate show an absolute reduction, since about 14 percent less coal is consumed. That is, emissions from the repowered 527 MW coal and gas unit are less than from the original 452 MW coal unit. Boiler repowering results in a reduction in annual emissions of SO_2 (1,060 tons), TSP (50 tons), sludge (53,310 tons), and ash (19,040 tons). The increase in NO_x emissions is 110 tons per year, only 15 percent of 750 tons per year increase attributable to a new coal unit. (It should be noted that the potential NO_x reduction available via boiler repowering by adjusting flame temperatures and oxygen content, as alluded to in the "Scenario Description" above, is *not* included in this analysis because test results which would allow quantification of this benefit are not available at this time. The reduction presented herein is attributable to the substitution of gas for coal, and a higher unit efficiency.) The coal unit would also increase annual emissions of SO_2 (1,290 tons), TSP (65 tons), sludge (52,500 tons), and ash (18,750 tons).

CONCLUSION

Repowering with natural gas provides electric utilities with a clean, economical, efficient and reliable means to generate electricity. An estimate has not been made of the potential capacity available through repowering; however, the following table, based on the Energy Information Administration publication *Inventory of Power Plants in the United States*,[12] breaks out installed gas turbine capacity in the U.S. by Census region. If all installed turbine capacity were converted to combined-cycle operation with results similar to those estimated in the peaking turbine repowering scenarios, an additional 23 megawatts of capacity could be available—equivalent to 3.5 percent of total installed U.S. generating capacity. In addition, the 46 gigawatts now used primarily for peaking could be run in an efficient baseload mode. This number should be viewed as a maximum for illustrative purposes, as an engineering analysis of the existing equipment stock was not performed.

Potential New Capacity from Conversions to Peaking Turbine Repowering by Census Region

Census Region	Installed Gas Turbine Capacity (Gigawatt)	Potential New Capacity (Gigawatt)
New England	1.3	0.6
Middle Atlantic	11.0	5.5
East North Central	6.3	3.1
West North Central	5.8	2.9
South Atlantic	9.8	4.9
East South Central	3.0	1.5
West South Central	2.2	1.1
Mountain	2.9	1.4
Pacific	4.1	2.0
Total U.S.	46.4	23.0

FOOTNOTES

[1] North American Electric Reliability Council, *1985 Reliability Review*, Princeton, New Jersey: 1985) p.2.

[2] *Ibid.*, p. 4.

[3] E. Stephen Miliaras, "Conversion of Utility Gas Turbines to Combined Cycle Plants," (Presented at Energy Technology Conference XIII, Washington, DC: March 1986) p. 257.

[4] Electric Power Research Institute, *Technical Assessment Guide*, (Palo Alto, CA: May 1982) p. B-55.

[5] E. Stephen Miliaras, *op. cit.*, p. 163.

[6] Electric Power Research Institute, *op. cit.*, pp. B-55, B-79.

[7] American Gas Association, *Historical and Projected Natural Gas Prices: 1986 Update*, (Arlington, VA: June 10, 1986) p. 23.

[8] U.S. Environmental Protection Agency, *Compilation of Air Pollutant Emission Factors*, (Research Triangle Park, NC: August 1982) supplement 13.

[9] U.S. Environmental Protection Agency, *Ibid*, p. 3.3.1-2.

[10] Radian Corp., *Emissions from Stationary Gas Turbines*, (Research Triangle Park, NC: November 1984) p. 4-78.

[11] U.S. Environmental Protection Agency, *Energy/Environment Fact Book*, (Washington, DC: March 1978) p. 24.

[12] U.S. Department of Energy, Energy Information Administration, *Inventory of Power Plants in the United States, 1981 Annual*, (Washington, DC: September 1982).

Section V

Regulatory Considerations

SELECTED AUTHORS

Chapter 23

Repeal of the Fuel Use Act and Incremental Pricing

DON SCHELLHARDT

*Special Counsel and Executive Assistant
to the Vice President, Government Relations
American Gas Association*

On May 21, 1987, in what marked only the third formal signing ceremony of the year, President Reagan signed into law H.R. 1941 (which is now known as Public Law 100-42). The new law repeals several Sections of the Powerplant and Industrial Fuel Use Act (FUA), including all of the Act's outright prohibitions on the use of natural gas or oil. The new law also repeals all of the incremental pricing program that was originally established by Title II of the Natural Gas Policy Act (NGPA).

The House version of this legislation, which reached the President's desk, was sponsored by Representative Billy Tauzin (D-LA). The Senate version, S. 85, was only marginally different from H.R. 1941 and enjoyed the joint sponsorship of Senator Bennett Johnston (D-LA) and Senator Pete Domenici (R-NM). Key roles were also played by numerous other legislators, including Representatives Jack Bryant (D-TX), Jim Slattery (D-KS) and Dan Coats (R-IN). A pivotal role was also played by Senator Wendell Ford (D-KY), a coal state legislator who displayed a crucial willingness and ability to negotiate a fair compromise with Senators Johnston and Domenici. The resulting consensus language, which established "coal convertibility" requirements for new baseload powerplants, led to unanimous approval of this legislation in all of the reviewing Committees —and on the floor of both Houses of Congress.

Except for authorization of one last spurt of incremental pricing surcharges by those interstate pipelines which need to "clear" their incremental pricing accounts, all of the artificial gas demand constraints removed by H.R. 1941 were swept from the statute books on the date of the President's signature. This date—May 21, 1987— marks the end of one era and the beginning of another.

All of the industrial boilers who were once subject to incremental pricing surcharges have been freed from the rate distortions of this program. Now gas utility retailers of interstate pipeline gas are no longer legally barred from undercutting the price of residual oil— and/or the prices offered by certain competing gas suppliers—in the "swing" industrial boiler fuel market.

In the case of the powerplant and Industrial Fuel Use Act, the situation is somewhat more complicated.

The following customers are no longer affected in any way by the FUA:

- Existing Major Fuel Burning Installations (i.e., large industrial facilities). (Existing *powerplants* were already freed from the FUA's gas use prohibitions by language in the Omnibus Budget Reconciliation Act of 1981.)

- New Major Fuel Burning Installations (*except* for certain large new industrial cogenerators, which are discussed below).

- New outdoor gas light customers. (*Existing* outdoor gas lights were originally subject to a FUA mandate for termination of service in January of 1985, but this FUA directive was removed by the Omnibus Budget Reconciliation Act of 1981.)

- New powerplants that are used for peakload or intermediate load only (i.e., new powerplants whose total annual power generation never exceeds rated design capacity times 3,500 hours).

The following customers are free to use natural gas or oil IF they can meet the new FUA "coal convertibility" requirements:

- New baseload powerplants (i.e., new powerplants whose total annual power generation exceeds rated design capacity times 3,500 hours).

- New industrial cogenerators which meet *all* of the following criteria: (1) a rated design capacity of more than 100 million Btus per hour; *and* (2) total annual power generation which exceeds rated design capacity times 3,500 hours; *and* (3) sale or exchange, for *resale*, of at least 50 percent of the electricity generated.

If a facility that wishes to use natural gas is subject to the "coal convertibility" requirements in the new version of FUA Section 201, the facility can choose to meet these requirements through a simple self-certification procedure instead of the previously required, time-consuming and cumbersome FUA exemption process. *A facility can certify compliance with the "coal convertibility" requirements through one of two actions:*

- The facility can use a gas combustion turbine (*unless*, for some unusual reason, the turbine cannot be easily retrofitted for the use of coal gas); OR

- The facility can install a boiler that has been suitably expanded and modified to: (1) "enable it to handle coal," *and* (2) leave "adequate space for the addition of all necessary pollution control equipment."

In neither case is it necessary for an affected facility to acquire extra land (for possible future coal storage and handling facilities), to demonstrate the availability of coal transportation connections, or to focus on any other matter besides the "coal convertibility" of the on-site power generation equipment.

Functionally, the revised FUA provisions have the following practical effects:

(1) Free market competition between energy sources is established in all end use markets *except* new baseload powerplants and certain large new industrial cogenerators.

(2) The latter customer classes, when and if they choose natural gas, are guided toward use of energy efficient gas combustion turbines, which can easily shift to coal gas, and are strongly discouraged from installing new gas-fired boilers.

SECTION-BY-SECTION ANALYSIS

H.R. 1941 is divided into two Sections. Section 1 of H.R. 1941 repeals several specific portions of the Powerplant and Industrial Fuel Use Act (FUA), and rewrites FUA Section 201, with the net effect of removing all of the FUA's outright prohibitions on the use of natural gas or oil. Section 2 of H.R. 1941 repeals, in its entirety, the incremental pricing program that was established by Title II of the Natural Gas Policy Act (NGPA). This program had operated to prevent most gas utilities, under most circumstances, from retailing natural gas to industrial boilers at a price below Btu parity with high sulfur No. 6 oil.

Section 1—Repeal of Certain Sections of the Powerplant and Industrial Fuel Use Act of 1978

Section 1(a) (1)

- This provision repeals Section 202 of the FUA. In the process, it lifts all FUA restrictions on use of natural gas or oil by new Major Fuel Burning Installations. Under Section 103 (a) (11) of the FUA, such facilities are generally considered "new" if construction or acquisition began after April 20, 1977. These Major Fuel Burning Installations, or MFBIs, are basically large industrial facilities.

Comment: The repeal of FUA Section 202 shifts all new MFBIs to pure free market competition between rival energy sources. However, some large new industrial *cogeneration* facilities might be subject to the "coal convertibility" requirements that remain in the FUA. (For an outline of these requirements, see the Legislative Analysis discussion of Section 1 (c)(4) of H.R. 1941.)

The possible coverage of large new industrial cogenerators can be traced to the fact that, under certain FUA provisions which remain in force, some cogenerators are considered to be "electric powerplants." Since new baseload "electric powerplants" remain subject to "coal convertibility" requirements under the FUA, new industrial cogenerators are subject to the same requirements IF these cogenerators meet the statutory criteria for qualifying as "electric powerplants."

However, when remaining provisions of the FUA are combined with certain provisions of H.R. 1941, an industrial cogenerator must meet *all* of the following statutory criteria before it can be classified as an "electric powerplant":

(1) The cogenerator must have a *rated design capacity* (which should not be confused with actual energy consumption) of at least 100 million Btus per hour. (Section 103 (a)(7)(A) of the FUA)

(2) In at least one 12-month calendar period, the *actual amount of electricity generated* by the cogenerator (not counting emergency power generation) must exceed the rated design capacity times 3,500 hours (an average of rated design capacity times 9.86 hours per day). (New Section 201 (c) of the FUA, as established by Section (c)(4)(A) of H.R. 1941, when read in conjunction with Section 103 (a)(18) of the FUA)

(3) At least half of the annual power generation (which should not be confused with the total energy output) must be sold or exchanged for *resale*. (Section 103 (a)(7)(B) of the FUA)

As noted earlier, unless an industrial cogenerator meets *all three* of the above criteria, it is not affected by the FUA at all. For a potential cogenerator who wishes to avoid the "coal convertibility" requirements, the omission of any of these three elements will be sufficient.

Section 1 (a) (2)

- This provision repeals Section 302 of the FUA. The stricken Section had given the Secretary of Energy some discretionary authority to ban use of natural gas or oil by existing MFBIs, on a case-by-case basis, where DOE could affirmatively prove that a particular facility had the capability to switch to a different energy source.

Comment: The repealed Section was rarely used in practice. Consequently, its removal from the statute books will have little practical effect.

Section 1 (a) (3)

- This provision repeals Section 401 of the FUA. The stricken Section had provided discretionary authority for the Secretary of Energy to: (1) ban the use of natural gas in any new boilers which are "capable of consuming 300 Mcf or more of natural gas per day" (the equivalent of 12.5 million Btus per hour); and (2) ban the use of natural gas in any "oil capable" existing boilers which have an actual record of consuming 300 Mcf or more on a *peak* day.

Comment: This discretionary DOE authority was never actually exercised.

Section 1 (a) (4)

- This provision repeals Section 402 of the FUA. The stricken Section had barred gas utilities from hooking up any new (post-1978) outdoor gas lights.

Comment: An earlier version of FUA Section 402 had also set a January 1985 deadline for termination of utility service to *existing* outdoor gas lights. However, this earlier portion of FUA Section 402 was removed by Congress as one component of the Omnibus Budget Reconciliation Act of 1981.

Please note that the outdoor gas light restrictions were being administered, in most cases, by the individual States. Now that the statutory mandate for these gas light regulations has been eliminated, affected gas utilities may find it prudent to make sure that the implementing State agencies are aware of this change in the law.

Section 1 (a) (5)

- This provision repeals Section 405 of the FUA. The stricken Section had required that, for any existing powerplant which used coal (or another alternative to natural gas or oil) during 1977, special regulatory permission was needed before the powerplant could increase its consumption of oil above the 1977 use levels.

Section 1 (a) (6)

- This provision repeals Title V of the FUA. The stricken Title had prescribed the conditions under which individual electric utilities could choose a System Compliance Option (for meeting FUA requirements) in place of certain otherwise applicable restrictions on the fuel choices of individual powerplants.

Section 1 (a) (7)

- This provision repeals Section 801 of the FUA. The stricken Section had required the Secretary of Energy to conduct annual collections of information on the extent, characteristics, and productive capability of coal reserves in the United States.

Comment: With all the FUA Sections repealed by Section 1 (a) of H.R. 1941, the current text of FUA Section 301 is left untouched.

This preservation of FUA Section 301 protects from repeal certain policies that Congress wrote into the Section during deliberations in 1981. As originally written, FUA Section 301 had barred all *existing electric powerplants* from increasing their use of natural gas above the use levels that prevailed in the mid-1970s. Further, the original version of FUA Section 301 had required all existing electric powerplants to be "off gas" completely by 1990. However, this earlier version of FUA Section 301 was extensively rewritten by Congress during action on the Omnibus Budget Reconciliation Act of 1981.

This process produced two major changes. First, both the 1990 gas use ban and the pre-1990 gas use restrictions were removed. Thus, existing electric powerplants are now free to use whatever energy source they choose. Second, at the specific request of certain combination companies and electric utilities, the amended FUA Section 301 was structured to leave electric utilities the option of *requesting* DOE to prohibit one or more of their existing powerplants from using natural gas or oil. Because such a DOE prohibition order (even when it is *requested)* can automatically trigger a relaxation of certain Clean Air Act emission standards, some companies consider it desirable to retain the option of volunteering for "mandatory" coal conversion. *Nothing* in H.R. 1941 closes the door on this option.

Section 1 (b)

- This clerical amendment changes the FUA's Table of Contents in order to reflect repeal of several sections of the FUA.

Section 1 (c) (1) and (2)

- These are various conforming amendments.

Section 1 (c) (3)

- This language amends FUA Section 104 in order to provide that whatever FUA provisions remain in force shall apply only to facilities in "the contiguous 48 States and the District of Columbia." The practical effect is to take a previously applicable *partial* exclusion from FUA coverage for Hawaii and the U.S. Territories, and expand that into a total exclusion which now covers *Alaska* as well.

Section 1 (c) (4)

This Section of H.R. 1941 eliminates the current text of FUA Section 201 and replaces that text with an entirely new FUA Section 201. Both the previous and current text address FUA requirements for *new* powerplants (which Section 103 (a) (8) of the FUA generally defines as those on which construction or acquisition began after April 20, 1977).

- This Section creates a new FUA Section 201 (a). The new FUA Section 201 (a) states that, unless an exemption is affirmatively obtained under Subtitle B of Title II of the FUA, a new power-plant may not be "constructed *or operated*" as baseload pow-erplant unless it has "the capability to use coal or another alternate fuel as a primary energy source."

Comment: The previous text of FUA Section 201 had made *all* new powerplants subject to an outright ban on the use of natural gas or oil.

The new text of FUA Section 201 (a) shrinks this FUA coverage in two ways. First, instead of covering all new powerplants, it covers *baseload* powerplants only (although, as was noted in the discussion of Section 1 (a)(1) of H.R. 1941, large new industrial cogenerators can be treated as "new baseload powerplants" under certain specific circumstances). Second, instead of imposing an outright ban on use of natural gas or oil, the new FUA Section 201 (a) requires only the *"capability"* to use coal or another "alternate fuel"; there is no re-quirement that such coal or "alternate fuel" must actually be used in the powerplant.

If a facility wishes to be free from even this modified FUA man-date, exemptions are still available in Subtitle B of Title II of the FUA. FUA Section 212 contains the permanent exemptions, while Section 211 contains the temporary exemptions.

- The new FUA Section 201 (b) provides some definitional guid-ance on compliance with the "coal convertibility" mandate in new FUA Section 201 (a). First, under new FUA Section 201 (b)(1), the new baseload powerplant must have *"sufficient in-herent design characteristics to permit the addition of equipment* (including all necessary pollution control devices)" that would be required for use of coal, or another alternate fuel, as a pri-mary energy source. Second, under new FUA Section 201 (b)(2), the facility *must not be "physically, structurally, or tech-noligically precluded" from using coal* (or another alternate fuel) as a primary energy source. The new FUA Section 201 (b) adds that these statutory requirements for an affected facility "shall not be interpreted to require any such powerplant to be immediately able to use coal or another alternate fuel as its primary energy source on its initial day of operation."

Comment: Several points bear emphasis.

(1) *The "backup" fuel can be coal gas as well as the solid form of coal.* A new powerplant that can be converted to coal gas meets the new FUA requirements because FUA Section 103(a)(5) defines "coal" as "anthracite and bituminous coal, lignite, and any fuel derivative thereof."

In addition, coal is not the only energy source which qualifies as an "alternate fuel" for purposes of FUA compliance. FUA Section 103(a)(6) defines "alternate fuel" as "electricity or any fuel, other than natural gas, or petroleum." Specific examples listed in this statutory definition include petroleum coke; shale oil; uranium; biomass; municipal, industrial or agricultural wastes; wood; "renewable and geothermal energy sources"; certain commercially unmarketable "waste products of refinery or industrial operations"; and "waste gases from industrial operations." Propane is not considered to be an "alternate fuel," since "liquid petroleum gas" is included in the FUA Section 103(a)(3) definition of "natural gas."

(2) *The compliance standard is retrofittability.* As noted above, the new FUA Section 201(b) speaks of "inherent design characteristics" which "permit the addition of equipment" needed for the use of coal, coal gas, or some other "alternate fuel." The same Section then adds that a facility does not have to be "immediately able" to use coal or another alternate fuel "on its initial day of operation." Consequently, it is clear that the equipment in question needs only to be easily *retrofittable.* This is why, although the statutory language itself speaks of "capability" for coal use, the legislative sponsors of H.R. 1941 more frequently spoke of *"convertibility"* for coal use. "Coal convertibility" captures more clearly the actual spirit of the new FUA requirements.

(3) *"The coal convertibility" requirements only apply to the on-site power generation equipment.* The relevant Senate Energy and Natural Resources Committee Report mentions that the Committee specifically rejected statutory language that "would have required an

electric powerplant to have adequate land space available to enable it to be converted to coal." (S. REP. NO. 30, 100th Congress, 1st Session at 7 (1987)). In discussing new powerplants that utilize gas combustion turbines, the same Report adds that *"it is only the initial power generation equipment itself* that must not be physically, structurally, or technologically precluded from using coal." *(Supra.* at 8.)

(4) *Gas combustion turbines, because they are easily retrofittable for the use of coal gas, are generally presumed to meet the new "coal convertibility" requirements.* The relevant Senate Energy Committee Report offers the following commentary: "According to information submitted to the Committee, no initial, additional cost would be involved by requiring a new combustion turbine to be capable of being converted to coal after it is installed. For example, combustion turbine powerplants currently being manufactured in the United States meet this coal convertibility requirement . . . Initially, such a turbine could use natural gas. The turbine, once installed, could later be retrofitted into an integrated gasification combined-cycle powerplant . . ." *(Supra.* at 7.)

Both the Senate Energy Committee Report, and its House Energy and Commerce Committee counterpart, mention the experimental Coolwater powerplant as an example of such a combined-cycle coal-gas facility. The 100 megawatt Coolwater powerplant produces medium Btu coal gas from a Texaco gasifier and burns it in General Electric combined-cycle turbines. The powerplant is located near Barstow, California.

The Senate Energy Committee Report also notes that, in the case of gas combustion turbines, the new FUA Section 201 "does not require the owner or operator to have adequate land space availability for the coal gasifier or other coal handling facilities. If a powerplant owner or operator should decide in the future that it is economic to utilize coal, the owner or operator could obtain off-site acreage for its coal gasification facilities and could connect the two sites by pipeline." *(Supra.* at 8.)

(5) *The "coal convertibility" requirements for new gas-fired boilers are more stringent.* The Senate Energy Committee Report

mentions the following criteria: "In the case of a baseload powerplant employing a boiler, the boiler must have characteristics (e.g., size) enabling it to handle coal, and there must be adequate space for addition of all necessary pollution control equipment. Space for actual coal handling facilities would not be required." *(Supra.* at 7.)

- The new FUA Section 201 (c) totally excludes peakload powerplants and intermediate load powerplants from any "coal convertibility" requirements under the Act.

Comment: As was noted in the discussion of Section 1 (a) (1) of H.R. 1941, this language—when combined with Section 103 (a) (18) of the FUA—operates to exclude from the "coal convertibility" requirements any power generation facility whose annual power generation consistently falls below rated design capacity times 3,500 hours.

- The new FUA Section 201 (d) permits affected facilities to choose to establish "coal convertibility" compliance through an optional self-certification procedure instead of the previously required, time-consuming and cumbersome FUA exemption process.

Comment: A.G.A. anticipates that, within the next few months, DOE will be issuing proposed regulations which define the format and criteria for self-certification.

The Senate Energy Committee Report notes that "The exemption process is still available." *(Supra.* at 8.) The Report adds, however, that "Nothing in the new self-certification procedure precludes the owner or operator of a new, baseload powerplant that already has an exemption pending from filing under the new self-certification provision." *(Supra.)*

Section 1 (c) (5)

- This provision amends FUA Section 211 (a) to permit petitioners for a cost exemption to compare the projected cost of coal (or another "alternate fuel") against the projected cost of "the fuel that would be used" if the exemption were granted. FUA Section

211 (a) had previously required cost exemption petitioners to use the projected cost of "imported petroleum" as a yardstick in all cases.

Comment: The previous FUA cost exemption language had sparked regulatory debates over whether the price of No. 6 oil or the price of No. 2 oil should be used as the "yardstick" for measuring the net projected cost of coal (or another alternate fuel). With the new statutory language, cost exemption petitioners that want to use natural gas can compare projected gas costs directly against projected coal costs. They will no longer be required to employ an often inaccurate and controversial fuel oil "proxy" for natural gas.

Section 1 (c) (6) through (24)

- These provisions embody a long string of conforming amendments.

Section 2—Repeal of Incremental Pricing Requirements

Section 2 (a)

- This is the specific provision that actually repeals the entire incremental pricing program, as established by Title II of the NGPA.

Comment: The incremental pricing program had basically used a very complicated mechanism to achieve the relatively simple result of price parity between natural gas and high sulfur No. 6 oil in the industrial boiler fuel market. This artificially imposed price parity served to distort interfuel competition, *in favor of imported oil,* by precluding otherwise available oil displacement opportunities in those areas where natural gas could have been sold for a lower price than high sulfur No. 6 oil. In addition, because the program only applied to natural gas that was "acquired by interstate pipelines," incremental pricing saddled interstate pipeline system supply with an artificial disadvantage in its competition with: (1) intrastate market gas; and (2) "carriage gas" purchased directly from producers by either industrial users or gas utilities. Thus, the incremental pricing program was distorting both interfuel competition and gas-to-gas competition. The long overdue repeal of the program should lead to higher gas sales in general and more competitively priced interstate pipeline supplies in particular.

Section 2(b)

- This provision allows incremental pricing surcharges to continue just long enough to permit interstate pipelines to recover costs which were incurred, but not yet recovered, under the incremental pricing scheme. Therefore, where it is necessary, the program may remain in force long enough to allow affected interstate pipelines to "clear their accounts."

Section 2(c)

- This provision authorizes the Federal Energy Regulatory Commission to "take appropriate action to implement this section" (i.e., to dismantle the incremental pricing mechanism).

Chapter 24

Public Utilities Regulatory Policy Act: Actual versus Avoided Costs of Purchased Cogenerated Electricity

PAUL L. WILKINSON
Manager, Policy Analysis

BRUCE P. McDOWELL
Policy Analyst
American Gas Association

Cogeneration offers a highly efficient and potentially economic alternative to conventional energy systems. While the typical electric powerplant releases two-thirds of its input energy to the environment, cogeneration systems capture and use a substantial portion of this wasted heating, cooling, or meeting industrial process needs. Cogeneration has also become an important supplement and alternative to traditional baseload electric powerplants.

These efficiency and competitive factors motivated Congress to include provisions in the Public Utility Regulatory Policies Act of 1978 (PURPA), designed to encourage the use of cogeneration in nonutility applications. One of the major provisions of PURPA and its subsequent federal implementation rules requires electric utilities to purchase cogenerated electricity at a rate which can equal, but not exceed, the utility's avoided cost of generating or purchasing additional electricity. There is considerable debate today regarding the level at which these "buyback" rates have been set before state commissions.

This chapter examines the actual cost of cogenerated electricity relative to the respective "avoided costs" as set by the state regulatory commissions, and the cost of other utility purchased power. The chapter is based on 87 cogeneration facilities which have filed for qualifying status under PURPA (QFs). These 87 facilities, located throughout the U.S., have a combined capcity of nearly 5,700 megawatts—about one-fourth of the total filed with the Federal Energy Regulatory Commission (FERC), and a much higher proportion of the capacity *on line* today.

It should be noted that this chapter focuses on the avoided cost rate as established by the individual state PUCs relative to the rate paid for cogenerated electricity. No attempt was made to determine whether or not the avoided cost rates established accurately reflect the "true" avoided cost of the individual utilities. States which set artificially high avoided cost rates subsidize excessive cogeneration to the detriment of electric utilities and their ratepayers. Conversely, avoided cost rates set too low discourage worthwhile cogeneration projects, forcing utilities to pursue more expensive options thereby penalizing ratepayers. Thus, where cogenerated electricity is properly priced, both gas and electric utilities, as well as their customers, benefit. On the other hand, when it is improperly priced (either high or low), all parties suffer in the long run.

EXECUTIVE SUMMARY OF
ANALYSIS RESULTS

Based on the sample QFs, virtually all cogenerated electricity purchased in 1985 was at a rate at or below the electric utilities' avoided cost. In addition, over two-thirds of cogenerated electricity was purchased at rates lower than the utilities pay for other purchased electricity (see Table 24-1).

- The "avoided cost" is the theoretical cost which an electric utility would incur for the purchase or generation of an incremental unit of electricity absent cogeneration power purchases from QFs. Of the electricity sold by the 87 QFs to electric utilities in 1985:

Table 24-1. The Actual Cost to Electric Utilities of Cogenerated Power Purchased from Qualified Facilities[1]

	Percent of Purchased Electricity
A. Actual Cost of Cogenerated Electricity Relative to the "Avoided Cost"[2]	
Above the Avoided Cost Range[3]	0.6
Below the Avoided Cost Range	5.4
Within the Avoided Cost Range[4]	93.0
Data Not Available	1.0
	100.0
B. Actual Cost of Cogenerated Electricity Relative to the "Standard Rate"[5]	
Above the Standard Rate Range	0.6
Below the Standard Rate Range	7.0
Within the Standard Rate Range[4]	91.4
Data Not Available	1.0
	100.0
C. Actual Cost of Cogenerated Electricity Relative to Other Purchased Power[6]	
Above Cost of Other Purchased Power	29.1
Below Cost of Other Purchased Power	70.9
Equals Cost of Other Purchased Power	—
Data Not Available	—
	100.0

[1] Based on sample of 87 qualified facilities which account for one-fourth of the QF capacity filed with FERC.

[2] Cost of generating or purchasing additional electricity, as set by the state regulatory commission.

[3] Results from sales governed under contracts signed before 1985 when the applicable avoided costs were higher.

[4] Seventy percent of subtotal from state of Texas, which did not calculate avoided costs in 1985, but all contracts were reviewed by the regulatory commission and the rates are considered to average below the cost of building a new powerplant. See discussion on page 262.

[5] Approved rate which utilities must pay qualified cogenerators for electricity, absent negotiation. Usually applied to smaller units.

[6] Excludes cogeneration from qualified facilities.

Note: Does not reflect effect of waste heat recovery savings.

Sources: 1985 Form 1s filed with the Federal Energy Regulatory Commission by major privately-owned electric utilities; *Energy User News*, Fairchild Publications (New York, NY: various issues); phone survey of various state utility regulatory commissions.

— Only 0.6 percent was above the 1985 avoided cost range. The cost of electricity purchased may exceed the current avoided cost in a given year if the contracted rate was determined in a prior year when the avoided cost was higher.

— 93 percent was within the established avoided cost range.

— 5.4 percent was below the avoided cost range.

— The above comparisons do not imply that state determined avoided cost rates have been set at the correct level, rather this is merely one yardstick for examining buyback rates.

• The "standard offer" is a rate required by FERC to be offered to small QFs with capacities less than 100 kilowatts. This rate may serve as a starting point for the negotiations of larger QFs.

— Similar to the results in the avoided cost comparison, 0.6 percent of the cogenerated electricity purchased was above the standard offer level in 1985, 91.4 percent was within the standard offer range, and 7.0 percent was below the standard offer.

— Approximately 70 percent of the cogenerated electricity purchased in 1985 was in the State of Texas which did not set avoided costs in 1985, and for which standard offer data was incomplete. However, all power purchase contracts had to pass commission review, and the QF rates in Texas are generally regarded to be below the utilities' avoided costs.[1] Thus, the cost of purchased power in Texas was considered to be within the avoided cost and standard rate ranges as set out in Table 24-1.

• The cost of cogenerated power from QFs was also compared to the cost of other electricity purchased by utilities.

— Approximately 70 percent of the cogenerated electricity was less costly than the average cost of other power purchased by the utilities, while 30 percent was more costly.

— Baseload cogeneration may reduce a utility's flexibility to purchase economical energy by requiring utilities to take QF power when they may have less expensive alternatives. Due to data limitations, this report could not quantify the difference between the cost of power which could have been purchased at particular points in time absent cogeneration and the cost of cogenerated power.

• Of the 16 million megawatt hours of electricity sold by the 87 QFs to electric utilities in 1985, 87 percent was generated with natural gas.

— Due to the declining price of natural gas, the cost of electricity purchased from new gas-fired QFs may be more attractive relative to the avoided cost than older gas-fired units without fuel adjustment clauses. However, older units coming on-line today without fuel adjustment clauses may produce cogenerated power which is more costly than power available from other sources, since fuel costs are likely lower than when the cogeneration agreement was executed. (This situation has presented a problem in some parts of California; see discussion on pages 266–267).

— Nearly 50 percent of all cogeneration capacity in the U.S. which is on line, under construction or on order is gas-fueled. The fact that 87 percent of the electricity generated by the sample in this analysis is gas-fueled suggests that the lower capital costs and shorter lead times of gas units makes them more likely to be completed than cogenerators fueled by other energy sources.

BACKGROUND

At the time of the passage of PURPA, U.S. energy markets were in a state of turmoil. PURPA was one of several pieces of energy legislation passed in an effort to respond to the market uncertainties following the Arab oil embargo. The goals of PURPA were to conserve electric energy, encourage the efficient use of utility plant and fuel, and to promote equitable rates to electric utility customers.

Fostering the growth of privately-owned cogeneration facilities was considered to be one way to meet these goals. PURPA and its regulations, as established by the FERC, removed some of the barriers to cogeneration growth by forcing utilities to purchase electricity from, and provide back-up electricity to, cogenerators at a "just and reasonable" rate. In addition, cogenerators were exempted from certain federal and state regulations that pertain to electric utilities.

In order to get these benefits, cogeneration facilities must meet certain standards, such as minimum efficiency requirements. Facilities that have filed applications with the FERC and that meet the standards are granted qualifying status.

METHODOLOGY

This analysis entailed two major steps—calculating each utility's average cost of purchased electricity from both qualified cogenerators and other outside sources of electricity, and determining the avoided cost for each utility as set by the respective state regulatory commission. Over 140 utilities were examined in the analysis. Several utilities were excluded because they served states from which no facilities had filed with the FERC for qualifying status (Nevada, South Dakota, Nebraska, Vermont, and Rhode Island). Only cogeneration facilities were included in this analysis, not small power producers which may also be QFs.

The 1985 Form 1 that each major, privately-owned electric utility filed with FERC served as the source for the amount and cost of cogenerated and other purchased power. Account number 555 of the Form 1 lists purchased power, except interchange power, from associated and nonassociated utilities, associated nonutilities, other nonutilities, municipalities, cooperatives, and other public authorities. Cogenerators are listed in the other nonutilities category.

The list of purchases from other nonutilities was then compared to the list of facilities that have filed with FERC for qualifying status under PURPA.[2] Only those that matched were included in the analysis sample. Facility owner, location, and, where possible, size were used to identify those cogeneration facilities that had filed with the FERC and were selling electricity to utilities.

The sample listing of cogenerators that sell electricity to utilities is not all inclusive. It is likely that many such facilities are not included, for a variety of reasons.

- Not all utilities detailed their cogeneration purchases by each source, instead grouping them into a generic category such as "Misc. cogenerators." Thus, those filing with FERC could not be identified for these utilities.

- Some facilities filed with FERC under one name, such as the builder's or financer's name, and were listed in the utility's Form 1 under a different name, such as the facility's owner's name, and thus could not be matched.

- Some facilities were listed in the interchange power account, but the cost per kilowatt of energy sold to the utility could not be precisely determined when listed in this account, and thus they were not included. (Interchange power refers to a facility that sells electricity to, and purchases electricity from, a utility.)

- Not all cogeneration facilities have or will file with the FERC for PURPA benefits, and a significant amount of cogenerated electricity is being sold to utilities by facilities that have not filed with the FERC. Since these facilities are not covered under PURPA, they were not included in the analysis.

- Only large privately-owned electric utilities were examined, and thus cogenerators that sell to small private or any public electric utility were not included. However, the vast majority of QFs sell electricity to the major utilities.

Despite these limitations, 87 QFs were included in the sample. Specifically, the database includes all operating cogeneration facilities which have qualified for PURPA benefits or have QF applications pending, and for which facility-specific data are publicly available in the FERC Forms 1. These 87 facilities, located throughout the U.S., have a combined capacity of 5,689 megawatts—about one-fourth of the total filed with the Federal Energy Regulatory Commission (FERC) prior to 1986, and a much higher proportion of the capacity *on line* today—a significant portion of filed capacity has not yet been constructed.

Once a utility's cogeneration purchases from filed facilities were identified, the total purchased power cost paid to those facilities was divided by the total kilowatt hours provided by those facilities. The result was the utility's average cost per kilowatt hour for purchasing cogenerated electricity from facilities that had filed with the FERC. The utility's cost per kilowatt hour for the remaining (net or filed cogeneration) electricity purchases was calculated by dividing the remaining purchased electricity expense by the remaining kilowatt hours received (excluding interchange power).

The standard rates for 1985 offered by electric utilities to cogenerators were obtained from a survey done by the *Energy User News*.[3] Standard rates are required under FERC rules for purchases from qualifying facilities with a design capacity of 100 kilowatts or less. Also, standard rates for facilities greater than 100 kilowatts may be set. The rates listed in the survey are usually for systems 100 kilowatts or less, but these rates often apply to larger systems or serve as a starting point for the larger facilities' rate negotiations. For the states in this study (those with filed cogenerators which sell electricity to utilities), about half have ruled that the standard rate applies to all units.

The standard rates are usually shown as a range expressed in cents per kilowatt hour. For those utilities that do not offer a capacity credit, the range reflects the seasonal variation in the energy credit. For those utilities that offer a capacity credit, the low end of the range reflects the lowest energy credit offered, and the high end represents the highest energy credit plus the highest capacity credit offered. If the capacity credit is listed in dollars per kilowatt per year, a cents per kilowatt hour figure is calculated by dividing the dollar per kilowatt figure by a 8760 hours per year. This yields a relatively conservative result, because few cogeneration units operate every hour of the year.

As stated above, PURPA does not require either long-term offers or any kind of standard offer for QFs above 100 kW. However, California adopted a long-term offer for capacity and another long-term offer for both capacity and energy. Both offers established prices which, given the subsequent significant declines in oil and gas prices, now appear to have been above market levels in some cases and the California Public Utilities Commission prospectively suspended them. The methodology of this study does not capture such contracts which will supply increasing quantities of power to the grid.

One of the California utilities has estimated that even if only some of those projects come on line, overpayments could increase by nearly 500 percent and reach $840 million annually as early as 1991. Problems such as this do not indicate a fundamental flaw in PURPA; rather, they indicate the need for appropriate guidance at the state level.

A phone survey of state utility regulatory commissions was performed to determine the avoided costs for each utility in the sample. Avoided costs are "the incremental costs to an electric utility of electric energy or capacity or both which, but for the purchases from the qualifying facility, or qualifying facilities, such utility would generate itself or purchase from another source."[4] Avoided costs can vary from standard rates. The state can set standard rates lower than avoided costs if the state feels that this lower standard rate can encourage cogeneration development. In addition, the standard rates may vary according to unit size. However, most of the standard rates for utilities in the sample were equal to the avoided costs.

As stated in the Executive Summary, the state approved avoided cost rate may not accurately reflect the true cost of available incremental power in all cases. For example, the California experience demonstrates overpayments to some QFs even though the buyback rates are not explicitly set above avoided costs. Buyback rates are set by multiplying an "incremental energy rate" (IER) by an avoided fuel price. The California Public Utilities Commission has set both parts of the equation above the actual avoided cost. For one California utility, error due to the CPUC's forecast of IERs higher than those which actually occurred cost ratepayers $102 million in overpayments from 1983 to 1985. In addition, cogenerators reap a benefit in the second term of the equation, the avoided fuel prices, because the rate at which they can buy gas is less than the gas price used for avoided cost calculation. This cost ratepayers of the utility $161 million from 1983 to 1985. Thus, the utility's ratepayers overpaid QFs by $263 million even though the QF prices allegedly represented avoided costs. Similarly, while Table 24-2 indicates that the avoided costs of the listed California utilities are in the 5–8 cents range, in fact, at least one of those utilities has a current incremental cost of energy less than 3 cents. Again, such issues must be addressed at the state level.

Table 24-2. Summary of Findings by Utility
(1985 Data)

State	Electric Utility	Average QF Electric Buyback Rate[1] (¢/kWh)	Avoided Cost Range (¢/kWh)	Standard Offering Range (¢/kWh)	Average Cost for Utility's Purchased Electricity[2] (¢/kWh)	Amount of Cogenerated QF Electricity Purchased (MWhs)
Alabama	AL Power Corp.	2.275	2.178-2.850	2.330-3.050	2.720	8,950
California	Pacific Gas & Electric	7.493	4.978-7.964	4.978-7.964	2.915	1,628,115
	Southern CA Edison	6.239	4.000-7.527	4.000-7.527	2.618	1,158,074
Colorado	Pub. Serv. Co. of CO	4.301	1.567-4.322	1.567-4.322	2.879	402
Florida	FL Power & Light	3.173	3.511-4.878	3.511-4.732	5.440	49,768
Georgia	Savannah Elec & Pwr Co.	2.529	2.480-4.058	2.480-4.058	3.911	12,082
Idaho	Idaho Power Co.	7.312	4.300-5.300	4.300-5.300	3.844	38,363
Louisiana	Gulf States Utilities	2.545	2.545[3]	4.334-4.983	3.985	137,211
	New Orleans P.S. Co.	2.974	2.974[3]	3.683-3.730	5.294	5,765
Maine	Central Maine Power	6.273	6.800	6.800	3.390	791,250
Maryland	Baltimore Gas & Electric	2.255	1.740-4.824	1.740-4.824	2.626	1,604
Massachusetts	Western Mass. Elec.	4.454	[4]	4.164-4.978	3.315	13
	New England Power Co.	3.806	[4]	3.090-4.910	3.892	1,747
Michigan	Consumers Power	3.020	[4]	2.720-6.480	2.96-5.15	17,631
	Detroit Edison	1.998	[4]	2.030-5.300	10.130	1,965
New Jersey	Jersey Central Pwr. & Lt.	4.594	3.050-5.864	3.050-5.964	4.714	21,887
	Pub. Ser. Elec. & Gas	5.634	2.730-6.134	2.730-6.134	2.902	89,406

State	Utility					
New York	Niagara Mohawk Poer	4.067	3.377–7.172	3.377–7.172	2.015	85,765
N. Carolina	Duke Power Co.	2.560	2.160–6.140	2.160–6.140	5.930	87,530
	Carolina Power & Lt.	6.183	2.800–7.160	2.800–7.160	4.032	154,014
Oklahoma	Pub. Ser. Co. of OK	4.113	4.113[3]	N/A[5]	2.040	26,592
Pennsylvania	Penn. Elec. Co.	2.850	4.020	4.020	3.740	15,728
S. Carolina	SC Elec & Gas Co.	2.414	2.184–2.759	2.184–2.759	3.082	67,309
Texas	Central Pwr. & Lt. Co.	2.685	N/A[5]	3.103	2.444	79,229
	Houston Ltg. & Pwr.	4.630	N/A	N/A	9.330	8,481,617
	S.W. Elec. Pwr. Co.	2.098	N/A	3.032–3.511	2.147	7,570
	Gulf States Utilities	2.506	N/A	3.515–6.397[6]	3.985	28,744
	TX-NM Power Co.	3.825	N/A	N/A	4.877	2,376,631
	Texas Utilities Co.	5.405	N/A	N/A	3.119	507,649
Washington	Puget Sound Pwr. & Lt.	5.585	1.220–1.600	1.220–1.600	5.411	56,452

[1] Includes facilities that have filed with the FERC and sell cogenerated electricity to a utility.

[2] Purchased power net of QF power.

[3] Negotiated contracts between QF and utility generally reflect avoided costs, according to commission staff.

[4] Commission staff estimates that avoided costs are at least as great as standard rates.

[5] Not available—either rates have not been determined by the commission or were not included in *Energy User News* survey.

[6] Represents range of energy cost (actual for April 1985) to the energy cost times an energy payment multiplier as listed in the tariff.

Sources: 1985 Form 1s filed with the Federal Energy Regulatory Commission by major privately-owned electric utilities; *Energy User News*, Fairchild Publications (New York, NY: various issues); phone survey of various state utility regulatory commissions; Tariff filings by Texas utilities for standard offerings.

DISCUSSION OF THE RESULTS

The findings of this analysis indicate that qualified cogenerators are being paid appropriately based on PURPA guidelines for their electricity with respect to avoided costs, and that these electricity sales in many cases are helping to lower utilities purchased electricity costs. (It should be noted that this analysis assumes that the avoided costs, as calculated or approved by each state regulatory commission, are correct. That is, they accurately reflect the avoided cost of the utilities.)

For utilities with approved avoided costs, less than one percent of the cogeneration sales averaged above the respective utility's 1985 avoided cost and standard offerings (Table 24-1). Only two utilities, one in Idaho and one in Washington, purchased electricity from QFs above the approved 1985 avoided cost rate (Table 24-2). Discussions with the staff of the respective state utility regulatory commissions revealed that both of these cogenerators had signed contracts with the utilities in the early 1980s, when avoided costs were higher. At that time, then, those cogenerators were paid rates within the proper avoided cost range, and thus are considered as being compensated at the proper avoided cost. A limited number of other individual cogenerators were found to be paid at rates above 1985 avoided costs and standard offerings (when averaged with the other cogenerators in the utility's service area; however, the utility's average purchase rate was within the proper 1985 range), and discussions with appropriate commission staffs revealed that these facilities also had signed contracts prior to 1985 when avoided costs were higher.

Ninety-three percent of the QF cogeneration sales were at a cost equal to or within the range of the 1985 avoided costs and standard offerings. These sales accounted for the majority of data for which 1985 avoided cost and standard offering rates were available. Most of these sales were within a range for avoided costs or standard offerings —the range spans from the lowest energy credit offered to the highest energy credit offered plus the highest capacity credit offered (where applicable—not all utilities have a capacity credit in their avoided costs). No attempt was made to verify that the cogenerator was in the proper section of the range, since that would require information that was not readily obtainable, such as the specific times at which the power was sold. It was assumed that if the cost was within the range, then the cost was equal to the avoided cost or standard rate.

Between five and six percent of the QF cogeneration sales were at rates lower than the 1985 avoided cost and standard rates. The FERC implementation rules of PURPA allow states to set rates below avoided costs, and the rules also allow negotiated rates between utilities and QFs. Thus cogenerators can, and do, sell electricity at rates below avoided costs. Cogenerators will often trade higher rates for other contract concessions, such as a guaranteed long-term purchase commitment.

Over 70 percent of the cogenerated electricity in this sample is being sold in Texas, where avoided costs were not calculated or approved in 1985. Despite a lack of definitive avoided costs, these sales are perceived to be at rates which are at or below the theoretical avoided costs. A 1986 Department of Energy study, using data as current as December 1985, stated that in Texas ". . . utilities have negotiated with qualifying facilities on contract terms and conditions in such a way as to result in rates that appear to be lower than the cost to the utility of building additional capacity."[5] In addition, all contracts are reviewed by the state public utility commission.

On average, then, virtually all of the cogenerated electricity sold to utilities in 1985 was at or below the applicable avoided cost rate. Some critics of PURPA contend that this rate is too high, that these sales will increase consumer electricity costs.

In reality, in many cases these cogeneration purchases have likely helped to lower the average U.S. utility purchased electricity costs. Over 70 percent of the cogenerated power purchased by utilities was at a rate below their average purchased cost of electricity (net of the cogeneration sales in this sample) obtained from outside sources. Of the 30 utilities in the sample, 17 of them, in 12 different states, were purchasing cogenerated electricity at rates lower than their average purchase electricity costs. Future cogenerated electricity purchases should continue to be attractive to most utilities, since the recent decline in energy prices has caused many utilities to lower their 1986 avoided costs.

The problems which have occurred in the setting of avoided cost rates, especially in parts of California, are largely problems which need to be worked out at the state level. It is up to the individual states to see that buyback rates truly reflect the avoided cost of the utilities within their boundaries. If this is accomplished, cogeneration will be neither over- nor underutilized as a source of electricity, and gas and electric ratepayers will be the ultimate benefactors.

REFERENCES

1. Pfeffer, Lindsay, & Associates, Inc. for the U.S. Department of Energy, *Emerging Policy Issues in PURPA Implementation* (Washington, DC: March 1986).
2. Federal Energy Regulatory Commission, *The Qualifying Facilities Report* (Washington, DC: January 1, 1986).
3. "State's Cogeneration Rate-Setting Under PURPA, Parts 1-10" *Energy User News,* Fairchild Publications (New York, NY: various issues).
4. *Code of Federal Regulations* Federal Energy Regulatory Commission, 18 CFR Ch 1. Part 292, "Regulations Under Section 201 and 210 of the Public Utility Regulatory Policies Act of 1978 With Regard to Small Power Production and Cogeneration." National Archives and Record Administration (Washington, DC: April 1, 1986).
5. Pfeffer, Lindsay & Associates, *Emerging Policy Issues . . . ,* pg. vi.

Section VI

Marketing Cogeneration: Natural Gas Distribution and Pipeline Company Perspectives

SELECTED AUTHORS

Chapter 25

The Risks and Rewards of Gas Company Cogeneration Development

RALBERN H. MURRAY
Consolidated Natural Gas Company

The rewards of cogeneration for a gas company can be substantial. More importantly, these rewards are multiplied by the allied cogeneration benefits that fall to other parties. These include: cost savings to industry and commercial institutions; to local communities, in which these employers and institutions provide a substantial number of jobs and services; and also, to electric utilities.

Figure 25-1 shows an estimate by the Gas Research Institute of this potential cogeneration benefit. This includes a potential cogeneration market of some 55,000 megawatts, of which about two-thirds is industrial. The total represents a potential gas load of 3.5 trillion cubic feet (Tcf) each year with attendant savings for energy uses and gas consumers. Considering that the nation's current gas demand stands at under 17 Tcf, one can see why cogeneration cannot be ignored.

NATIONAL COGENERATION MARKET POTENTIAL...

● Market (in Megawatts)

Sector	Ownership			
	Utility	Third Party	Self	Total
Multifamily	210	430	210	850
Commercial	7.960	8.460	2.330	18.750
Industrial	11,950	9.230	15.000	36.180*
	20.120	18.120	17.540	55.780

*DOE 43,000 in Total
*GRI 11,000 by 1995

● Potential Gas Sales Are 3500+ Bcf Per Year
● Seventy-Five Percent Are Currently Third-Party Ownership

Source GKCO-ARGONNE-GRI

Figure 25-1

Thus, potential gas sales are clearly an important "reward" of gas industry involvement in cogeneration. But what level of cogeneration gas sales increase can actually be achieved? Why is this important to anyone but the gas industry? A quick way of answering these questions is to look at a typical cogeneration project.

A hypothetical, but reasonably real-world, application is shown in Figure 25-2. As shown at the top of the diagram, this industrial customer requires about 44,000 pounds of steam per hour, which is produced from a conventional gas boiler. To raise this steam he consumes 65 million Btus (MMBtu) of gas an hour. His seven megawatt electric load requires an additional 70 MMBtu of fuel in an electric utility power plant.

Now let's look at a cogeneration plant sized to meet his needs. The resulting cogeneration facility would cost at most on the order of $7 million and would require 97 million Btu per hour. Using this example, we can get a feeling as to the overall impact of cogeneration.

The first immediate point, as summarized in Figure 25-3, is that

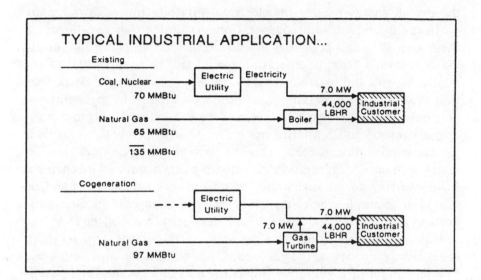

Figure 25-2

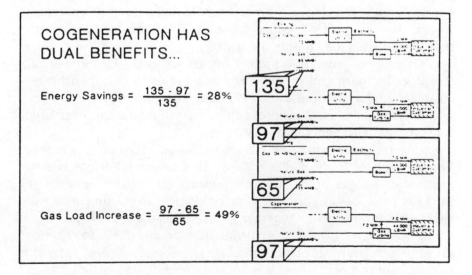

Figure 25-3

the overall energy use for the electric utility plus the industrial plant declines 28 percent with cogeneration. Assuming that two-thirds of these savings would go to pay for the cogeneration plant, the resulting cost savings from cogeneration would still be over $400,000 annually for this industrial customer. Indeed, it is this type of savings that drives both the nation's and industry's interest in cogeneration. Secondly, the actual gas volumes sold to the industrial customer increase from 65 to 97 MMBtu per hour. This 49 percent increase is a fundamental, inescapable fact of thermodynamics. Moreover, in many instances, cogeneration can also provide a built-in mechanism for converting an industrial customer from oil to natural gas and/or coal. For example, replacing oil loads with cogeneration plants co-burning coal and natural gas may be a very attractive concept.

However, factors other than gas load are important. One of these is clearly the quality of the gas load. Cogeneration can also help here. The left graph of Figure 25-4 shows the monthly gas use for a potential 4 megawatt cogeneration project at a paper mill in our region. As shown on the right chart, for the gas industry to acquire the same load from residential use would require us to serve over 4000 new homes. That's a string of homes over 75 miles long. Moreover, it would also require us to store substantial volumes of gas to serve such a residential load during the winter.

Clearly the cogeneration project on the left is the far more attractive load for both the gas industry and indirectly for its other gas customers. This is because such a stable monthly cogeneration load enables the gas industry to spread fixed cost of operations over larger throughputs. This benefits all customers.

Cogeneration provides an attractive, stable, long-term gas load. That, in and of itself, makes cogeneration an attractive gas sale opportunity to a gas producer. Acknowledgment of this opportunity by a gas producer can also be a major element in causing such a cogeneration project to go together. This is because a gas-fired cogeneration project is a major capital undertaking and fuel can represent up to 65 percent of its operating costs. Thus, a stable gas supply with a constantly competitive price is not only nice; it is essential!

This is the case because cogeneration investments and savings are for all practical intents and purposes funded by the difference

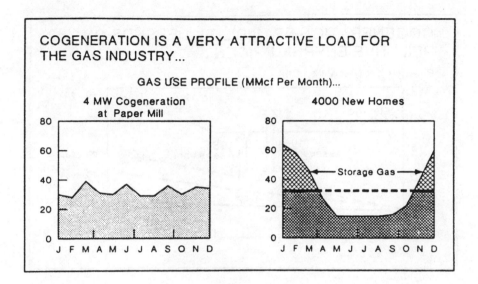

Figure 25-4

between electric prices and the gas fuel cost. The electric price is generally known and fixed over time. This is because it is typically a contractual function of the long-range avoided cost of the local electric utility. Figure 25-5 shows a representative series of avoided costs for a New York State utility. These are the upper series of lines with the dashed line showing their avoided costs as they were set in early 1984. The related solid line shows the most recent series of numbers which have been proposed by that state's public service commission.

Also shown on Figure 25-5 are a series of Chase Econometrics gas price estimates. Again the fainter, less solid dashed lines are the earlier estimates; the solid line, the latest estimate. This is not to say that the most recent estimate is, in itself, any more accurate—but it does at least represent Chase's latest thoughts on the subject.

There are several messages in this chart. First and most obvious, is that future perceptions of gas and electric prices do indeed vary over time.

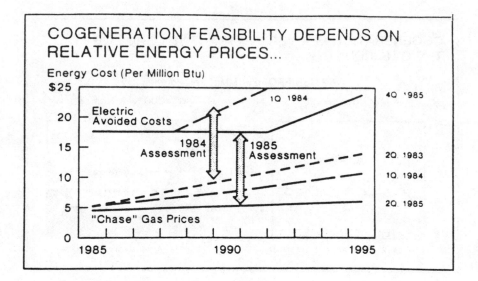

Figure 25-5

Second, and more subtle, is that although both electric purchase pricing and gas pricing have changed, the spread between them has remained relatively constant for at least the 1990 cogeneration feasibility test shown in this example. These test results are shown by the energy price differential arrows which connote hypothetical cogeneration feasibility assessments, which were undertaken in the years of 1984 and 1985 respectively. Their equal length denotes a relatively constant level of cogeneration attractiveness. This is really no surprise as long as the utility commission that sets avoided costs and the gas producer both use identical crystal balls.

However, the third point is that both the cogenerator and the bank financing the project won't trust anyone's crystal ball. This is shown in Figure 25-6 which uses the same energy price curves. Even though we may all concur that the 1985 assessment arrow does represent the most likely gauge of future cogeneration attractiveness and gas prices, no crystal ball is perfect. Gas prices might return to earlier forecasts even though we may all view such an event to be extremely

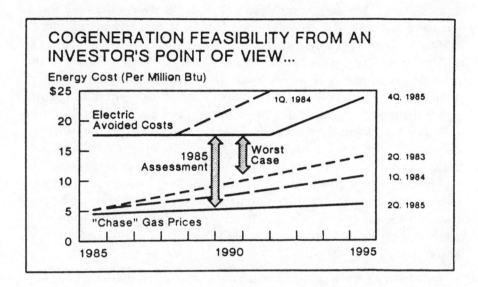

Figure 25-6

unlikely. However, from a practical point of view, the third-party or industrial owner will not enter—and a bank will clearly not finance—a cogeneration project without a relatively well-defined and certain relationship between electric revenues and fuel costs. This does not necessarily mean that the gas price needs to be a function of electric price. To our knowledge the gas industry has not approached this degree of innovation—attractive as it may well be in the future. Since the electric purchase price is most often a string of contractual avoided costs, some type of gas pricing having an acceptable fixed escalation is all that is required. This allows all the PC punchers to pronounce favorable prognostications from cash flow pro formas and debt coverage tests.

What does all this mean from a practical point of view relative to our tabulation of gas company cogeneration rewards? Put quite simply, it means that this reward is only likely to occur if one can secure a gas supply under a long-term contract which has known future pricing or which is indexed to a project variable like electric avoided costs. We do not mean to imply this is always the case and a hard-and-fast rule. But, it will be the case in almost any major cogeneration project with a meaningful gas sale.

The background data contained in Figure 25-1 also surfaces an additional gas company cogeneration potential. This is the possibility of gaining revenues from "owning and operating" the cogeneration plant. Its data suggests that as much as 70 percent of cogeneration facilities may not be owned by the thermal user. This projection is eminently consistent with current experience. About 70 percent of the cogeneration projects undertaken last year were by third-party owners.

Thus, revenues from "owning and operating" cogeneration facilities represent yet another potential reward from gas company involvement in cogeneration. There are a number of reasons why cogeneration by third-parties has proven so attractive to institutional and industrial end-users:

- Industry's own capital resources, in many instances, are already strained.

- For much of industry, the first priority is to modernize aging plant and production facilities to meet growing foreign competition.

- Cogeneration requires a large up-front outlay. This outlay can equal 25 percent of the industrial plant's overall book value.

- Many customers feel they lack the energy, financial, and regulatory expertise to develop cogeneration.

Indeed, depending on the activity—or rather lack of activity—of third-party developers in your region, at least some level of "ownership and operation" is essential.

Entering such a business is not easy. It requires a high degree of dedication and experience. Cogeneration projects are invariably complex. They require a mix of business, engineering, regulatory, financing, and marketing skills to put together. In a region where cogeneration is still emerging, one must spend a great deal of initial effort just to convince a customer that cogeneration is a viable alternative. This is an area where a marketing group can be of substantial help.

Given that this first step is accomplished, one must then complete: conceptual engineering; operating and capital cost estimates; pro forma projections; thermal sale negotiations; electric sale negotiations; enviromental assessments; contract development; gas purchase and transport negotiations; engineering and construction procurement; plant start-up; and a whole range of concurrent management, technical, financing, and regulatory checks and decisions. This is obviously no small task!

It is also important, in most instances, to find a suitable outside partner. This goes well beyond the simple financial desirability of a joint venture. A synergistic partner can provide a valuable combination of knowledge, skills, and attitudes.

It is critical to recognize that gas production and distribution operations, legal and financial departments, marketing and planning departments—while important resources—are not the customer. The customer is the thermal user and in many instances also the electric utility. This means that any effective "own and operate" function must be placed high enough in the gas company so as not to be biased by being lodged in a special interest group. It must also be relatively free to cross divisional lines in order to secure needed inputs and constructive assistance in a timely manner. It must have top management support and the capability to manage and resolve the inevitable conflicting interdepartmental viewpoints which will inevitably occur toward such a cogeneration endeavor.

From a practical point of view in being able to deal meaningfully with industrial users and cogeneration venture partners, such an "owning and operating" subsidiary must also be relatively free to choose between fuel type—including coal—and between suppliers. Otherwise, such an operation may be viewed by many thermal users to be suspiciously biased. If the gas utility runs cost-efficient production, transport, and also distribution operations, it will likely get what cogeneration gas load is to be gotten in the end in any event. If it isn't that efficient relative to its competitors, then it might as well find out now.

Again this is based on the presumption that the utility is willing to recognize that its customer is the cogeneration thermal user, not in-house departments and affiliates. If it can't accept this, then, quite honestly, it has no business being in the business. Such an unrestricted arm's length arrangement can also provide valuable insight as to how outside customers likely assess gas sales efforts, follow-up, and pricing, all of which have important implications for other markets.

In developing cogeneration, it is also important to strike a good balance between thermal and electric output. As a prospective owner-operator, it is not appropriate to turn a "good" 25 to 50 megawatt cogeneration project into a 200 to 300 megawatt "PURPA Machine." The promoters of such machines are one or two suppliers of large electric utility type gas turbines and a handful of developers, constructors, and financers seeking large capital projects. In some instances they have been encouraged by customers seeking "free" steam or by fuel suppliers seeking vastly increased loads or market share.

Not many such projects will actually get off the ground. One reason is that these types of projects typically run into a lot of opposition from electric utilities, public service commissions, and other parties. However, such abortive "PURPA Machine" takeoffs are already leaving sound industrial and institutional gas-fired cogeneration projects having important customer benefits to reap the whirlwind. Actions in states as diverse as California, Texas, and New York already include:

- Avoided cost biases in favor of coal versus gas-fueled cogeneration.
- Lowering of state avoided costs to prevent "windfall profits" to cogenerators while simultaneously grandfathering hydroelectric projects.
- Lowering cogeneration project eligibility to a 20 megawatt cap.
- Discussions in Washington and in various states about reopening PURPA or related state rules.

"PURPA Machines" carry with them a complex set of issues. Attempting to fly "PURPA Machines" works to the long-range detriment of cogeneration in general, and the securing of cogeneration benefits for industrial and institutional energy consumers in particular. The gas utility industry should work to alleviate, rather than aggravate, this issue.

Thus, "owning and operating" cogeneration facilities can be an additional gas company "reward" under some circumstances. Whether this will be attractive to a given gas utility is really dependent on the commitment, talent, and dedication that it is willing to apply to such an effort. Frankly, gas utilities shouldn't plan on attempting to "own and operate" cogeneration unless they can assemble a dedicated team that has both an entrepreneurial spirit and a related corporate commitment of support.

What has been discussed so far in this chapter has had a relatively near-term benefit horizon. In thinking through these various possibilities, it is important to assess what niches are particularly important for each company and for each region's cogeneration development. This really becomes a matter of matching the company's strengths and goals with the needs and status of cogeneration development in each particular region.

Now let's step into the longer-term. Consolidated Natural Gas Company largely serves a mature industrial region in the United States. We serve directly or indirectly portions of New York State, Ohio, Pennsylvania, West Virginia, and also the Northeast. Figure 25-7 shows gas sales of key industrial states in our region over the last decade. As you can see, overall gas use has declined over 20 percent due to conservation. As a result of this plus plant closings, industrial sales have been particularly hard hit.

The same gas sales are also shown in Figure 25-8, which includes an overlay of the region's nonagricultural employment over the same period. Our region, like many other mature portions of the United States, has been hard hit by recession—and that weakness has been exploited by growing foreign competition. Nonagricultural employment in the states of New York, Ohio, Pennsylvania, and West Virginia, for example, has grown only one percent in total over this entire ten-year period. In comparison, the rest of the nation as a whole grew 24 percent. By failing to keep pace with the dashed line representing employment growth elsewhere in the nation, these states have, in effect, "lost" three million jobs.

Clearly, if these jobs had not been lost, the region's gas sales would look far different. This also has an important side effect. Fixed costs associated with our entire industry's multibillion dollar

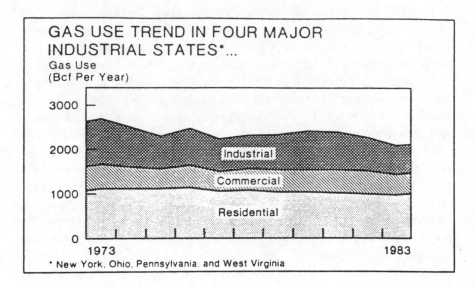

Figure 25-7

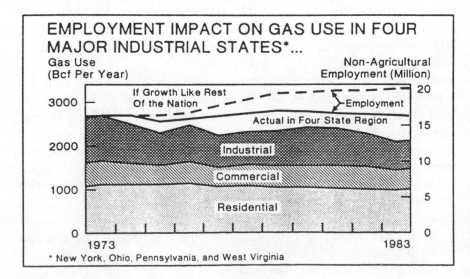

Figure 25-8

investment in transmission and distribution systems represent more than 30 percent of the delivered cost of gas. This type of a gas sales decline tends to increase the gas bills of residential, commercial, and remaining industrial consumers.

Cogeneration savings are important to industry. They can represent as much as 15 to 18 percent of a company's thermal bill. In many energy intensive industries, which are the key markets for cogeneration, energy represents 30 percent of the cost of their materials. According to DOE, potential industrial cogeneration savings in these four states alone total over $1.3 billion dollars on an annual basis. Not counted are the savings to hospitals, educational institutions, and other commercial businesses—which are also struggling to control rising costs. Thus, we feel cogeneration can play a key role in preserving and adding many jobs to the region, by strengthening old and attracting new industry.

Like many gas companies, we are a vertically integrated operation. We have Appalachian and Gulf Coast production, a large transmission network in our region, and a number of our own and other retail distribution companies to whom we supply gas. A great deal has been said within our industry about the importance of gas company efficient integration into gas transport and production operations. We feel that this is a sound way to keep our prices low and to continue to be an attractive investment for our thousands of stockholders, many of whom are also our gas customers. But, frankly, all of this production and transport is meaningless without a viable community and series of customers on the end of the pipe. After all, it is self-evident, or it should be, that the gas industry can only produce and transport what it can sell—and industry and its local worker-homeowners can only buy if they are there to consume. As gas companies, we cannot follow our industrial customers and their gas loads to the South. Nor can we follow the steel and automotive industry to Japan and Europe.

Let us give an example of how this can work. We and HYDRA-CO recently visited a paper mill in the Midwest that has just installed a cogeneration system. Mill executives told us that they face growing foreign competition, and that energy savings are critical to that plant's future viability. Not only did cogeneration preserve over 200 jobs at that mill, it also preserved over 300 indirect jobs in the community that relies on that mill as part of its employment. This particular system used gas turbines. However, such a success story in preserving local jobs is clearly independent of the particular fuel used in the cogeneration plant.

This example can be translated into the industrial 7 MW cogeneration plant case study used earlier in Figure 25-2. This is a real-world case study. It is a typical industrial plant based on data from the Census of Manufacturers. Figure 25-9 shows the impact of that plant on its local electric utility, its local gas company, and on the community where it is situated. The plant consumes on an overall basis some 57 million kilowatt hours of power each year and seven-tenths of a Bcf of natural gas. The plant directly employs 700 workers. Moreover, based on U.S. Department of Commerce data, this plant will provide at least 1.6 jobs in satellite industry and commerce for each job at the industrial facility. This related energy use has also been priced out using regional energy data. This is shown on the second line. The third line reports the energy use at the homes of these 1800 direct and indirect employees.

ANNUAL ELECTRIC LOAD AND JOBS PRESERVED BY COGENERATION*...

	Electric	Gas	Jobs
Industrial Plant (57 Mil Kwh 0.7 Bcf)	$2 9 Million	$3 3 Million **	700
Satellite Business (15 Mil Kwh, 0.08 Bcf)	1.0	0 4	1100
Residential (11 Mil Kwh. 0 1 Bcf)	1.0	0.6	
	$4 9 Million	$4 3 Million	$50 4 Million ⓐ $28 000/Job

*Based on a 7 MW Industrial Cogeneration Plus Regional Energy Use Patterns and Prices Also Includes Present Gas Company Loads

**Excludes Cogeneration Fuel Use Which Dependent on the Fuel Type Would Either Add 0 26 Bcf or Subtract 0 53 Bcf of Gas Use

Figure 25-9

When the numbers are totaled, an interesting—if not startling—fact emerges: by keeping local industry viable, cogeneration actually preserves more electric than gas load. Moreover, this does not count the costs that the electric utility avoids by not being required to build new capacity for its other customers. At only a 1.7 cent per kWh savings for cogeneration, this would add yet another one million dollars of benefits each year to the electric utility's ledger. For example, in our four-state region, at the same time that gas loads decreased 21 percent, electric use increased 6 percent. Clearly, the electric utility and the local community can win big from cogeneration in terms of preserved loads and jobs.

This is a message that needs to be told to the electric utility industry, which cannot move its service area elsewhere. This is also a message that needs to be delivered to state legislators and regulators. One vehicle for this could be state-by-state Cogeneration Coalitions. The gas industry can band together with industrial and institutional energy users to be an effective delivery tool. These types of communications and coalitions have been largely bypassed by the gas industry and by prospective cogenerators in many states. This is unfortunate for the long-range health of both the gas and electric utility industry, for their industrial and institutional customers, and for the local economies they serve.

The benefits to a gas utility of becoming involved in cogeneration are:

—Increased gas load

—Improved gas system load factor

—Increased gas production sales

—The advantages of "owning and operating" diversification, and finally, and perhaps most importantly,

—The chance to enhance the economic vitality and prospects of the service area

It is abundantly clear that cogeneration represents an opportunity, not only for what it does for the gas company in the short run, but for what it does for others, and for what it does for the gas company's service territories as well.

Chapter 26

Gas Utilities and Cogeneration: A Partnership in Success

VICTOR M. RICHEL
Elizabethtown Gas Company

Many cogeneration systems are in large industrial-scale plants that produce power measured in megawatts, with large requirements for process heat, usually in the form of steam. These energy requirements are usually used in making paper and chemicals, or refining crude oil, manufacturing fabric and many other industrial processes.

It is not unusual that such industrial facilities would utilize cogeneration, because cogeneration has been around since the late 1800s when on-site power generation was commonly used in industrial sites in the United States and Europe. As a matter of fact, in 1900, 50 percent of the total United States electric generating capacity was located at industrial sites, and over half of that was the result of cogeneration. By 1950 cogeneration had dropped to about 15 percent and by 1974 on-site generation accounted for only 5 percent of the total generating capacity here in the United States. Today, it is estimated by many, that cogeneration will account for 15 to 20 percent of generating capacity by 1995, thereby contributing more to the electrical needs of large consumers than nuclear power.

The National Energy Act of 1978, and the Public Utility Regulatory Policies Act, also known as PURPA, contained legislative incentives to promote the use of cogeneration. Before PURPA, we were unable to convince consumers that on-site generation made economic sense. We sold some jobs, but most of those "Total Energy" systems required equipment redundancy and oversizing which almost never allowed for reasonable returns on invested capital or equity. There was no system support, no utilization of excess power to make the systems viable. There were no incentives for either the customer or the electric utility to pursue on-site generation.

Along came PURPA and most of the problems began to go away. Electric utilities were in most cases required to buy back excess power at their avoided cost, which immediately turned total energy systems into cogeneration systems. Supplemental service rates and grid interconnection immediately made installations economic.

Obviously, we are now in a post-PURPA environment, with great opportunities to develop new business for gas utilities, as well as entrepreneural opportunities for those companies that seek nonutility investments. Increasingly, large-scale industrial plants throughout the country are turning to cogeneration. While large industrial cogeneration facilities will continue to take full advantage of cogeneration potential, an enormous opportunity exists for the commercial marketplace to take advantage of the great savings that cogeneration presents.

This chapter describes some exciting new packaged cogeneration systems that are currently available, and are in the process of being field-tested at locations throughout the United States. These kinds of packaged systems will allow small commercial customers who have a relatively good balance of thermal and electrical demand to take full advantage of the significant benefits of cogeneration.

Our market analysis data indicates that the retail energy price structure in most American cities makes gas-fueled cogeneration competitive with systems with installed costs at or below a thousand dollars per kW or engine/generator capacity. At that level, and with high energy utilization on-site, the market for cogeneration in the commercial sector could be as large as that identified in the industrial sector.

The Gas Research Institute (GRI) initially targeted the three following commercial applications for cogeneration packages: *hospitals, restaurants and supermarkets.* The three development systems cover a diverse mix of sizes and applications and each incorporates evolutionary technology advances. In each of these applications a large number of buildings share common energy use characteristics making them ideal candidates for a standardized package.

It is important to note that custom designed systems are currently available, and offer the ability to develop a system uniquely able to meet the needs of a specific use. However, packaged cogeneration systems offer the obvious advantages of:

- Integration of all system components

- Optimization of system performance

- Reduction in total installed cost

- Reduction in O&M costs

- Integrated service responsibility

- Reduced delivery times

Packaged systems offer some tremendous benefits for hospitals:

- These systems provide about 70 percent of heating and cooling thermal needs

- They satisfy approximately 60 percent of electric energy requirements

- Overall system efficiency is 75 percent

- Depending on local utility rates and facility size, annual energy savings range from $90,000 to $250,000

- First cost premium payback is usually 3 to 4 years

A packaged cogeneration system designed for restaurants and developed by Waukahau Engine Company could save between $15,000 and $30,000 a year in operating costs depending on local electric and gas prices.

Packaged cogeneration for restaurants can provide 100 percent of heating and cooling needs:

- Provides in excess of 70 percent of electrical requirements
- Overall system efficiency exceeds 70 percent
- Annual energy savings are estimated to be $15,000–$30,000 depending on climate and energy prices
- First cost premium payback is 3–5 years

Supermarket managers are highly motivated to save energy since their energy bills account for a greater percentage of sales than their profit does. For a cogeneration system costing only $50,000 more than conventional supermarket refrigeration, system payback should be less than two years with annual energy saving of $25,000 to $30,000.

Packaged systems offer these benefits for supermarket cogeneration systems:

- Provides nearly 100 percent of refrigeration needs and augments the following electrical and thermal loads:
- — Store dehumidification
- — Condenser subcooling
- — Lighting
- Space and domestic water heating
- Overall system efficiency of 75 percent
- Annual energy savings expected to be about $25,000 to $30,000
- First cost premium payback is 2–3 years

In an attempt to provide industrial and commercial customers with the opportunity to determine their cogeneration potential, we subscribe to a program offering feasibility studies for cogeneration that determine the economic benefits these energy systems can provide. The computer software system is called "Cogeneration and Energy Planning Program" (CEPP) and is offered by Engotech, Inc. of Schenectady, New York.

We offer a preliminary screening analysis for customers in our franchised area to determine if a cogeneration system is economically feasible. The analysis is free, and enables evaluation of cogeneration applications and alternatives.

A large number of customers have used the program to do preliminary analysis and then have decided to proceed with a more advanced analysis of their individual requirements.

We feel strongly that cogeneration offers enormous potential for not only our customers who can conserve energy and operating expenses but also for the State of New Jersey as we seek to maximize energy and utilize it in its most efficient manner.

Cogeneration has arrived at just the right time. The days of new large generating facilities being planned and built have passed. It is obvious that most electric utilities are following a policy of capital minimization. The siting and environmental problems associated with nuclear and central generating stations seems to present potential financial and operating difficulties that electric utilities can easily avoid by supporting cogeneration.

Accordingly, adequately structured cogeneration ventures have the potential to allow electric utilities to indirectly bring new capacity on line. If the electric utilities do not show willingness to participate and promote cogeneration, the alternative is that gas utilities and industrial and commercial customers may install cogeneration facilities and retain all the benefits, or participate in third-party organized ventures.

The current PURPA regulations restrict electric utilities to a 50 percent ownership position in cogeneration facilities. There is no such restraint on gas utilities and, accordingly, many gas utilities are actively interested in promoting and developing cogeneration facilities.

There are many finance options available to consumers considering cogeneration. The primary financing opportunity is an alternative which allows for the local gas and electric utilities to jointly co-venture a cogeneration installation with the individual customer participating to some degree. Additionally, the third-party ownership and operation is an option which is being pursued by many customers around the country. Such systems allow for firms with extensive cogeneration design and installation experience to own, operate and maintain systems while providing the customer at the host location with significant operating savings. Third-party financing is an option which can allow utilities and customers to take full advantage of many of the benefits of cogeneration while minimizing capital requirement and maximizing energy savings.

Chapter 27

Marketing Equipment at the Leading Edge of Technology

KEN RAPP
The Brooklyn Union Gas Company

Previous chapters have explained the benefits of cogeneration gas sales to utilities and to the *entire* gas industry. Chapter 26 described some of the accommodations gas distribution utilities are willing to make for these high-value customers. The Brooklyn Union Gas Company has reached a similar corporate consensus. On January 1, 1985, Brooklyn Union formed an Energy Systems Department within the utility structure.

Basically, this department has received three mandates:

- To develop the market for the TOTEM 15 kW module from BIKLIM. BIKLIM is an Italian firm related to Fiat which is marketing a small cogeneration module. It has sold more than 1500 in Europe.

- To ensure an orderly market and to assist in market development and sales of small to mid-sized cogeneration systems (as we will see later, these are two quite distinct tasks).

- Finally, essentially to act as Project Managers (in our area) for large design installations. We have several large real estate developments and redevelopments that we have been squiring toward cogeneration.

Of course, Brooklyn Union is willing to invest money in projects of any size. We see several benefits:

- Addition of gas load.

- Return on investment.

- Market development. Our participation reassures the customer, who is frequently initially unsure about the project. It is also reassuring to the developer to have the support of the local gas utility.

Actual equity participation in small projects will probably be minimal, but benefits may be substantial.

BACKGROUND—BROOKLYN UNION

We operate in a densely populated urban environment that experiences little new construction; we add only about 2,500 new small houses to our network each year. We do have the benefit of having fairly complete coverage of our territory; we rarely have to extend mains, although we do have to reinforce lines for certain large jobs.

Our sendout last year was:

- 66,472 million decatherms to small residential sites

- 21,463 million decatherms to multi-family apartment housing

- 19,108 million decatherms to commercial/industrial sites

- 107,043 million decatherms of total load

Of this about 20,315 million decatherms was sold on an interruptible basis.

The future potential market that Brooklyn Union faces is essentially limited to conversions from oil to gas, if we ignore cogeneration (for the moment). We have substantial opportunity in what I call the conventional market:

- 15,800 million decatherms remaining in the small residential area
- 15,810 million decatherms to multi-family apartment housing
- 4,130 million decatherms to industrial sites
- 9,050 million decatherms of commercial load
- For a total of 44,790 million decatherms of total load.

This potential can provide several years of sustained load addition. However, the prospects for adding all, or a major share, of this conventional market load are not bright considering the fact that this market has not yet succumbed to better gas prices and to sustained marketing and advertising activities in the past. Our sendout has remained the same each year since 1980:

	Total Sendout	Degree Days	Mdth per Degree Day
1980	105 Bcf	4,172	22
1981	114 Bcf	4,471	25
1982	117 Bcf	4,907	24
1983	109 Bcf	4,999	22
1984	117 Bcf	4,317	27
1985	113 Bcf	4,894	23

With the narrowing of the gas price advantage and at current sales levels, there will be very little opportunity left ten years from now, or even sooner. That is one reason why we have decided to emphasize cogeneration.

Although we are developing new cogeneration sales efforts, we are no stranger to the technology. Today we serve 12 sites using gas-fired cogeneration. The total installed capacity is nearly 100 mW. If we segment it by market we have:

11,200 kW of commercial load	343,450 Mdth
23,600 kW of industrial load	2,646,798 Mdth
53,100 kW of residential load	2,984,031 Mdth

For a total of 87,900 kW and 5,974,279 Mdth.

All of these units, with the exception of one site at 100 kW, are total energy systems and are completely divorced from the grid.

But what about the future potential cogeneration market? We initially asked ourselves if we could pick up a substantial amount of cogeneration load. With commercial electric rates in the range of 12¢–39¢/kWh and with an incentive gas rate of 63¢/therm, we felt that we could make a go of the market—maybe!

Through surveys and market research, we estimated the size of the economic market. Our conclusions are based on a *very specifically* assumed installation configuration:

- Operation in parallel with the utility for *purchase* of back-up and supplemental electricity, but *not* usually for sales. Con Ed, our local electric utility, has been forgiven the mandate to purchase electricity. However, they must sell it to cogenerators under the *present firm commercial rate*. This means there are no ratchets, no contract demand charges, and no reliability penalties. PURPA and our local P.S.C. require this.

- Thus, since all energy is consumed on site, we recommend a base-loaded electrical system. The energy prices in New York make our economics less sensitive to thermal use than in other cities. However, economic and qualified operation of a system (even in New York) requires that a substantial amount of thermal energy be used.

Taking this into account, we see a retrofit market requiring cogeneration units in the following sizes:

For 15–40 kW units	— 119,280 kW of total load	or	47%
For 40–60 kW units	— 23,000 kW of load	or	9%
For 60–100 kW units	— 10,480 kW of load	or	4%
For 100–200 kW units	— 9,450 kW of load	or	4% 30%
For 200–500 kW units	— 15,400 kW total	or	6%
For 500–1000 kW units	— 24,750 kW	or	10%
And for larger units	— 49,500 kW	or	20%
Total	252,000 kW		

The corresponding gas load is:

For 15–40 kW units	—	8,946 Mdth	or 51%	
For 40–60 kW units	—	1,380 Mdth	or 7%	
For 60–100 kW units	—	628 Mdth	or 3%	
For 100–200 kW units	—	567 Mdth	or 3%	29%
For 200–500 kW units	—	924 Mdth	or 5%	
For 500–1000 kW units	—	2,104 Mdth	or 11%	
And for larger units	—	4,206 Mdth	or 20%	
Total		18,755		

This description of the market does not include residential load or central cogeneration plants in real estate developments. The development of the residential market is tied to rent control regulations and submetering issues and does not figure in our immediate plans. Meanwhile, we are pursuing the condominium sector. Since most buildings have electric meters for each apartment, we will be serving common space usage only. This usually requires units of 20–60 kW even for very large buildings of 200–300 apartments.

The large-scale redevelopment market requires installation of multi-megawatt equipment (the largest presently 35 mW). While these installations are not as large as many industrial cogenerators, they do represent the upper end of Brooklyn Union's market.

It is evident, based on the above, why we are particularly interested in the 15 kilowatt TOTEM unit. Even Thermo Electron and Waukesha units cannot serve the lower end of our market, and there lies nearly 50 percent of our total cogeneration potential. In many areas of the country energy costs and economies of scale would preclude small-scale cogeneration hardware; however, we feel TOTEMs can be widely distributed in New York and other selected areas in the country.

THE TOTEM 15 KW MODULE

The TOTEM unit comes in three standard versions:

- a grid independent version
- a parallel version, and
- a version that combines both.

We have tested the fifth generation unit; the current model is the sixth generation.

All have an electric output of 15 kW. Input is 2.13 therms per hour and thermal output is 1.34 therms per hour. This represents an overall efficiency of 87 percent. The engine generator runs at 3,600 rpm.

The unit is completely enclosed and is 44 inches by 35 inches by 34 inches and weighs 1,080 pounds. The enclosure is almost completely soundproof and noise level is only 70 decibels at one meter. Ambient noise levels frequently drown out the noise from TOTEM; sometimes you must put your hand on a TOTEM to determine if it is actually running. Water is delivered at 187°F with a flow of 8.8 gallons per minute. Power will be 60 Hz, 3 phase, at 120/208 volts. An option allows 480 volts. The parallel version uses the generator as a starter.

The units are modular. They can be stacked. They normally run on natural gas, but can be run on any gaseous fuel. In fact, many of the units in Europe (about 1500 at present) operate on biogas, either from sewage or agricultural waste.

Brooklyn Union has been involved in investigating TOTEM since 1978. The program now underway has three concurrent thrusts:

- Further technical testing of two E-5 versions in our general office. We now have one E-6 also in operation.

- Field testing and market development of new E-6 version.

- Economic investigation of entering TOTEM sales.

The E-5 model has been under test since 1979. This European machine was modified by us for parallel operation with the North American grid. This required speeding up the engine and minor modification of the electric system. To date we probably have the most complete and best data and success in the world on TOTEM operation specifically and small-sized cogenerators generally. We have 100,000 hours of operating data. Unit availability has exceeded 99 percent on a monthly basis. An E-6 machine was added in August 1985 and, to date, we have not had problems with any major component of these units. The only nonperiodic maintenance that has been needed is infrequent: head changes, timing chain replacement.

The unit has operated well beyond any expected level of performance. We are very pleased with this test unit and are presently arranging to import seven E-6 units that have been modified at the factory for operation in North America. One of these units will join its predecessor in our general office. The others will be placed in the field. Three will go to a restaurant and the other three will be distributed to two other small commercial and industrial sites. One of these sites will be a residential care facility and the other will be a small metal plater (the business type that presently "enjoys" the highest electric rate in the country—an average of 25¢–39¢ per kWh). While we are very satisfied with TOTEM operations in a controlled, carefully monitored environment, these field tests will allow us better to gauge needed service in the field. These units should be in place in early 1987.

Several regional utilities have expressed an interest in concurrent testing and marketing of TOTEMs: Connecticut Natural Gas, Central Hudson Gas and Long Island Lighting.

We are considering distribution of a second set of TOTEMs to these and perhaps other utilities for field testing purposes. The testing will be mainly for gauging technical performance, service costs and institutional barriers. This will help to more nearly define installation and service costs outside the hyper-costly New York City environment.

The business assessment side of the coin has involved contact between Brooklyn Union and a variety of entities who have controlled TOTEM production. From 1978 to 1979 we dealt directly with Fiat. From that point until 1980, we dealt with Thermotec, a subsidiary set up to market TOTEM. In 1984, however, the automotive parent sent its child out on its own as BIKLIM. This new acronym is partially owned by Fiat and partially owned by a variety of other semi-official Italian companies. They are now building a factory near Naples to begin full production of these modules. They are including their new, state-of-the-art automobile engine in the module (the FIRE 1000 model). This engine should be even more fuel efficient than the present model.

Our approach to the TOTEM at the start in 1979 was to look at sales of the unit by a subsidiary, much as our present subsidiary—Gas

Energy Incorporated—controls North American sales of the Hitachi Chiller-Heater. However, it was determined that an entity seeking a profit totally from equipment sales was not, in fact, going to receive a profit. The technology was not ready for that. Interest in the TOTEM marketing then subsided for nearly four years, although we did continue the testing of the units.

With the successful performance of TOTEM hardware testing and new cogeneration awareness of the 1980s we looked at TOTEM again. However, our view of ourselves had changed radically. No longer are we looking at TOTEM as an equipment marketing opportunity for a subsidiary. We have considered the total impact of TOTEM on the company if we operate an energy sales and service organization in the utility.

The benefits are:

- Equipment sales profits
- Profits from the service organization
- Gas sales
- Development of the market

The last item is by no means the least important of the four. As we said at the beginning of the chapter, we are interested in sales *and* market development. Sales and installation of TOTEMs will serve to familiarize our customers with the concept of cogeneration. This will help us prime the market for the entry of other suppliers and developers. Since we benefit from TOTEM in several ways, our distribution of TOTEM in no way makes us competitors of other cogeneration suppliers. This microniche is not presently a competitive one.

The exact business relationship between Brooklyn Union and BIKLIM is being finalized. We are the exclusive North American distributors of the module. We have reached agreement on all but one issue—the one of liability. Having seen how litigious Americans are, the Italians are unwilling to expose themselves to huge lawsuits involving consequential issues from disgruntled American users. They have thus asked that Brooklyn Union accept *all* liability for consequential damages arising from TOTEM.

Once this is achieved, we will be ready to proceed with establishment of a TOTEM sales and service organization. We have examined the effects of sales on the utility by making assumptions about:

- Total market penetration
- Necessary investment for equipment and inventory
- Investment in manpower
- Income from sales and service

We project a profitable operation for both customers and Brooklyn Union.

But the market will not be limited, or completely served, by TOTEM. We have already indicated how TOTEM sales will aid in the synergy of the modular cogeneration market. How do we plan to approach cogeneration projects in which we do not have a financial stake? A brief history of the last 12 months should give a clear picture of that.

As former head of the total marketing and advertising efforts of Brooklyn Union, it was obvious to me that the development of cogeneration would never occur with less than full specialization, focus, and attention. Formation of a new separate energy services department exclusively devoted to cogeneration was needed. This, in fact, occurred and Brooklyn Union's Energy Systems Department was born on January 1, 1985. It is a small group that is totally devoted (in all ways) to making cogeneration work. Rabid customer interest in the concept of cogeneration does not always translate easily or automatically into a rapidly developed cogeneration site.

Our first order of business was to get attractive gas rates for cogeneration. The steady load that these customers provide made it relatively easy to approach the PSC with the idea of lower cogeneration gas rates. Our presentation could be based almost totally on the lower cost of service to these high-load-factor customers. We succeeded and developed two rates:

- A subset of our present interruptible temperature controlled rate that allows a lower minimum input for cogeneration customers—1 million Btu/hr vs. 2.5 million Btu/hr. The present cost of this is 49¢/therm.

- A totally new rate that provides firm gas to these customers at a rate substantially below our regular firm service rate: 59¢/-therm vs. 79¢/therm.

Once these rates were in place, we then examined the relationship of our customers with Con Ed. It is apparent that qualified facilities should receive regular firm service as backup.

Our next concern was how to approach a customer, or customer class, when promoting modular cogeneration. Here we shall only describe our sales process to retrofit customers who can use a mid-sized, packaged module. Later we shall discuss our sales approach to larger projects. By using an example, we can outline the problems that arise in selling high-technology gas equipment in a market not quite ready to receive it.

Promotion of the cogeneration concept is not a problem; response to our mailings has been above 50 percent, *if* we can contact the right person in the organization. In a city with energy costs as high as New York, nearly everyone is willing to respond to an offer by Brooklyn Union of lower energy costs. We emphasized the "Brooklyn Union" because our presence in the market as part of the sales process has produced almost all progress to date. We believe any progress in modular cogeneration will need the complete, and whole-hearted, participation of the gas utility. Vague offers of support are not enough.

Our response to the customer is an economic evaluation of cogeneration at its site. We recently had a software package designed for us that handles all the parameters of cogeneration in New York City. It has all energy rates built in and steps the energy uses through all the complicated blocks of local utility rates. They are very complex and demand charges make it inadvisable to use average electric prices.

We next prepare a presentation package of economic results—providing, of course, that we feel cogeneration to be economic. As we all know, even in New York, cogeneration is not for everyone. In this presentation we:

- Briefly outline the benefits of the project in a letter.
- Describe cogeneration and small, packaged modules.

- Give a complete outline of a cogeneration development schedule, as we anticipate it at his site. We describe the various ownership options, too.

- Outline economics of a system on a shared savings, or a full ownership plan. This is accompanied by graphic interpretation.

- Finally, invite the customer to continue discussion with us and a cogeneration developer with whom we will team up. We ask the customer to sign a letter indicating his intention to proceed to this next step.

The cogeneration developer will also get a copy of the Brooklyn Union cogeneration presentation. He can then decide if he wants to enter the sales process with us. The number of developers willing to enter this market is still small, for three reasons:

- Concern about economies of scale of small units—units which typically are not operated 24 hours a day.

- Lack of contractor experience in installing and bidding on these jobs. The contractor is frequently unable to estimate his costs when asked for a bid by a developer. They are usually unaware of:

  — how fast modules are to install, and
  — what electrical equipment and work is necessary.

  This last problem is exacerbated by lack of standard interface requirements for small units.

- Lack of marketing commitment by package/module manufacturers. There is frequently a breakdown between an enthusiastic manufacturer and terrified distributors.

We have, however, located several local developers who are willing to bid on small to mid-sized jobs. They enter the sales process along with Brooklyn Union. We have discovered, much to our pleasure, that our presence in the sales process is the *sine qua non* of the process. Most customers would not even *respond* to a solicitation by a cogeneration developer *without* the presence of Brooklyn Union.

They have had too many lighting ballast salesmen in the past ten years and are so tired of energy-related solicitations that it takes Brooklyn Union to break the barrier.

After the first meeting between customer, developer and Brooklyn Union, a proposal and letter of intent is forwarded to the customer. We are usually a signatory on the letter of intent.

What exactly is Brooklyn Union's commitment (in terms of money) to cogeneration projects? We have said that we participate in selling and lend our good name to the process. Do we go farther?

You all realize the constraints a utility must operate under when spending its money. We are in cogeneration primarily for gas load growth.

We feel this goal is served by entering into cogeneration investment. We have, therefore, petitioned the PSC to allow us a blanket authorization to invest $15 million in cogeneration projects—without regard to the form of the investment.

We anticipate using this money only to prime the pump on small projects, with the preponderance of investment going to multi-megawatt projects. On the small end, we plan only to enter as a co-venturer in the first few cogeneration projects of small developers. This will show our willingness to both the customer and developer to support the project and to deliver gas at the best possible rate.

For the larger projects, the same assurance is sometimes necessary. We have found some customers unwilling to enter projects *unless* Brooklyn Union takes the first step financially. This attitude is perfectly understandable, and we are willing to cooperate.

While the larger installations are a small amount of the total economic cogeneration sites, they represent 20 percent of the total load. This is especially evident when we include several large redevelopment or rehabilitation projects in Brooklyn—a new phenomenon for our area.

In all of these projects, we are heavily involved in the promotion and we are considering an equity position. These projects require a lot more time than do small projects. We have had to prove to the customers that cogeneration is economical. To do this we have:

- Engaged engineering firms to perform preliminary analyses of energy equipment and economics at proposed building sites.

- Assisted in preliminary negotiations with Con Ed
- Worked with and provided assistance to the cogeneration developers

But most important, we have been engaged in impassioned sales presentations to the customers who frequently seem remarkably unimpressed by cogeneration savings. They seem to see only the problems and not the opportunities. Keeping the sales process active has been one of our major jobs.

This chapter has outlined our cogeneration sales process and our rationale for entering the market. We are totally convinced that the cogeneration market will expand rapidly, even with total on-site use of both electric and thermal use.

Chapter 28

The California Cogeneration Success Story

MICHAEL F. NEIGGEMANN
Southern California Gas Company

In the 1960s, Southern California Gas Company (SoCalGas) was one of the pioneers in the field of cogeneration development with a program known as "Total Energy." We promoted cogeneration in commercial applications throughout southern California. Several facilities were built, including one at our Anaheim base and one at our Woodland Hills facility.

The invaluable experience of the 1960s contributed to new regulations in the 1970s. The establishment of PURPA in 1978 and the development of favorable "avoided cost" buyback rates and electric utility standard offer contracts set the stage for successful development of cogeneration in California.

SoCalGas was an active participant in these early regulatory and legislative actions and played a major role in establishing the strong regulatory and legislative support enjoyed by California cogenerators.

By demonstrating the benefits of cogeneration through our program to our state government and regulators, California has become one of the most progressive states in the promotion of cogeneration.

MARKET STRATEGY

As a utility regulated by the CPUC, SoCalGas must be sensitive to the needs of several diverse constituencies. Those include the needs of our customers, both individually and collectively, for reasonably priced, safe, and reliable service. We must also provide strong financial support to our parent company, Pacific Lighting Corporation, and satisfy the profit motives of our shareholders. To accomplish this, SoCalGas must maintain a competitive position in the marketplace and retain market share by promoting energy efficiency while encouraging gas use.

Cogeneration ideally fits this market strategy. Promoting energy efficiency through cogeneration satisfies:

- Individual customer needs by providing efficient energy use, thereby reducing the customer's overall energy costs

- Ratepayer needs by spreading utility system fixed costs over a larger customer base, thereby reducing individual ratepayer costs

- Stockholders appreciate the new revenues and greater stability in gas sales

- Company receives stable gas demand and stable revenue

The major benefit of this strategy was the ability to receive funding for this program. SoCalGas' Cogeneration Program received full funding for its $2.4 million annual budget in the Company's last rate case.

COGENERATION PROGRAM OBJECTIVES

In line with and beyond the Company's general market strategy an ambitious set of program objectives have been established for the Cogeneration Program. These are:

- Encourage the design and installation of cost-effective, energy-efficient cogeneration facilities.

- Expand our market share in previously uncompetitive markets.

- Develop first those market segments which are the best and largest applications.
- Pursue a variety of smaller cogeneration projects on a much larger scale.
- Use cogeneration to reduce the likelihood of fuel switching by our larger customers.
- Expand the awareness of cogeneration as a viable means of improving energy efficiency.

COGENERATION PROGRAM

The success of SoCalGas' Cogeneration Program is primarily a result of the successful execution of the following program elements or tactics:

1. Incentive cofunding
2. Special gas rate
3. Special service priority
4. Special gas pressure and main options
5. Advertising/publicity
6. Promotional brochures/handbooks
7. Technical support

INCENTIVE COFUNDING

To encourage customers to evaluate their cogeneration potential, SoCalGas offers incentive cofunding. SoCalGas will cofund half the cost of a feasibility study up to $10,000 and half the cost of an engineering design up to $40,000 for a total of $50,000. Prior to 1983, SoCalGas would also cofund up to $50,000 towards project construction for a total of $100,000.

The Company will also cofund the temporary installation of a packaged cogeneration unit to prove its feasibility. SoCalGas' 1985 cofunding budget is $970,000.

SPECIAL GAS RATE

SoCalGas has developed a natural gas rate to maintain parity between electric utilities and cogenerators. It is lower than most of the Company's other rates. The cogeneration gas rate is adjusted monthly and is based on the weighted average rate charged to the electric utilities for the previous recorded month.

By billing cogenerators at gas rates equivalent to what the electric utilities would pay for natural gas, cogenerators can be assured of economic viability. This is because cogenerators in California are paid electric "avoided cost" buyback rates which correspond to electric utility gas rates. So if natural gas rates rise, then the "avoided cost" payment revenues to cogenerators will also rise, offsetting increased gas cost. Another benefit of this special cogeneration rate is that it reduces the economic risk of the energy project, making it easier for cogenerators to obtain financing.

SPECIAL SERVICE PRIORITY

Another element of SoCalGas' program is its higher service priority. Through our program, cogenerators receive a higher service priority than other customers of similar size with alternate fuel capability. This special priority provides cogenerators with preferred service in the event of curtailment during peak winter seasons. The higher service priority has helped alleviate investor and financier concerns over gas supply availability.

SPECIAL GAS PRESSURE
AND MAIN OPTIONS

SoCalGas will also provide a dedicated main to those applications which require high-delivery pressures. The customer pays for the main and has the option to pay the cost over an extended period. Providing a dedicated main with high-delivery pressure can allow the customer to eliminate the need for a natural gas compressor. This reduces the cogenerator's capital costs and associated compressor operating costs.

ADVERTISING

SoCalGas cogeneration advertising efforts have focused on reaching audiences and potential cogeneration customers not normally contacted by company sales people. Cogeneration advertising has generally been directed through the print media and designed to complement field sales force activities.

PROMOTIONAL BROCHURES/HANDBOOKS

An important element of our program has been the development of informational brochures and generalized feasibility study handbooks. These promotional materials have helped to support company sales force activities and increased customer understanding of the value of cogenerating.

Through the use of generalized feasibility study handbooks, SoCalGas has been able to market cogeneration on a much broader basis, using less technical marketing personnel. Handbooks developed for large swimming pool applications and small medical facilities have proven to be successful tools for marketing small package cogeneration systems. SoCalGas is currently developing cogeneration handbooks to promote cogeneration in high-rise multi-family market and metal-plating industry.

TECHNICAL SUPPORT

Our incentive funding may be invaluable but our technical assistance can be priceless. The Company's team of cogeneration sales engineers represent the backbone of the program. They, more than any other element of our program, are responsible for the program's success. The collective expertise of this team is second to none and they have been instrumental in identifying potential sites, suggesting design innovations, helping to secure financing, trouble-shooting project impediments and convincing customer company officials about the benefits of cogeneration. Our cogeneration team offers technical assistance from project's inception to the project's successful completion.

PROGRAM ACCOMPLISHMENTS

Cogeneration has truly been a success story in southern California. Since 1980, over 650 megawatts of cogenerated power has been installed in SoCalGas' service territory. This represents increased sales and revenue of 21 Bcf/yr and $90 million, respectively.

In addition to these installations, another 59 projects representing over 1000 megawatts are currently under design and construction. Beyond this, another 203 projects representing over 880 megawatts are currently being evaluated for feasibility.

This success, we believe, is due in large part to our cogeneration program.

COGENERATION OUTLOOK

The climate for cogeneration in California is changing and cogeneration's future is cloudy. Impediments to future cogeneration growth include:

- *Suspension of Standard Offer No. 4:* This standard electric utility buyback contract provided for fixed energy payments to cogenerators for a period of five years. Essentially, this provides a guaranteed fixed income stream to the cogenerator for his electric production. The CPUC has restricted electric utilities from offering this form of contract for at least one year. Without this predictable revenue stream, obtaining project financing will be difficult.

- *Air Quality Requirement:* Southern California emission controls standards require selective catalytic reduction and a carbon monoxide catalyst on most cogeneration systems. Naturally this adds significantly to the initial cost of the cogeneration project. The cost of both of these systems ranges from approximately $400,000 for a 5 megawatt project to $2.75 million for a 50 megawatt project.

- *Excess Electric Capacity:* At least for the short-term there will be excess electric generating capacity in the western United States. This excess capacity will hold down avoided costs, reducing the attractiveness of selling power back to the utility grid.

On the positive side, a number of factors will contribute to the continued growth of cogeneration in California, including:

- *Rising Retail Electric Rates:* Industrial electric rates in California are some of the highest in the country. With several costly nuclear power plants, totaling over $30 billion, now entering electric utility rate bases, these retail rates will rise even though the marginal cost to generate electricity may fall. As a result, industrial customers may find cogeneration an attractive option for avoiding electric retail purchases.

- *Enhanced Oil Recovery (EOR) Opportunities:* California has large ongoing EOR operations in the central California San Joaquin Valley near Bakersfield. This market is ideally suited for large cogeneration development. The cogeneration potential for this market is estimated at 3000 megawatts. We expect the development of over 1000 megawatts of EOR cogeneration over the next three years. We recently began service to a 300 mW project jointly owned by Texaco and Edison in the area.

- *Availability of Packaged Systems:* Pre-engineered packaged cogeneration systems should greatly reduce capital costs for cogeneration making it much more attractive, especially in the commercial market.

- *Air Quality Utility Offset Credit Benefits:* Cogenerated power is generally cleaner than that generated by electric utilities. To the extent cogeneration displaces electric utility generation, air quality is improved. In essence, cogeneration growth helps improve air quality and as a result will, to a limited extent, be encouraged by state government and regulators.

CONCLUSIONS

The Southern California Gas Company is very proud of its cogeneration program and plans to continue aggressively marketing in cogeneration. The success of our program can be attributed to:

1. Establishing a highly-trained competent cogeneration sales "team"

2. Obtaining regulatory support for both the concept of cogeneration and utility cogeneration marketing efforts

3. Innovation in promoting cogeneration through special rates, special services, and imaginative advertising

4. A company-wide commitment on the part of all the people and departments of Southern California Gas Company

Cogeneration is a smart, inherently-efficient method, that makes good economic sense to the customers.

Chapter 29

Key Issues in Industrial Cogeneration Development

GARY D. HOOVER
Northern Natural Resources Company
Division of ENRON

InterNorth merged with Houston Natural Gas in 1985. The combined companies' pipeline extends some 37,000 miles from "coast to coast and border to border." The HNG system which includes Florida Gas Transmission Company and TransWestern spans the entire width of the United States along its Southern States; the InterNorth system, known as the Northern Natural Gas pipeline travels from West Texas up through the Central portion of the United States with northern terminations in Wisconsin, Iowa, Minnesota, Nebraska and South Dakota. The merger has created a great deal of synergy in the combined company, not only in pipeline systems but in cogeneration efforts as well.

HNG-InterNorth is also involved in liquid fuels, petrochemicals, exploration and production of oil and natural gas through our Bel-North and HNG Oil Company units, as well as international trading activities in energy products. Northern Natural Resources Company was founded about four years ago to act as a separate operating company within InterNorth with the sole mission of analyzing and developing new energy-related business opportunities. Cogeneration happens to be one of these new businesses, and, in fact, it is the largest business of those in the NNR array currently.

It is necessary to distinguish between three main types of cogeneration projects and hence three types of cogeneration development paths. The first type of cogeneration includes the very large facilities which sell the majority of their electrical output to an electric utility. We are joint-venture partners in a gas-fired facility located in Bayport, Texas which supplies about 300 megawatts of electricity to Houston Lighting and Power Company, along with steam to Big Three Industries there.

We are also constructing a plant in Texas City, Texas which will provide about 400 megawatts of electricity to Texas Utilities Company as well as 25 megawatts of electricity and 300,000 pounds of steam per hour to a Union Carbide chemical refinery there. Obviously, the range and depth of development activities on these types of projects are quite lengthy and detailed and require special expertise in a multitude of business disciplines.

The second type of cogeneration facility is one which is located at a large industrial facility where the total cogeneration plant output—both electrical and thermal—is consumed totally by the local industrial processes. There may be some small selloff of electricity to the local utility as a load-balancing technique but, in general, the facility is built to provide only the energy needs of that specific site. This type of development is still very detailed and complex but the need for careful negotiations with an electric utility is not required in this case which simplifies things considerably. Some of these projects will require complete ownership and operation of the cogeneration facility by our Company; others will involve partial ownership through a joint venture with the industrial host facility or an equipment supplier.

The third type of cogeneration plant consists of those which may be located in commercial applications such as large hotels, airports, hospitals, and universities. Again, these plants are typically designed so that all of the output is consumed locally although in some cases there has been some selloff of electricity to the local electric utility.

Regarding cogeneration plants that supply large quantities of electricity to electric utility companies, we do not feel that cogeneration, even large-scale cogeneration, will replace the large central electric generating stations which are owned and operated by electric utilities. Large output cogeneration does have a place in the overall electric energy market, however, and is changing the way in which utilities think of themselves. We expect cogeneration will become a significant source of electrical energy, but not the dominant source.

DEVELOPING A LARGE COGENERATION PROJECT

Obviously, the location of the plant site is of prime importance during the early phase of development. Locating an industrial plant with a fairly significant need for steam is crucial. This plant must also be located within the territory of (or at least fairly close to) an electric utility which needs the electrical output from the plant.

It seems to us that several other developers have undervalued the latter part of this siting issue. A large portion of cogeneration development has concerned itself with the needs of the industrial host facility leaving the electricity selloff to be controlled by the regulations of PURPA, which states that electric utilities *must* purchase electricity provided by cogenerators at their full avoided cost.

Eliminating the electric utility's needs from the cogeneration development equation can lead to serious pitfalls, probably resulting in either poor project economics or required abandonment of a specific project. This process requires a fair amount of careful analysis before any potential project is even discussed. If an electric utility does not need power now and can't forecast any need into the future, there is really no reason to develop or start to develop a plant in that area unless there is a good deal of assurance that the electricity can be "wheeled" to an adjacent utility that needs additional power.

During this period of initial project development (which may take several months) we rely on a fairly sophisticated economic model to continually evaluate the attractiveness of the project. This model requires us to look at the three main project ingredients: capital cost, electric buyback rate from the utility, and fuel, in addition to many other issues which can have significant bearing on project feasibility. This is an iterative process with many assumptions changing throughout the development process. As a result, we may find ourselves renegotiating certain items so that the project is a good deal for all parties concerned.

Once we have settled on the initial overall project structure with a host industrial facility and an electric utility, it is often necessary to convince the state Public Utilities Commission of the need for the plant and, of course, application for a qualifying facility must be filed with the FERC. Many regulatory issues need to be addressed, especially in states like California where emissions and structural design must be carefully documented with the California Energy Commission.

Once the permitting issues are surrounded, we begin to address the financing needs of the project and probably start to meet with several investment bankers and financial advisors to arrive at an optimal financing system. In some cases, we may decide to use our own straight equity to finance the project.

Concurrently with this activity, our technical staff will have been looking at optimal plant design and will have been involved in the early phases of equipment selection and procurement. We have contracted with outside engineering services by this time as well. Any joint venture agreements probably have been worked out or at least they will have been well along in the negotiation process.

The early stages of project development require us to look at fuel availability and approximate pricing in a general way. Before contracts are signed, we will have carefully explored fuel pricing and availability and will have negotiated carefully for a long-term fuel commitment. Obviously, the fuel price and availability (whether it be natural gas, coal, oil, or municipal solid waste) is a major factor in deciding whether each project will proceed.

If all areas and disciplines are in good order at this point, we are ready to execute the main contracts and proceed with construction. Building the plant requires a whole new set of skills and special services. Besides hiring a general contractor with construction management services, we also will deploy some of our own technical and construction personnel to the site to monitor progress.

During the latter construction phases, we will move through a number of equipment start-up procedures and address general plant debugging. Maintenance and operating issues will be addressed along with overall billing and administration. The plant will be staffed adequately for smooth long-term operation. Obviously, it is in our best interests to design the plant and staff it so that we can provide maximum availability. We have been aiming at this since the inception of our cogeneration business line.

DEVELOPING A SMALLER INDUSTRIAL PROJECT

Next, let's consider the second type of cogeneration facility mentioned earlier: the smaller scale industrial project where the total electrical and thermal output of these plants is consumed locally. A facility with 5 to 50 megawatt output is considered to be in this category. Locating a plant, or in the case of these smaller facilities, identifying potential hosts, is again the first step.

Several criteria are used to select large market growth areas. Most obviously, we look for areas with high electric rates or areas where rates are predicted to increase. Equally important is the need for self-sufficient power generation by local industries. After this search, a closer evaluation of an individual site interested in hosting a facility is warranted. Initially, all costs involved in designing, installing and operating the plant must be addressed. The system cost itself, including engineering, purchase, and construction, can be obtained by tying in with a reliable equipment supplier, perhaps as a joint-venture partner.

Fuel supplies can be 90 percent of the cost of operations—and it is advisable at this point to begin work on a long-term fuel supply contract. Operations and maintenance as well as financing costs must also be considered.

These costs are fitted into our economic model for analysis. Of course, as work on the project progresses, these inputs are constantly varying and being revised. This model currently allows for varying tax structures as well.

Once the host facility is convinced that cogeneration makes

economic sense, and all the parties concerned are satisfied with the results, we ask the host to sign a Letter of Intent so that a larger investment of time and money can be made in the form of additional engineering and financial analysis with more and more refinement.

Power, steam, and fuel contracts must be signed: negotiations will proceed concurrently with the more complex engineering work. With these "inside-the-fence" facilities, the power contract is signed with the host facility instead of a utility. Thus, the normal buyback rate negotiations do not take place. Since both steam and power contracts are with local industries the development time frame of the overall project is shortened, allowing the plant to be operating and generating revenue more rapidly.

As work progresses on these contracts, there will again have been meetings with several financial advisors to arrive at the optimal financing arrangement. A properly constructed project will be highly credit worthy.

At this point, the various permit applications are filed, similar to the larger projects. The remaining steps of procuring equipment, constructing, and operating the plant follow much the same series of events as previously outlined. If we have entered into the project with an equipment manufacturer, these stages may be carried out by this firm.

Fuel price and availability is key to attaining good economics for a cogeneration project. Fuel comprises more than two-thirds of the annual costs for a large cogeneration facility. In certain cases, it can become an even larger percentage. Obviously, fuel pricing along with assurance of its supply over the life of the cogeneration project is the main economic factor in determining feasibility.

When thinking in terms of the large cogeneration facilities that sell a large portion of their electrical output to the electric utility system, it is important to keep in mind that these plants may be displacing generation which is fired by fuels other than natural gas. In Texas, for example, the avoided cost calculations in certain areas are based on avoiding lignite-fired generation with long-term fixed escalation on lignite prices. Furthermore, the lignite reserves may be owned by the utility so that the marginal price of mining the fuel is predictable and relatively inexpensive over long periods of time.

In other cases, the avoided cost calculations are based on a combination of coal and gas fuels. We are facing much more here than gas vs. gas competition, and this is inherent in the nature of gas-fired cogeneration in most areas of the country today excluding Southern California and parts of the Southwest, Middle South, and Florida. If such plants are to look economically attractive as sources of electrical supply to utilities and to developers such as ourselves, natural gas must be priced attractively on a long-term basis.

If we are avoiding coal-fired central station generation with utility-owned reserves, we must secure gas on long-term basis with pricing that matches the solid-fuel recovery costs. Many readers may think that gas-fired cogeneration plants are much less expensive to build compared to solid fuel plants, and that should be taken into account. The major point, however, is that the fuel price is such a large portion of the plant economics that capital differential effects are minimized, and the price of fuel to the cogeneration facility must be competitive with the fuel avoided by the electric utility. If we are all looking for continued growth and opportunities in gas-fired cogeneration, this fact must be addressed.

Now, let's reconsider the fuel supply for a medium-output cogeneration facility. Again, the fuel price must be competitive and must be tied down long-term with fixed escalators or by index to local electric price movement. Not only is this important for consideration during the development phase, but it is crucial when financing is being considered. Through long-term fuel contracts, we reduce project risk and provide support for project financing. Local distributing companies play a major role here. We need local gas transportation to our cogeneration facilities; it must be reliable and economically priced, especially in view of the large gas quantities used in these plants. LDCs are also a major component of the fueling equation.

What can we expect in the future for gas-fired cogeneration? Identification of electric utilities who need power and industrial host sites who need both heat and electricity will become more difficult. We expect, however, that there will be more opportunities for the second type of cogeneration facility, involving total consumption of output by the industrial host itself; what we call "in-the-fence" cogeneration.

This is best confirmed by looking at some recent statistics. Eighty percent of QF applications (over 350 projects) in the last year have been in the medium or small segment of the market. Estimates of total potential ranges up to 300,000 sites or a $65 billion investment for facilities under 10 mW alone. As utilities increase rates, more industries will look to cogeneration. Also, 70 percent of the electric load growth in the U.S. will occur in the Northeast and sunbelt states—areas to watch for cogeneration. As the potential number of sites for cogeneration increases, however, so does the competition in the business. This will be from equipment manufacturers, independent developers, or even electric utilities, themselves.

We also expect that those facilities which do provide electricity to the utility grid will have to be more responsive to the particular utility's load schedule with operation at peak-load periods only or perhaps partial-load operation at off-peak hours. It may be important that utilities have some form of dispatching control over many of the future cogeneration plants built in the United States.

If the fuel pricing issues can be addressed, gas-fired plants will continue to be the most popular type of cogeneration. Besides requiring a lower capital investment, they are constructed in much shorter time frames and some of the air pollution control issues which need to be addressed so carefully with solid fuel plants can be eliminated.

We do not believe that cogeneration will become the dominant source of electrical supply but we do believe that it has a proper, significant role in providing electrical energy to our nation. Electricity will play a larger and larger role as the preferred source of energy and some utilities will be short of conventional generating capacity in the near future—even within the next five years.

Large cost overruns in nuclear projects have had devastating effects on some utilities leaving them without the capability of meeting near-term load requirements and in poor financial condition as well. Some utilities, such as Atlantic Electric and those in New England are actively searching for long-term contracts with cogenerators and small power producers. San Diego Gas and Electric Company has gone as far as saying that they will no longer build their own central generating facilities and will rely on power generated by others.

The future is bright but we must work together to reap the benefits of these opportunities and make them work to the advantage of all interested parties. Economically priced, long-term fuel supply is critical to success, and successful gas-fired cogeneration development leads to a very significant market addition for natural gas suppliers, pipelines, and local distribution companies.

What is ENRON doing to insure a market position? Again our gas sourcing, with the ability to match alternative power costs, is all important. Our pipelines service 90 percent of predicted high growth areas for cogeneration. Secondly, our responsiveness and quick development turnaround times are impressive, as is our flexibility in offering creative business deals suited to specific project factors. Similarly, we are willing to take equity positions in these ventures. Finally, we feel it is important to maintain our contacts with manufacturers, engineering firms, small developers, and utilities, and to remain open to all new projects and ideas.

Chapter 30

Cogeneration:
A Marketing Opportunity for
Pipelines

JACOB S. ULRICH
MidCon Corp.

Marketing "cogeneration" presents both new opportunities and new challenges to pipeline company management. These opportunities and challenges include not only the chance to sell more gas but the ability to develop new businesses and new methods of thinking about marketing that will benefit the entire pipeline industry.

Theodore Levitt, in his now-classic article "Marketing Myopia," stressed the importance of analyzing customer needs. His conclusion was that

> The entire corporation must be viewed as a customer-creating and customer-satisfying organism. Management must think of itself not as producing products, but as providing customer-creating value satisfactions. It must push this idea (and everything else it means and requires) into every nook and cranny of the organization. It has to do this with the flair that excites and stimulates the people in it. Otherwise the company will be merely a series of pigeonholed parts, with no consolidating sense of purpose or direction.[1]

---

[1] Theodore Levitt, "Marketing Myopia," *Harvard Business Review* (Boston: Harvard University, September-October 1975), p. 10.

Unfortunately, many industry veterans think of marketing, at the pipeline level, only in terms of a commodity product, with pricing through rate design the only marketing tool. Load retention is the operative strategy, with much more thought given to how the FERC can protect market-share than to what the long-term marketing strategy should be.

This is not the case with cogeneration. Although certain characteristic factors, such as the FERC regulations and inter-fuel competition affect the marketing of either gas or cogeneration, the possibility of product differentiation is much greater with cogeneration.

Marketing cogeneration will require a different approach than marketing gas, with an imaginative and well-reasoned strategy necessary for success.

Many people are aware of the demise of the railroad industry, and the near-bankrupt state of many major railroads in the 1960s and 1970s. Yet only thirty years before, the railroads were considered, along with the steel industry, to be among the strongest, most successful businesses in the country, if not the world.

In the years following, the total number of people and tons of freight moving throughout the country increased dramatically, yet the railroad's fortunes declined. Why? Part of the reason was the *inability* of railroad management to think of their business not as running a railroad, but as providing transportation.

The above illustrates the single most important issue the pipeline faces when entering the cogeneration market:

Do we want to sell gas or do we want to be in the energy business?

The answer to this question, more than anything else, will dictate marketing strategy. Before examining how this issue and others will shape the marketing strategy, consider four scenarios:

Case A:

The plant manager of company A, a steel producer, seeks your assistance in developing and implementing a cogeneration project at the plant. You currently sell company A 20 mmcf/day for a variety of uses.

After an exhaustive study of the facility, your engineers tell you that, due to the large quantities of by-product gas available, installation of an extraction steam turbine/generator set is clearly the best design. A combined-cycle or gas turbine is not economical. By installing the steam turbine/generator set and new boilers to efficiently use the available by-product gases company A will save $5 million per year in electric costs and reduce total gas consumption by 10 percent. What do you recommend?

Case B:

The corporate energy director of company B, which buys electricity and gas from LDC B, a good customer of your pipeline, tells you his company is considering cogeneration. You examine the proposed system and calculate that by installing a gas turbine/generator system he will increase his gas purchases from LDC B, and you, by 15 percent, but decrease electric purchases, which are significantly greater dollar-wise, by 50 percent. LDC B has been a valuable customer for many years. What do you recommend?

Case C:

Plant C is located one mile from your pipeline. C is purchasing gas from your customer, LDC C, which adds $1.50/mmBtu to the delivered cost. Your marketing rep tells you that plant C wants to cogenerate but with the current gas/electric price differential it is not quite economical. A quick analysis reveals that you could sell an additional 25 mmcf/day if you built a direct line to plant C, which would facilitate the decision to proceed with a cogeneration project. What do you recommend?

Case D:

A large industrial food processor, company D, reveals to you that they want to install a coal-fired fluidized-bed boiler/generator system with the support of a third party. Currently, your customer, LDC D supplies all of the gas and electricity to plant D. If the fluidized-bed system is built, LDC D will lose virtually all of the sales, which are a significant portion of its overall revenues. Your company will subsequently lose 50 mmcf/day sales to LDC D. What do you recommend?

Your recommendations to the above, and more importantly, top management's response, are critical in shaping your company's cogeneration strategy.

Either approach, increasing gas sales, or creating a new business, can be accomplished successfully if the proper marketing approach is used. Remember that selecting a strategy and striving to implement that strategy will almost always produce better results than proceeding without any plan.

INDUSTRIAL MARKETING

To become a significant player in this game requires a commitment to the concept of industrial marketing in the truest sense. We in the pipeline industry are at a distinct disadvantage compared to seasoned industrial marketing players such as equipment manufacturers, architectural/engineering firms and some local distribution companies. Accordingly, some time and effort should be devoted to reviewing what others have developed regarding industrial marketing.

As an initial point of review, we should examine the four key concepts that differentiate industrial marketing from other types, as B. Charles Ames set forth fifteen years ago.[2]

(1) *Focus on improved profit performance,* not sales volume or market share per se. Typically, as numerous studies have shown, the return on investment for consumer and commodity producing organizations is directly correlated to market share. Industrial, noncommodity products are less price-sensitive, often one-of-a-kind, making market share or volume per se less important. An arguable conclusion.

---

[2] B. Charles Ames, "Trappings vs. Substance in Industrial Marketing," *Harvard Business Review,* 48, 4 (July-August, 1970), p. 93–102.

(2) *Identify customer's needs* through analysis and understanding of the customer's business. This means not only how the customer operates but who his competitors are, how their actions affect his, industry economics, etc.

(3) *Segment customers.* A key concept in any marketing strategy, but especially important where consequences of decisions have much greater inertia and relationships with customers last long after the sale.

(4) *Design a product/service package.* Very few cogeneration projects are the same. Parameters vary with state, electric utility, plant generations, financing, personnel, etc. The total package is more important than the physical system.

Finally, Ames noted that industrial marketing requires a greater commitment from top management than traditional consumer marketing.

The reasons for this are readily apparent. A decision to market likely involves a significant commitment of resources, both capital and human. Not only is the marketing department affected but areas such as gas supply, transportation, engineering, and planning will be involved. This functional interdependence requires a clear commitment from top management. Without it, individual functional areas can prevent the implementation of a successful marketing plan, e.g.:

- "Our accounting system isn't set up for it."
- "What do you mean it's too expensive, we designed it to pipeline standards."
- "But we've never committed to supply gas adjusted to avoided cost before."
- ad nauseum

An important point to consider at this juncture is establishment of a subsidiary where the only responsibility is cogeneration. One advantage is that the personnel are more likely to advocate the cogeneration concept within the company if they are freed from conflicting responsibilities.

Just as the typical cogeneration project does not exist, the typical or correct marketing approach has not been defined. The approach will vary from company to company and region to region, and is influenced by both internal and external constraints.

We need to examine a number of choices to determine the marketing strategy that makes the most sense for us. The problem is clear. We have a new product/service package, different market(s) and limited resources. How can we transform the challenge into an opportunity?

The following framework can be used as an aid in the selection of an appropriate marketing approach. Don't attach undue significance to the terms, but rather focus on the concepts.

Company Strategy   Implementation Options

Internal Resource

Channels

Company Strategy

Three basic strategies exist for pipelines to follow:

(a)   Promotion

(b)   Merchandising

(c)   Energy Services

Promotion includes activities designed to encourage others to use cogeneration. This is an appropriate strategy if your goal is to increase gas sales. Basic promotional tools include:

(1)   Special gas rates for cogenerators

(2)   Increasing customer awareness of cogeneration

(3)   Subsidizing cogeneration feasibility studies

(4)   Offering technical assistance to potential cogenerators

(5)   Maintaining a "clearinghouse" for disseminating cogeneration information and supplying names of local experts

(6)   Other services to facilitate cogeneration

A promotional strategy requires a lower level of commitment than other strategies. The risk level is also lower and subsequently returns are limited.

[Note: One aspect that should be part of any promotional campaign is educating customers as to the differences between "Total Energy" and cogeneration. A large number of people remember "Total Energy" as a fiasco. Make sure they know the positive impacts of PURPA and advances in control technology over the last twenty years.]

Another surprising objection to gas cogeneration is the possibility of curtailment. Many plant personnel remember the 1970s and wonder if it can happen again. Cogeneration promotion needs to stress the availability of gas and the changes that have occurred since the NGPA.

You would be surprised at the misleading information regarding gas prices that is distributed by industry opponents. In the last year certain electric utility spokesman have projected double-digit percentage increases in gas prices through the 1980s. Counter this with more recent, objective forecasts.

Merchandising requires a greater commitment and carries a greater level of risk and potential return than promoting. Not only is the pipeline promoting cogeneration, but is also selling the hardware, e.g. gas turbine/generator packages. The revenues from equipment sales will supplement gas sales revenues.

An eastern distributor is following this strategy through distribution of Hitachi absorption chillers. (An example of a "push-pull" strategy would be to discount these units to potential cogenerators and increase demand for cogeneration, thereby increasing chiller sales and gas sales.)

Energy Services, the sale of electricity and thermal energy to a customer, instead of equipment or gas, is the highest risk/return alternative. Correspondingly, the commitment of pipeline management is also the highest of the three strategies. The pipeline must integrate those areas not typically involved in marketing, such as finance, engineering and gas supply, into a cohesive project unit. Gas supply, engineering, procurement, construction and energy sales contracts must be linked to support the project. Top management's support is critical in decisions involving the apportionment of risk among internal and external parties to the project.

Implementation Options

- Passive/Active
- Internal/External
- Channels

Very simply, three basic decisions can be made as to how a cogeneration program is to be implemented.

The first choice is whether to be *passive* or *active* players, react to the markets, or act to shape the markets. Many pipelines have had passive marketing strategies in the past without harm. (Actually the argument can be made that an active philosophy does not make sense in a regulatory environment.) Many pipelines are currently active at the federal level, but passive at the state.

Cogeneration is influenced as much, if not more, by state regulations, and requires monitoring, if not action, at that level. If a state passes a law that obstructs third-party financing, such as Georgia or Indiana, do you avoid that state or lobby to have the rules changed?

If an active implementation philosophy is chosen the question of using *internal* and *external* resources arises. This is primarily a function of your corporate resources, the strategy you've chosen, and the channels available.

Different companies have taken different approaches. HNG/Inter-North has an engineering arm that can provide engineering design and construction services for cogeneration projects. MidCon uses independent engineers on major projects. Both approaches have pluses and minuses. The choice is based on your strengths and priorities.

Given an active, external choice, or even a passive choice, various cogeneration *channels* can be used.

Local distribution companies, especially gas-only, are logical partners in promotion, merchandising, or energy service strategies. The LDC should have a better relationship with and more knowledge of the industrial customers in its service area. Often the LDC is capital constrained but can contribute manpower to the marketing effort. Obviously, this is one channel that will make sense in many situations.

Equipment manufacturerers are a source of external resources and an existing channel. They can often provide free information regarding technical and economic feasibility. The manufacturer's sales force may also be able to provide leads and knowledge of the industrial market in certain segments.

In many areas, independent developers have located the most attractive cogeneration prospects and have invested considerable time "selling" the idea. However, very few industrial companies are comfortable with a small, limited resource developer, without a successful track record. Projects typically take much longer to develop than people realize, so developers who are not well along by now are not likely to be successful, especially with large projects. The pipeline may be able to turn this into an advantage, by bringing financial staying power and long-term stability to a project in the development phase. The project may even be bought as it nears completion or is operating. This would make sense for a company with a strong operating background. (Energy Factors has been very successful with this approach.)

Another major channel is the architectural/engineering business. Whether large national or regional firms are involved, they usually have excellent knowledge of the industrial base and how client industry plants are operated. Often they have completed feasibility studies in the past that can provide a shortcut to identifying or eliminating prospective cogenerators.

Remember that the main reason for bringing in an external resource is that they can or should be able to do a specific task better than you. The cost of doing so must be weighed against the decreased risk that results from a more limited role.

A *caveat:* The goals of the pipeline are often different than those of equipment manufacturers, developers and engineers. Typically a pipeline company looks at the project as a long-term revenue source, while incurring corresponding long-term obligations to match. Manufacturers, developers, and engineers want to get out as soon as possible. Their at-risk period is much shorter and much more predictable than the long-term energy supplier. *The returns should be structured to reflect this fact!*

Rule-of-thumb (rough)—on a net present value basis, the cost of gas accounts for approximately two-thirds of total project costs, while capital equipment is about one-quarter. (The remainder is O&M, insurance and other "soft" costs.)

The above choices are all interdependent, and many combinations are possible. How does a pipeline go about choosing the strategy, implementation options and channels that maximize return? The key is focusing on the optimal market segment(s).

SEGMENTATION

> The most important decision an industrial firm makes is the selection of its customers. The successful firm adapts its product offering, prices, and promotional strategies to the needs and buying practices of its customers.[3]

Segmenting is simply selecting customer groups that are likely to respond in a similar manner to a marketing strategy. It requires careful analysis of the cogeneration market and of a company's internal resources, as well as creativity in matching the company's capabilities with the customer's needs.

Three basic criteria exist for segmentation variables:

- *Measurable*—among potential customer base
- *Relevant*—for a substantial group of customers
- *Operational*—responsive to a specific marketing strategy

A number of variables exist that may be used to segment the cogeneration market. The first broad differentiation would likely be between commercial and industrial applications.

Commercial applications lend themselves to segmentation by size and line of business, e.g. 200 bed hospitals.

Within the industrial segment a number of variables, including the following may be used in the "macro" segmenting phase:

---

[3] Frederick Webster, *Industrial Marketing Strategy* (John Wiley and Sons, 1979), p. 73.

- undifferentiated
- location
- electric utility
- electric demand
- energy cycle
- thermal load
- business line, S.I.C. group

Using combinations of the above and/or other macrosegmentation variables may be sufficient for defining an operational segment. However, in many instances, a more complex segmentation, known as microsegmentation is beneficial. An example of microsegmentation would be to further segment according to where the purchasing decision is made.

One of the services that MidCon, as well as other major pipelines, e.g. HNG/InterNorth, has used to help segment is Dun and Bradstreet Technical Economic Services. The D&B database consists of over one hundred observable characteristics for over 20,000 industrial plants in S.I.C. codes 20 through 39. The service can be used as both a marketing tool and a model for screening project economics.

The distinctions between segmentation and economic feasibility are especially important. Both must be analyzed, but the order is important. For instance, the primary economic variables are:

(a) *Energy price differential*, i.e. the spread between the cost of gas and the retail rate or avoided cost of electricity

(b) *Capital cost* of cogeneration system

(c) *Capacity factors*, i.e. the number of hours per year the cogeneration system can be "profitably" used

(d) *Energy load profiles*, i.e. the steam/electric ratio relationship

None of these would be a good segmentation variable, even though they are observable and relevant. They are not operational, and would not permit isolating customers with a similar market response. However, after a segmentation strategy has been designed, the above economic variables can be modeled to reduce the number of targeted projects. Generally, the segments with the greater potential return are the most attractive, if the segment is operational. (However, be aware that misleading results can be obtained by defining segments too broadly. The returns from targeting all plants using over 100 mW would be very attractive, but it would be difficult to develop an appropriate operational strategy.)

The integration of the various components of the marketing framework with operational segments is a difficult, time-consuming process. However, the resources invested will bear directly on the quality of the cogeneration marketing plan. A well-reasoned marketing plan will greatly enhance the probability of successfully landing a cogeneration project.

Exhibit 30-1

Primary Reasons for Failure

- *Not segmenting correctly*

  Diluting resources through pursuit of too many projects

  Successfully landing low-return projects

  Reacting to market changes—lost time

  Not competitive in area targeted

- *Underestimating required resources*

  Diluting resources as development time stretches

  Superficial research leading to poor strategy

  Not competitive—not understanding customer, not maintaining marketing support

- *Not recognizing current relationships*

  Inter-company relationships with competitors

  Intra-company decisionmaking—customer, pipeline

Exhibit 30-2

Other Notes

- The decisions regarding the purchase of gas, the desirability of cogeneration, and how cogeneration will be implemented are often made in different areas. Target your marketing accordingly.

- Relationship building is more important for large-ticket, one-time sales than for sales of periodic, interchangeable items. Cogeneration systems fall into the "relationship marketing" category while gas marketing is usually more "transaction marketing" oriented.

- Advertising by pipelines is usually targeted towards the investment community. Many executives see little, if any, point in general advertising. However, the concept of company image is important in industrial marketing, especially where a sales force is used.

- In many parts of the country, gas cogeneration systems are facing increasing competition from solid-fuel systems. A coal supplier or broker will guarantee delivery, a twenty year term, initial prices and escalation index. Many potential cogenerators want the pipelines to provide the same. To be competitive, pipelines will need to be more flexible in the future.

Chapter 31

Cogeneration Energy Services:
A Utility Marketing Opportunity

JOHN H. WILLIAMS
Energy Factors, Inc.

BACKGROUND

With a surplus of natural gas from producers relative to the lowered demand for product as a result of conservation efforts, the Gas Industry is faced with an upset of the historic supply/demand picture. The Industry has responded with a re-emphasis on marketing its product wisely and with an efficient utilization emphasis. Just adding volume does not make any more sense in the gas utility business than it does in a manufacturing business if the product pricing is insufficient to increase or maintain profit margins along with volume. In the natural gas distribution business, obtaining a fair return is directly related to the seasonal load factor on the system, the ability to utilize the capital investment in the ground. Thus, as a general rule, gas usage that helps to levelize the annual usage pattern, fill summer valleys, or shift winter peaks to other seasons are preferable and profitable.

Programs have been used over the years to promote gas air conditioning, agricultural water pumping, compressed natural gas for vehicles, and more recently, cogeneration—thus seeking more profitable business income that a flattened annual load pattern offers. These programs were stifled during the 1970s and early 1980s by the knee-jerk reaction of regulatory bodies in denying expenditures for what looked to them as unjustifiable sales promotion at a time of shortage. The over-correction of supply versus demand has now put the Gas Industry in a position in which new markets are vital to the financial stability of the utility industry on behalf of ratepayers and shareholders alike.

Cogeneration, with its relatively flat annual usage pattern, seems an ideal target market for natural gas sales. The best applications are (by the very nature of the cogeneration concept) those in which thermal and electric needs are a 24-hour, 7-day a week operational requirement. Whether small engine-generator packages or relatively large gas turbine power stations are involved, cogeneration facilities can add significantly to the long-term baseload for a gas or dual-utility. These attributes are in the best interests of the ratepayer in stabilizing the basis for the utility ratebase and for the shareholder in assuring a profitable usage of the sunk costs in the piping system.

"Cogeneration Energy Services" represent the methodology that will take cogeneration to the marketplace on a broad commercial basis.

COGENERATION ENERGY SERVICES
—A DEFINITION

Cogeneration Ownership Responsibilities

The concept of cogeneration itself is now generally understood by most plant engineers, energy managers, and the architectural and engineering community nationally. Many have read technical and business journal articles or have conducted studies of a cogeneration application. Few, however, truly understand "Cogeneration Energy Services" (or Energy Services Ownership).

Cogeneration Energy Services are often referred to as "third-party" cogeneration, a misnomer that confuses and has a negative connotation. To truly provide cogeneration energy services, companies such as Energy Factors cannot be "third-parties," but must instead be rather intimate partners of a "host" company. The "third-party" label was first used in describing the sometimes suede-shoe financing offers that filled the marketplace just a few short years ago. Many an ambitious investment banker, stockbroker account executive, lawyer, and/or CPA claimed the sudden capability to put together multi-million dollar financing packages to fund cogeneration projects. These entrepreneurs jumped at any project where a major industrial or commercial facility owner was willing to sign a contract for energy. They offered a "third-party" ownership built on tax credits and stretched to the limit the intent of the Public Utilities Regulatory Policy Act (PURPA). Customers and entrepreneurs alike soon found that there was a lot more to installing an onsite power plant than getting a few dentists to put money into limited partnerships. The survivors in the "third-party" financing world have been the major leasing companies, the investment banking houses, and the true "Energy Services Operators."

This chapter highlights the cogeneration Energy Services Operators (ESO) who offer to build, own/lease, operate, and maintain cogeneration systems over a 10, 15, or 20 year life. These Energy Services Operators offer to arrange the financing and the up-front engineering and permitting, to manage the construction project, to start up and debug the system, and to operate and maintain the equipment in a variety of modes of operation. We at Energy Factors believe that true cogeneration energy services are a partnership, not necessarily in the legal sense, but a commitment to live together for the term of the contract. The commitment involves a give and take on the part of both parties, a willingness to work for the mutual benefit, to ride with the bad times and to share the glory in the good times.

To understand cogeneration energy services ownership, we must first understand the responsibilities of any owner of a cogeneration system. The cogeneration system owner has the following obligations:

Capital Investment

The owner must provide for the investment capital competing with other uses for the monies, determine what impact it will have on the balance sheet of the corporation, and assess the downside risk that is associated with the investment. The investment estimate must include provision for engineering and permitting costs, staff time involved, interest during construction, working capital needs, legal and accounting overheads, contingencies, and, of course, the construction costs.

Permitting and Licensing

In addition to the general building and land-use permits that are generally required, cogeneration systems will normally require air emissions and sometimes water quality permitting under a variety of local, state, and federal standards. To connect to the local utility system for standby and surplus sales benefits, the system must qualify under a Federal Energy Regulatory Commission set of rules which include an energy utilization standard that must be met. The certification of the cogeneration system, although self-certifying in many cases, still takes owner's staff time and specialized knowledge and is subject to review and challenge by FERC.

Supplier Negotiations

The job of deciding which overall system configuration and specific equipment to use can only partially be delegated to a consulting engineer. The owner's interest in fair value, long life, low annual maintenance, delivery reliability, and overall quality must be protected by obtaining knowledgeable input from those who have been through the process.

Electric Power Sales/Grid-Connection Agreements

The negotiation of a Power Sales Agreement under the Public Utilities Regulatory Policy Act (PURPA) involves an in-depth understanding of "wheeling," "time-of-use" (TOU) sales contracts, "capacity credits" versus "energy credits," fuel escalation clauses, and electric load dispatching considerations. The technical aspects of grid-connection and the costs associated with the interconnection are used by many utilities as a negotiating barrier to discourage practical cogeneration. The cogeneration system owner must be prepared to refute these efforts and to negotiate the best possible terms for his installation.

Fuel Supply

The use of utility natural gas service versus "contract carriage" from gas brokers is a controversial issue throughout the energy industry. Playing the "self-help" fuel purchase game can mean a short-term windfall with serious mid-term consequences. The cogeneration system owner must become aware of the intricacies of these carriage arrangements before a wise decision is possible on the overall benefits. In addition, a solid knowledge of alternative fuel prices and availability is a key consideration for the owner as well.

Operation and Maintenance

Actually operating and maintaining a cogeneration system is an owner responsibility that cannot be completely subcontracted. The owner's monies are invested in major pieces of machinery for which on-line performance and reliability are the key to any operating savings. The owner has an obligation to his management to know when an engine or turbine will require downtime for overhaul, to know the amount of deterioration of performance to expect as the system ages, to know the actions that can be taken to bring back the rated performance, and to know if subcontractor maintenance charges are fair. Maintenance procedures must be established, subcontracts bid and let, and contract performance monitored and reviewed.

Energy Services Ownership

Cogeneration offers substantial saving to many facilities, but the owner obligations in building, operating, and maintaining such plants are significant and specialized in nature. Cogeneration "Energy Services Ownership" places most of the listed "Owner" responsibilities with an operating partner whose business it is to assume these roles.

Energy Services Ownership can be applied to a much broader area than cogeneration. Companies exist which will install and maintain small commercial air conditioners, heaters, insulation, solar panels, et al. We are limiting our discussion to energy services applied to cogeneration systems.

A true Cogeneration Energy Services company will offer to develop the environmental permitting, the in-depth economic feasibility analysis, the engineering, the construction management, and the financing. The Energy Service company will build, own/lease, and operate the system under a variety of contractual arrangements. In-plant personnel can be used for the day-to-day operation (on subcontract) or the operation can be completely independent of in-plant personnel—the choice is that of the "host" company. All investment capital required is supplied by the Energy Services Ownership.

The form of the contract relative to energy payments is very flexible. In an economically feasible cogeneration project, there are annual savings to be gained from a sizable capital investment. After the capital recovery has been accounted for to the party putting up the risk capital, there still remains a bottom-line savings to be divided. If the feasible savings are $500,000 per year, for example, a contract can be written that "shares" the savings on some percentage basis, or discounts the electricity price, or discounts the steam or chilled water price—but all still divide up that same $500,000. A "Discounted Utility Price" is the easiest to understand and is in broad usage. "Shared Savings" bases have been used, especially in building heating, air conditioning, and conservation applications. Shared savings contracts can be difficult to finalize due to the inability to establish "base energy usage" values. There are also tax implications to be examined closely under the Wallop Amendment to current tax law. As previously stated, however, there are many ways to contract for the savings.

Contract "Deal" Points

Of equal importance to the amount and method of tariff payment in the construction of an "energy services" contract are the following "deal" points:

(1) Definition of Baseline Value of Energy Services
(2) Termination Provisions
(3) Downside Guarantee (Floor)
(4) O&M Responsibilities (In-Plant)
(5) Extraordinary Improvements

Definition of the Baseline Value can be a difficult point of negotiation in shared savings contracts. How efficient is the current thermal system before cogeneration? How much in-plant O&M will be
saved by the cogeneration system? How much will be saved on electrical demand charges to plant needs not served by the cogeneration
system (when appropriate)? For the host company, all of these
points must be thought out and negotiated to a point of satisfaction
that the baseline is fair and equitable. Experienced energy services
companies will tailor their proposals to serve as starting points for
these negotiations. The matter is manageable, but not without some
major discussion.

Termination Provisions cover several key points of interest to both
parties. Should an unforeseen circumstance result in the closing of
the host company's facility either temporarily or permanently, some
provision must be made for the recapture of the Energy Service
Company's financial outlay. If the property is being sold to another
major energy user, the cogeneration system may be assignable as a
condition of sale.

Sometimes, however, corporations do not wish to have such conditions on the sale of their property and a cash settlement is needed.
Another termination provision will cover the case where the host
company seeks to purchase the system. Because of tax considerations,
this provision is usually available after the end of the first contract
term (e.g. 10 years) and the market value at that time as determined
by an independent appraiser.

Take-or-pay conditions are rarely used but may be found in some
termination agreements. Since cogeneration equipment has more
salvage value to a cogeneration company with multiple installations
in which to use the salvaged equipment, termination provisions for
plant closures will vary widely dependent on the service company's

asset base. If a cogeneration services company only has one or two small projects, the closure of any one might break them financially.

Downside Guarantees (Floor) refers to the risk acceptance by the energy services company if its projections of fuel and/or electricity prices are completely invalid. What assurance does the host company have that the system output will not cost more than conventional utility costs? Some form of guaranteed downside is a point for negotiation. As an example, Energy Factors usually guarantees that the host company will never pay more for the energy delivered than it would for comparable conventional utility rates for a customer of that class.

O&M Responsibilities (In-Plant) can usually be structured to the satisfaction of the host company. An energy services company must protect its capital investment by having knowledgeable personnel operating and maintaining its equipment. In many cases, in-plant personnel with training and established procedures to follow are ideal as the day-to-day operators with back-up maintenance from the energy services company. In large systems (20 megawatt), an on-site Energy Services Coordinator might be provided by the energy services company to establish procedures and oversee subcontracts to maintenance vendors. Good energy service companies have established maintenance procedures, supervisory personnel, and an engineering trouble-shooting capability to back up their contracts.

Extraordinary Improvements is often a section of the contract necessitated by today's world in which government regulations might be changed on a retroactive basis to require the retrofit of expensive equipment (e.g. environmental control technology). Some provisions can be placed in the contract to describe how the cost of these modifications will be shared between the parties. The energy services company should also retain the right to upgrade the system if newer equipment allows for an even more profitable operation.

Cogeneration Energy Services Company Example
—Energy Factors

Energy Factors is an example of the long-term, financially stable, highly-specialized cogeneration energy services companies that have grown out of the past few years of market development. Originally a division of a combination utility, San Diego Gas and Electric, Energy Factors was spun off as a private, over-the-counter stock company in 1983. Currently working to expand its sizeable experience in the San Diego area to a national scope, Energy Factors ranks at the top of national cogeneration firms.

With cogeneration energy services as its sole business, Energy Factors has current revenues exceeding $26 million per year, book assets of over $64 million, a credit line of $20 million, and cash on hand of over $10 million. On-line and running is 72 megawatts of cogeneration and over 5000 tons of air conditioning tonnage. Energy Factors operates a downtown central chilled water facility in San Diego serving over 54 blocks of commercial air conditioning needs. Energy Factors installations range from 800 kilowatts to 28000 kilowatts in the six plants now in operation. Three plants totaling around 20000 kilowatts are in the design and construct phase at present, one of which will use wood fuel. Over 140 megawatts are in the contract development stage with signed development agreements in hand.

Energy Factors operates reciprocating engines, gas turbines, and steam turbines in these facilities. Compressed air, steam, chilled water, hot water, hot exhaust gases, and, of course, electricity are the energy forms available. Consideration is being given to projects burning coal and biomass, and one major project will provide over 1000 tons of low-temperature refrigerated ammonia.

Because of its financial strength and its experience with multiple installations, Energy Factors is able to guarantee that host customers will pay no more than conventional utility rates under the worst scenario. With its background in utility negotiations, fuel supply, and power generation maintenance, Energy Factors is a prime example of the energy services companies that will develop in the next decade.

Energy Factors forms joint ventures with host companies or with local utilities. Whether a *legal* partnership is formed or not, successful energy services contracts are built on a *moral partnership and a dedication to excellence.*

THE UTILITY MARKETING OPPORTUNITY

The utility marketing opportunities are many and varied, but fall into four major categories:

Utility-Owned and Operated Cogeneration

Like the predecessor of Energy Factors, the Applied Energy subsidiary of San Diego Gas and Electric Company, a natural gas utility can own and operate a cogeneration subsidiary as long as it is willing to accept a utility rate of return on its investment. A limiting factor, however, is the current 18 percent before tax project rate of return sought by investment bankers for the debt evaluation of such projects. Cogeneration facilities are a higher risk project than conventional utility investments and investors properly seek a higher rate of return for money dedicated to such risks. Where the utility can mitigate this concern through internal financial strength, a utility-held subsidiary can be an acceptable mode of business operation. The pluses include the confidence and trust that most utilities have from their customer base. The negatives include the newness to the utility of the risk aspects of the cogeneration business and the potential confrontation with privately-owned cogeneration companies who may see such operations as a restraint of trade.

One major restriction on utility ownership is the Public Utilities Regulatory Power Act which limits *electric* utility ownership to no more than 50 percent of a qualified cogeneration facility. Combination electric and gas utilities are thus restricted in their ownership position.

Non-Utility Ownership

If the corporate structure is a holding company with the utility operation a part of the overall organization, a non-utility cogeneration subsidiary should be possible. Under this mode of operation, unregulated rates of return in the 25–35 percent and above on equity range could be targeted.

The benefits in this mode are many. Kept arms length, the non-utility subsidiary can seek higher returns than the utility while still benefiting from the association in the minds of the customer as a solid, conservative long-term entity with which to contract. The disadvantage in the real world is that utility holding companies usually have utility philosophies of business that inhibit their operation in the "deal-making" world of cogeneration. To the extent that this philosophy can be modified, the non-utility can do very well competitively (e.g. InterNorth).

Utility Partnerships with Private Cogeneration Companies

Whether restricted by the 50 percent rule or not, many utilities can take an active role in the support of cogeneration in their service area through participation with non-utility, private cogeneration companies. Brooklyn Union Gas Company is involved in such a role with at least three different private companies, including Energy Factors, in their service area. The degree of participation can range from an aggressive equity positioning to a support role in negotiating fuel supply, obtaining permits, and arranging power sales contracts. The pluses are the obvious ones of letting those who specialize in cogeneration carry the majority of the responsibilities while adding a local partner whose stability is apparent to the customer prospect. The negatives include the difficulty of maintaining a utility neutrality among competing cogeneration firms. Equity participation can be as little as 10 percent and still be meaningful to the partnership. If the tax appetite of the utility is high, the partnership benefits can constructively be used to benefit the cogeneration companies which may have more tax shelter than can be utilized.

Utility Marketing Support

If partnerships or direct participation in cogeneration is beyond the desired role of a particular utility's management, the company can still promote cogeneration in an aggressive and effective way through marketing programs which identify thermal users, educate the customer as to cogeneration benefits, and underwrite part of the costs of engineering and economic evaluations necessary for the customer to truly define the applicability of his application. Southern California Gas Company's cogeneration marketing is an outstanding example of what can be done and the results that can be obtained. Under Southern California Gas Company's Cogeneration Program, the gas company pays for up to half of the cost of a preliminary feasibility study by the customer's choice of engineering firm with an upper limit to the expenditure of $10,000. If the customer proceeds into hard engineering of a cogeneration system, additional monies are available up to a maximum of $50,000. The program has been in effect for several years and the revenue from gas sales added from cogeneration in Southern California Gas Company's service area is in the $100 million/year range.

CONCLUSIONS

At $4.00 per million Btu for natural gas, each 2500 kilowatts (2.5 megawatts) of cogeneration add $1 million/year in gas utility revenues for a 25-hour/day, 7-days/week load. Most of today's larger (over 1000 kilowatt) applications of cogeneration are operated around the clock with electricity sales to electric utilities even when the host plant is at a reduced night or weekend status. Target goals of 25 megawatts/year are possible representing $10,000,000 in natural gas revenues if 10-20 major industrial plants exist in the service area. Hospitals of 500 beds or more can also represent a $1-2 million/year revenue prospect for their 2-3 megawatt cogeneration systems.

Cogeneration Energy Services can represent one of the outstanding utility marketing opportunities for the 1980s. There is no need to wait for research and development—the equipment is commercially available now. The marketing opportunity lends itself to a variety of customized approaches in numerous utility service areas with promise that the cogeneration applications will broaden as electric rates rise and nuclear rate shock takes effect in more areas of the country.

Improving equipment might have an impact in making more applications feasible, but there is no need to wait.

What is needed is an understanding by utilities that a revised philosophy of business is necessary—a philosophy that understands that a part of a load is better than no load at all, and one that also understands that energy users in industry are trending to the subcontract of all plant services, including those of providing chilled water, steam, hot water, refrigerated ammonia, and electricity. American industry is learning to compete on a world scale and part of the lessons being learned are that a business must be lean for the tough times and flexible to expand in the good times. To industry this means minimum staff with the ability to add service personnel and with emphasis within that minimum staff on the production of the end-product. Energy is not the end-product for most gas industry customers—its supply in its many forms can be and is being subcontracted. Cogeneration Energy Services are a sign of the times—they are also a utility marketing opportunity for today.

Chapter 32

Cogeneration Organizational Aspects

MARTIN M. DUGGAN
HYDRA-CO Enterprises, Inc.

This chapter is presented in two parts. The first part will briefly describe why Niagara Mohawk elected to establish HYDRA-CO Enterprises, Inc.—an arms-length subsidiary. I will discuss what we have learned and what we have accomplished in our four years of existence, and review something about the benefits that have or will be obtained.

The second part will focus on the organizational structures that have evolved, and will start with a definition to insure that we are all working with a common understanding.

PART 1

Why did Niagara Mohawk elect to establish an arms-length subsidiary? What convinced them to take the risk? The answer lies in two areas. In 1981, the New York State legislature amended the public service law of the state to conform with PURPA. One of the unique modifications was that if any utility in New York State wanted to participate in the development of cogeneration, small hydroelectric or alternate energy facilities, they had to set up an arms-length subsidiary.

But the initial impetus came long before the energy crises and PURPA. Many years ago Theodore Levitt wrote a brilliant article for the *Harvard Business Review* entitled "Marketing Myopia." Levitt's thesis, and the basis for my subsequent actions, was that the unhappy fate the nation's railroads and their ancillary industries could have been avoided. In the early days the managers of the railroads could have expanded the options offered their customer, and the service mix associated with these options.

The constraints imposed on the railroads by regulatory agencies were largely after-the-fact reactions against predatory maneuvers designed by the railroads to destroy competing carriers.

Levitt suggested that if the then leaders of the railroad industry ever asked themselves what business they were in they must have said the railroad business. We know, with the benefit of hindsight, the answer should have been the transportation industry.

More recently, the *Economist* magazine had a series of articles on railroads. These articles discussed the need for railroads to redefine their roles. The author pointed out that railways are safer, more fuel efficient, less noisy than auto or air travel. However, the railroads cannot live on environmental godliness alone. They have to define the traffic they should be attracting—and which services to abandon. We should learn from the railroads.

For years I tried to convince Niagara Mohawk to break the utility version of the railroad mentality. But I wasn't very successful. Utilities benefitted from the economies of scale, increasing reliability of service and decreasing unit cost to the consumer. That has all changed. We've gone through a severe energy crisis. Congress, with the passage of PURPA, essentially deregulated the manufacture of electricity. Public Utility Commissions and state legislatures are reducing the barriers to utility energy subsidiaries. We've moved out of an era when it was *de rigeur* to construct ever-larger power plants in cathedral-like buildings to serve heavy industry.

We are still in an era where our utilities are beset by problems relating to the near-past spiral of interest rates, to energy conservation, and an incredible level of "econuttery" that has effectively stifled electric production progress. We are in an era that is seeing regulated entities like railroads, airlines and banks fall on very hard times. There is constant speculation as to which utility will be the industry equivalent of Penn Central, Braniff or Continental Illinois Bank.

If the utility industry is to avoid the fate of its transportation industry counterparts, I believe we must ask and answer four questions.

First: What business am I in? For most the answer probably is the utility business.

Second: What business do I want to be in? If others see things as I do and as does the management of Niagara Mohawk, then the answer will be the energy business. Who knows more about the safe, reliable and, in most cases, economic manufacture, transportation, delivery and billing for energy than does a utility? Nobody! Why, then, give up this edge?

Third: How will I get to where I want to be in the future? Obviously there are many viable options. To me, there is one option that is not viable. That is, within a regulatory framework. Regulation has a tendency to deaden incentive. Regulators have been known to assign risk and the attendant penalties to the shareholder, and, concomitantly, assign benefits and rewards to ratepayers. Such no win situations will not produce the initiative required.

Fourth: Why am I in business? I submit that the only proper answer is to make a profit or to have an excess of income after the expenses and taxes are paid. Any other answer will lead to trouble.

HYDRA-CO

HYDRA-CO Enterprises is a wholly-owned, arms-length, semi-autonomous subsidiary of Niagara Mohawk. Wholly-owned is self-explanatory. At the present time as we require funds, we, like most companies, raise money through the sale of stock— to Niagara Mohawk. Now, when you work in a corporate structure with a single stockholder it is imperative that you understand that the stockholder can call a meeting at any time and, as the owner, make any changes he sees fit to make. Thus, we say that we are semi-autonomous.

"Arms-length" means many things to many people. To us it means that we are in separate quarters. It means that any services we receive from the parent corporation we pay for. It means that we have our own pay scale and our own Board of Directors. Most important, it means that if we are not successful we do not have a job back at headquarters. Let me assure you that the lack of security is a powerful motivator and stimulator, and a condition I believe to be a prerequisite for a successful venture.

In October 1981 five of us left Niagara Mohawk. Of that group I was the only non-engineer. Among us were three professional engineering licenses, one masters degree in hydro-engineering, and two masters in business administration. Our backgrounds and work experience included marketing, marketing research, transmission field experience, generation planning, hydro operation, project management and cost of service.

We had stars in our eyes and a capitalization of thirty million dollars. Now we are seventeen and growing, having added secretarial, word processing, accounting, risk management and data processing capability. We are looking for additional personnel to assist in project management, project development, and project operations.

In our four years of existence we have been battered, bruised, occasionally bloodied—but unbowed. Of our original capitalization we have spent or committed about twenty million dollars. What have we learned?

First, most entrepreneurs, engineers, even the industry itself, don't have enough capital to develop a project. Besides, energy really isn't their business. Next, the typical developer doesn't have sufficient credibility to obtain non-recourse project financing. They don't have the requisite engineering and project management skills. Third, very, very few people in this country have the expertise necessary to negotiate with a utility and to develop an energy services contract.

We have learned that in order to move projects forward, we have to be a "bank." Some form of bridge financing is required in order to get all the paperwork in place that the financial community will require before it will commit to or let us draw down on construction financing and/or non-recourse project financing.

We have learned that, if we don't do our homework very carefully, we will get burned. The two main elements, the electric sales contract and the steam sales contract, are signed long before a spade of earth is turned over. Unlike those in the utility industry, we have no place to pass on cost overruns, changed tax rates, higher-than-planned interest rates, and the like.

In this regard, I pass on a *caveat*. Someday, in the not-so-distant future, utilities will be required to build, operate and maintain their facilities under the same contractual conditions they impose on developers, and equally important, within the confines of their estimate of avoided cost. Several states are already discussing and developing such legislation.

We and other developers have learned that while we may not be able to live with the utility's concept of avoided cost, we can exist by discounting retail electric rates. Utilities don't want to lose customers, and developers such as HYDRA-CO don't want to take them. To circumvent this option several utilities are contacting customers considering cogeneration. They ask simply, "What rate will it take to get you away from cogeneration?" The Federal Justice department is already investigating some utilities for possible anti-trust action. This tactic will only add fuel to the fire. Further, can you imagine the reaction of other customers of that utility? "Why not me?" Who is subsidizing who?

What have we accomplished?

Under PURPA, an electric utility, a subsidiary thereof, or any such combination cannot have more than a 50 percent equity interest in a qualified facility. Negotiating joint ventures is, has been, and probably will be, a painful activity.

To date we have negotiated and signed eight different joint venture agreements with four different entities to develop, own, operate and maintain hydroelectric facilities. We are currently negotiating three additional agreements. We have developed six sites, about 13 mW, with a total capital cost of some $29 million. We have two sites under construction—20 mW total capacity and a capital cost of $30 million. We are in the final stages of negotiations for some 70 mW of hydroelectric with a total capital cost of about $140 million.

In the area of what we describe as alternate energy projects we have a couple of things going. We own a geothermal site in Idaho and have signed yet another unique joint venture agreement with a developer. We have developed 6 mW at a location in Nevada. We have also negotiated the myriad agreements necessary to get a 30 mW wood fuel project under construction.

We are working with Air Products and Chemicals on some turbo expander projects. There is a large market available if one can harness the energy currently being dissipated when natural gas is reduced from high pressure to low pressure. The trick is to produce electricity and not violate the Fuel Use Act. We think we can do this with yet another joint venture partner and joint venture agreement.

In the context of more conventional projects, we have signed contracts for two 4 mW gas-fueled turbine cogeneration projects in upstate New York. We have submitted proposals to a tire manufacturer and a chemical manufacturer for 50 mW installations. It is doubtful if these projects will come to fruition because of the intransigence of the Public Services Commission staff.

What about benefits to Niagara Mohawk? Benefits like everything else come in a variety of sizes and shapes. We have financial benefits, for us and for our owners. In 1984 we provided Niagara Mohawk with almost $2 million in benefits. But in the long run I honestly believe the benefits to society will outweigh the financial ones. A successful cogeneration project means that the existing or new industry is going to be there for the next 10, 15, 20 years. Thus the community will have some employment stability and property tax stability. Think of the impact on a community when a major industry closes. For example, in Syracuse, Allied Chemical closed its Solvay works. Fourteen hundred people were laid off. The site is to be leveled, turned into a park. The cost to rehabilitate the land can reach $400 million. But the impact on suppliers, schools, hospitals, etc. is even greater.

With a good cogeneration contract the utility will keep an electrical customer, can double or triple a gas load, and have some assurance that residential and commercial customers will have the wherewithal to pay their bills. These are benefits that are often ignored when assessing the overall impact of a cogeneration project on a utility.

In a more substantive area, we have acted as project manager on all our major projects, to date. They have come in on time and under budget. In 1984 we, as a Development Company, had net income of some $400,000. We enjoyed project underruns in excess of $800,000. We did not have to raise any funds for our ongoing administrative and general expenses.

In the context of the corporate world, what we have accomplished and what we have underway is small potatoes. But our industry started out with small projects. To close, I'd like to share a thought that has helped me get over some very rough times. Thomas Watson, Jr., when he was Chairman of IBM, said:

> As you stand and are counted, you will first run into the group which equates newness with wrongness. If it's a new idea it is uncomfortable. Second, you are sure to meet cynics, people who believe anyone who sticks his neck out is a fool.
>
> If you stand up and are counted, from time to time you may get yourself knocked down. But, remember this: a man flattened by an opponent can get up again. A man flattened by conformity stays down for good.
>
> Follow the path of the unsafe, independent thinker. Expose your ideas to the danger of controversy. Speak your mind and fear less the label of crackpot than the stigma of conformity.

Chapter 33

Key Elements of a Cogeneration Marketing Plan

JAMES L. CRIST
The Peoples Natural Gas Company

Often in sales, a new product comes along and the orders come down from the top to "get this product sold"! All the sales people scramble around, start making sales calls, and after awhile, if sales are languishing, someone says, "Hold it, let's take a better look at what we are trying to do and determine who our prospects really are." Cogeneration can be a successful product, *if* the proper market planning is done in advance of the proper sales calls.

AN ILLUSTRATION

Consolidated Natural Gas Service is a fully integrated natural gas utility that serves at retail in Ohio, Pennsylvania, and West Virginia and wholesale in New York state. The cogeneration marketing work described in this example was conducted during the past three years in Ohio at The East Ohio Gas Company. There are many similarities between The East Ohio Gas Company and The Peoples Natural Gas Company. They are in older industrial areas that suffered greatly during the recent recession. Many of our customers have gone out of business or moved southbound, and the ones that are remaining are just now seeing their business improve and are very cautious as they look at new investments and the future. This is a very tough market area for cogeneration, as the market assessment results that we learned are reviewed.

There are two main reasons that The East Ohio Gas Company was interested in cogeneration marketing. The first reason for cogeneration interest was to improve the health of our customers. As mentioned earlier, the ones that are surviving are coming back rather lean and very aggressive toward cost cutting with very tight operating cost controls. The attitude with which we approached cogeneration project development was one of putting ourselves in our customers' shoes and wanting to develop the best project for the customer. The second was to preserve existing gas load and develop greater gas throughput. East Ohio saw considerable decrease in their sales volumes in the recent five-year period. The Peoples Natural Gas Company has seen even a greater percentage loss. So we are all very interested in retaining existing load and perhaps moving more gas through the system.

COGENERATION PROJECT DESCRIPTION

We are involved, basically, in two types of cogeneration projects: large ones, and small ones. A large project is the type we find in our industrial or very large commercial customers. They would normally use gas turbine technology, and would range between two and ten megawatts. They have a fairly lengthy development time period, usually 30 months or so. They have a large capital cost, greater than $1 million. These projects use a lot of gas also, anywhere from 200 to 600 million cubic feet annually for the projects that we have been working on the East Ohio service territory.

Small projects are not to be ignored. Small projects are under two megawatts; in fact, some are as small as 60 kilowatts. These would be found in very small industrial and commercial customers. They will use gas engine technology, and much of the gas engine technology is packaged modular systems. They have a shorter developmental timetable, roughly 18 months. They can be done for underneath

$1 million, and although each project individually sells lesser amounts of gas than large projects (for example, the gas load range is between 60 million cubic feet per year and 200 million cubic feet per year), there are many more of these projects than the large ones.

COGENERATION MARKET ANALYSIS

The first step is to analyze who the customers are by reviewing the revenue data available internally. Once we examined the types of gas usage that our large industrial customers had, we learned that we did not have very many 20-, 30-, 40-, and 50-megawatt projects. The bread and butter of our cogeneration market consists of projects under 10 megawatts. This knowledge is very important to guide the following steps in the market assessment.

The second step is to train the customer representatives. We were not trying to make cogeneration experts out of our field people. We were trying to give them enough information so they could be conversational in cogeneration with customers and learn how to identify good prospects from bad prospects.

The third step is using our trained sales representatives. We collected specific data from the 70 largest industrial accounts in the East Ohio area—data not just on thermal usage, but on electrical usage during the summer and winter months, and peak demand usage. We found this data to be fairly easy to collect as most of the customers were aware of cogeneration and interested in examining anything that promised to save them money.

From a customer's perspective, he wants to do projects that will give him some economic incentive. So we took a very detailed look at the economics of cogeneration. To do that, instead of relying on some of the outside vendors or outside consulting services, we developed our own computer model using the PC. This model, along with the next step, developing engineering data internally, has enabled us to conduct three years of cogenerating marketing for total outside expenses underneath $20,000. This represents a significant savings versus contracting outside the company for engineering and economic feasibility analyses on a project-by-project basis.

After conducting economic sensitivities, we learned that the most important variable governing the economics of cogeneration is not the gas price but indeed the electric price. Therefore, we contracted with a consulting firm to project electric rates for the specific utilities in our service territory.

As we looked at some project traits in other areas of the country where cogeneration is developing very rapidly, we could see that East Ohio Gas had some obstacles to be addressed. Projects in Ohio did not have three-year paybacks. Because the electric utilities in Ohio are engaged in several nuclear plant construction projects, there was no capacity component in the avoided costs of Ohio, and the regulations in the state were not established and were not very specific to provide the necessary guidance that cogenerators need.

If we looked at incremental electric costs, we saw electric prices between four and five cents per kilowatt-hour. This is an important point to remember. The incremental costs on a declining block rate schedule are less than the average cost that the customers are paying. As you do economic analyses, be very careful to use the incremental electric costs and not average costs.

The regulatory area was virgin territory in Ohio. Each project had to be negotiated with the electric utility prior to going to the Commission. Even though the electric utilities were overcapacity and had developed innovative sales programs to add new load without charging for demand, they were placing a stiff demand charge for backup power on cogenerators.

CUSTOMER SERVICES

With the market assessment complete, several customer services were developed internally prior to making sales calls on customers for cogeneration projects.

Analysis of a customer's energy data and collection of the proper data is the most important step of any cogeneration project. We assist a customer in collecting the necessary data, verifying the accuracy, and then analyze it and put it into a form that can be used for conceptual system design.

Equipment recommendations are fairly easy to make after the proper data is collected. There just are not that many competitive pieces of equipment in a certain project size range. Our engineering database consists of detailed cost and performance information on half a dozen gas turbines and about a dozen natural gas engines.

Our involvement in projects is not intended to supplant outside engineering firms and detailed outside engineering work; therefore, we developed a list of engineers with cogeneration expertise.

We make our economic evaluation model, COGENT, available to our customers to run various sensitivity cases on their projects.

Since cogeneration requires considerable capital, we investigated several mechanisms to assist our customers along those lines. The East Ohio Gas Company or The Peoples Natural Gas Company are not in the business of financing cogeneration projects. In fact, Consolidated Natural Gas and two other registered gas companies serving the region are prohibited from owning cogeneration plants due to a technical interpretation of the Holding Company Act. The two ownership options available to a customer are self-ownership, possibly using bank financing, or third-party ownership. There has not been much third-party ownership in our region.

Several presentations were developed for the technical community to stir up interest in cogeneration and inform them of our capabilities of working with them on project teams.

The most important issue today, in Ohio, is the regulatory area. We found it necessary to monitor and compile information from the Public Utility Commission, the Ohio Consumers Counsel, and the electric utilities in the area to assist our customers in proceeding through the myriad of regulatory activities and avoiding potentially lengthy project delays.

PROJECT DEVELOPMENT STEPS

The first step begins with an analysis conducted by us for the customer. We look at his energy use profiles, determine a rough, conceptual design with some suggestions of equipment that could be used, and conduct a computer economic evaluation to determine if the project is feasible. The objective at this stage, if we have a feasible project, is to give the customer enough information and incentive to move this project internally and approve funding for feasibility engineering.

Feasibility engineering is conducted by outside engineering firms at a cost to the customer of anywhere between $5,000 and $20,000. We have had very good results in encouraging customers to conduct feasibility engineering and, in fact, have over half a dozen projects with about 20 megawatts potential, which would result in annual sales of about two billion cubic feet.

The results of feasibility engineering usually confirm the analysis conducted by ourselves and oftentimes can suggest some improvements, different system configurations, or alternate equipment. The important point is that now an independent assessment of cogeneration has been made and presented to the customer.

The third step is a very important step: the decision the customer must make to proceed with his project. This usually requires considerable time for the internal decision-making process and consideration of several of the important regulatory and financial issues.

Once that decision to proceed has been made, steps four and five, the regulatory and financial steps, are conducted simultaneously. Several important regulatory issues, such as the interconnection equipment standards that the electric utility has, along with the back-up rates for cogenerators, must be clarified or negotiated. In the financial area, the customer must decide whether he is going to own his project or attempt to interest a third-party entity to own and operate. If the customer chooses to own his own project, then he must pursue financing arrangements with commercial banks.

Once regulatory and financial issues are resolved, then the project is free to proceed to detailed design and construction. Normally this is where the gas utility in traditional markets would first become involved. That is, the customer would walk into one of our offices, explain that he has a project under way in his plant (or he is building a new house or constructing an office building), and he wants to work with our engineering people to ensure the delivery of natural gas at the right time. You can see that cogeneration projects entail considerable front-end development that is not usually part of the natural gas marketing program of traditional areas.

The last step would be the final delivery and sale of natural gas to the cogeneration project. Delivery of natural gas as well as sale of natural gas is involved because capital-intensive cogeneration projects will not proceed unless perceived risk in the buyer's mind is reduced in regard to fuel supply. A buyer is looking for and should expect a long-term (five- to ten-year) gas supply contract with pricing escalators known in advance. Most of the cogeneration projects evolving in our market area will be through the use of transported gas, not tariff-sheet gas. Such increased gas throughput benefits our other existing gas customers.

Two projects have entered the regulatory and financial stage of project development. The first is the Beechwood Sheraton Hotel. It will use a 100-kilowatt packaged cogeneration module manufactured by Cogenics. This particular hotel is currently an all-electric hotel, having been built during the time period several years ago when there were only a few years of gas supply left in the country. The heat output of cogeneration will be used for domestic hot water, and the electricity will serve a portion of the hotel's electrical needs. Discussions are being conducted with the utility and the Commission on backup rates and interconnection requirements.

The second project that has proceeded to the regulatory and financial stages of development is Deaconess Hospital. This medium-sized hospital will be using an 800-kilowatt natural gas reciprocating engine system producing 15 psig steam. The steam will be used for food service, domestic heating, and domestic hot water. The electricity will serve a portion of the hospital's needs. This project is also engaged in discussions with the local electric utility concerning interconnection requirements and backup power rates.

There are several rules of thumb that will aid in a review of cogeneration prospects. The first has to do with baseload minimum consumption that will help determine if the user is large enough for cogeneration. We found that summer loads should be a minimum of 1,000 mcf per month before a user would be applicable for a 60-kilowatt package like the Thermo electron module. For larger-packaged units, such as the 500-kilowatt Martin cogeneration unit, users should consume 5,000 mcf per month in the summer.

There are also important thermal restrictions in how the heat from cogeneration is used. The very small systems, 60 to 200 kilowatts, normally produce only hot water; therefore, the customer must be using hot water, not steam. Larger engine systems, from

300 to 650 kilowatts, can produce low pressure, 15 psig steam, or they can produce a combination of higher pressure, 100 psig steam, and hot water. It is necessary to determine if the customer can also use the hot water if he has a high-pressure steam need. Otherwise, there would be considerable waste of valuable Btus.

RECOMMENDATIONS

The objectives of establishing a detailed market development plan and program are to establish credibility. Users need services that you can provide with a teamwork approach. Cogeneration projects require considerable up-front sales and marketing effort compared to traditional gas sales opportunities; therefore, it is very necessary to maximize your effectiveness, which can only be done after a thorough market analysis is conducted. The third objective is to provide management with better projections of future sales. Cogeneration is a very difficult market to project, and only when a market analysis has been conducted can you determine the estimated gas volume that can be sold for cogeneration. The fourth objective for developing such a comprehensive program is to identify and create a customer, either a new sale in an existing customer, or a new customer.

To reiterate:

(1) Good market analysis precedes a sales effort in cogeneration.

(2) It will be necessary to offer in-depth customer services if you expect to be a real project participant that can help influence decisions favorably.

(3) The marketing process for cogeneration is long and involved and must be approached by steps in the proper order.

Three simple steps must be followed:

(1) Start examining user data from your revenue system. That is the first step in determining what size of project potential there is in your particular service territory.

(2) At the same time, learn more about the electric utilities. Go and talk to them. Some of them might be very cooperative

and helpful. At least with some others, you might learn of their reasons for not wanting cogeneration and through better communication might learn of other sales opportunities that you can capitalize on.

(3) Go talk to the Public Utility Commission in your state to determine the degree of cooperativeness that you will receive should you get a project moving to the regulatory phases.

Section VII

Case Studies

J. A. ORLANDO, P.E.
GKCO, Incorporated

Chapter 34

Case 1:
Small Commercial Application

This case consists of a small nursing home where natural gas is used for domestic water heating and for space heating. This latter function is provided using a two-pipe heating or cooling system. Air conditioning is provided by electric chillers.

WALKTHROUGH

The objectives of an initial site inspection or walkthrough are to determine whether the site is technically compatible with cogeneration and whether a cogeneration system has the potential for economic viability. Technical compatibility between mechanical/electrical systems was reviewed in Chapter 8. However, in addition to the compatibility between the site's energy systems and the cogenerator's heat recovery capability, it also is necessary to determine whether there is adequate space for a cogeneration plant, including fuel processing equipment, whether the existing systems are adequate, and whether future energy requirements will be significantly different than historic requirements.

The walkthrough indicates that the site's mechanical systems, which consist of a two-pipe heating or cooling system, are compatible with the heat recoverable from a small reciprocating engine.

It also is necessary to develop a rudimentary measure of economic viability. Many such procedures can be used and one is illustrated in this chapter. Table 34-1 provides a summary of the site and the cogeneration system characteristics. In Table 34-2, operating costs, revenues and allowable installed cost are computed. In this example, average costs for purchased power, fuel and maintenance are used. Note that the value of the recovered heat is divided by 0.75 to reflect the 75 percent efficiency of a conventional boiler.

Table 34-1. Site and System Characteristics

Site

- Natural gas used for domestic water and space heating
- Annual boiler efficiency of 75%
- Average electric costs of 8¢ per kilowatt-hour
- Average natural gas cost of $5.00 per mmBtu
- Electric utility standby charge of $1.50 per kilowatt per month

Module

- 115 kW capacity
- 1,390,000 Btu per hour fuel input (12,100 Btu/kWh)
- 642,000 Btu per hour of recoverable heat available (5,500 Btu/kWh)
- 8,000 hours per year availability
- O&M cost of $1.25 per engine hour

Table 34-2. Preliminary Economic Analysis
Waukesha 115 kW Module

I   *Cogeneration System Costs*

   1. Fuel Cost
      (1,390,000 Btu/Hr) ($5.00 mmBtu) $ 6.95/Hr

   2. Maintenance Cost @   1.25/Hr

   3. Standby Costs @   .26/Hr

   4. Owning Costs @   .12/Hr

      TOTAL $ 8.58/Hr

II   *Cogeneration System Revenues*

   1. Displaced Electric Purchases
      (115 kWh @ $.08/kWh) $ 9.20/Hr

   2. Avoided Fuel Costs

      (642,000 Btu/Hr) ($5.00 mmBtu)

               (.75)   4.28/Hr

      TOTAL $ 13.48/Hr

III *Cogeneration System Costs*

   1. Operating Cost Reduction $ 4.90/Hr

   2. Annual Margin at 8000 Hr/Year $ 39,200/Year

   3. Allowed Installed Cost for 4-Year Payback $ 156,800

If all the recoverable heat can be utilized, then the module will produce $4.90 in cost reductions per operating hour and, at 8,000 hours a year, it will yield annual savings of $39,200. In this hypothetical case, it was assumed that the site owner required a four-year simple payback and the resulting annual savings would justify a capital investment of up to $156,800. The installed cost for a 115 kW factory-assembled cogeneration module can be less than $100,000 for simple installations, and the required investment is clearly less than the allowed investment.

The standby cost of $1.50 per kilowatt per month totals $172.50 per month for this module. Assuming an average availability of 667 hours per month (8,000 hours per year), standby charges total $0.26 per engine hour. Annual owning costs are estimated at 1 percent of the equipment cost of $100,000, or $.125 per engine hour.

In this case, a rather simple economic analysis requiring a few minutes indicates that a 115 kW installation may be viable. Based on this analysis, a site owner can decide whether to proceed to the next level of analysis, which may require a minimum of four to eight professional hours to as much as four days.

PRELIMINARY ECONOMIC SCREENING

The above analysis makes many assumptions, including one that the system is operated at full load for 8,000 hours per year and that all recoverable heat and all power could be used on-site. In addition, average utility costs were used rather than incremental costs. In the next, more comprehensive level of analysis, these assumptions are replaced by more realistic approximations. This procedure is illustrated by the following analysis of a small, hypothetical industrial facility, which requires low-temperature hot water for process use.

Figures 34-1 through 34-3 summarize monthly energy usage for the facility. The most significant difference between this analysis and the more general approach used in the walkthrough is the development of month-by-month data. Hourly data would be used at the next level of analytic detail. Table 34-3 provides a summary of the site's monthly energy usage.

Conventional energy characteristics by month as taken from utility bills are the basis for this analysis. At this point, it is important to have some understanding of the use of the natural gas in order to estimate what fraction can be displaced by recovered heat. Applications where gas is displaceable include service water heating, while such applications as cooking, some drying and process firing cannot use recovered heat in place of natural gas.

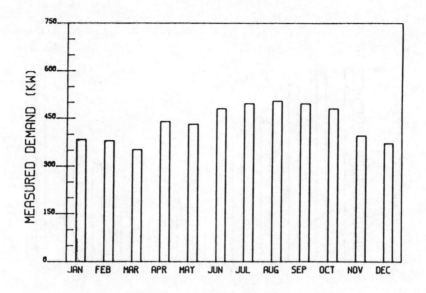

Figure 34-1. Monthly Electrical Demands

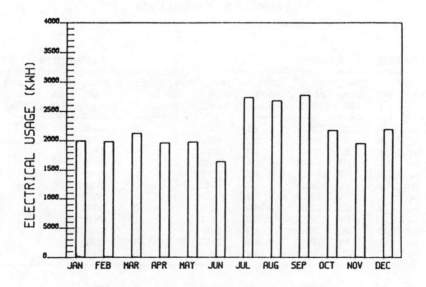

Figure 34-2. Electrical Requirements

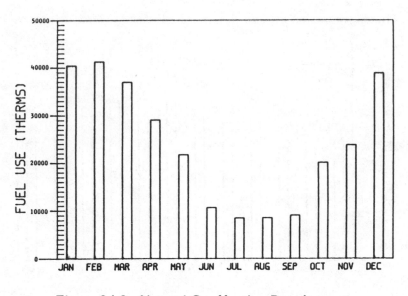

Figure 34-3. Natural Gas Heating Requirements

Table 34-3. Energy Data

CASE: 1

ALT HTG FUEL   0 BTU/GAL   BLR EFFICIENCY a   0.75

MONTH	DAYS	CONV'L DEMAND (kW)	CONV'L ENERGY (kWh)	CONV'L LOAD FACT (Hours)	CONV'L GAS USE (Therms)	NON-DISPL'BLE FIRM GAS (Therms)	DISPL'BLE FIRM GAS (Therms)	INT'BLE GAS (Therms)	CONV'L OIL USE (Gal)
Jan	31	384	200000	521	40470	0	40470	0	0
Feb	28	380	198800	523	41310	0	41310	0	0
Mar	31	352	212400	603	37040	0	37040	0	0
Apr	30	440	196000	445	29110	0	29110	0	0
May	31	432	197600	457	21790	0	21790	0	0
Jun	30	480	164600	343	10700	0	10700	0	0
Jul	31	496	274000	552	8500	0	8500	0	0
Aug	31	504	268400	533	8570	0	8570	0	0
Sep	30	496	278000	560	9110	0	9110	0	0
Oct	31	480	217600	453	20140	0	20140	0	0
Nov	30	396	195200	493	23850	0	23850	0	0
Dec	31	372	219600	590	38920	0	38920	0	0
TOTALS	365		2622200	506	289510	0	289510	0	0
MAX		504		603					
MIN		352		343					

The cogeneration module can be operated in any of three modes.

- Baseloaded with any excess heat rejected and excess power sold to the utility.

- Thermally loaded so that power is produced only when there is a need for heat. In this case, excess power, if any, is sold to the utility.

- Electrically loaded and tracking, with no power sales to the utility. Excess heat, if any, is rejected.

Each of these modes can be approximated in the calculation of the system's electrical output. In this example, the system is electrically baseloaded and output is limited only by the module's availability—8,000 hours per year. It is assumed that the site's minimum demand never drops below 115 kW. If hourly data are available, a load duration curve such as that of Figure 34-4 can be used for sizing.

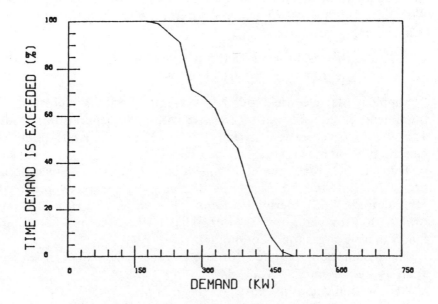

Figure 34-4. Load Duration Curve

In this example, the cogeneration system is sized so that it provides only a fraction of the site's thermal requirements and, therefore, all recoverable heat can be used. In cases where there are significant hourly changes in heating requirements, thermal storage may be required to smooth over these variations and reduce peaks.

The output of this set of procedures is an estimate of monthly supplemental power and fuel requirements, cogeneration system output and the module's fuel requirements.

As illustrated by Table 34-4, during a 31 day month the 115 kW system will produce 78,137 kilowatt hours (Column A).

(115 kW) × (31 days/month) × (24 hours/day)
× (8,000 hours availability/year)
÷ (8,760 hours per year) = 78,137 kWh

With recovered heat available at the rate of 5,580 Btu/kWh, a total of 4,360 therms of heat is available during the month (Column B). This total represents approximately 10 percent (Column C) of the site's required heat and it will all be used to displace boiler fuel. With an efficiency of 75 percent for the boiler, the fuel displaced (Column D) is calculated:

(4,360 therms, heat) ÷ (.75 therms, heat per
therm fuel to boiler) = 5,813 therms

Supplemental electrical purchases are the residual electrical requirement of 269 kilowatts (384 kW without cogeneration less the 115 kW of cogeneration capacity) and 121,863 kilowatt hours of electricity (Columns E and F, respectively).

The net fuel requirement is the 40,470 therms of conventional boiler fuel less the 5,813 therms displaced by the recovered heat, for a net of 34,657 therms (Column G). With an electrical heat rate of 12,100 Btu per kilowatt-hour (HHV), the system will require 9,455 therms of gas fuel (Column H).

The assumptions required for the next set of calculations in this analysis are presented in Table 34-5.

At this step, monthly costs for both conventional and cogeneration systems are computed. As shown in Table 34-6, billing demands (Column A) based on the rate ratchet must be calculated as a basis

Table 34-4. Energy Analysis

CASE: 1

CAPACITY	115 kw	HOURS PER YEAR a	8000
HEAT RATE	12100 Btu/kwh	HEAT RECOVERY a	5580 Btu/kw
ALT HTG FUEL	0 Btu/Gal	BLR EFFICIENCY a	0.75

MONTH	DAYS	CONV'L DEMAND (kw)	CONV'L ENERGY (kwh)	CONV'L LOAD FACTOR (HOURS)	CONV'L GAS USE (Therms)	NON-DISPL'BLE FIRM GAS (Therms)	DISPL'BLE FIRM GAS (Therms)	INT'BLE GAS (Therms)	CONV'L OIL USE (Gal)	COGEN INST CAP (kw)	(A) COGEN PROD'TION (kwh)	(B) REC'V HEAT AVAILABLE (Therms)	TOTAL DISPL'BLE FUEL REQ'D (Therms)	(C) AVAIL/REQ'D HEAT (%)	USED/AVAIL HEAT (%)	(D) ESTIMATED FUEL SAVING (Therms)	(E) SUPPL'L DEMAND (kw)
Jan	31	384	200000	521	40470	0	40470	0	0	115	78137	4360	40470	10.8	100.0	5813	269.0
Feb	28	386	198600	523	41310	0	41310	0	0	115	70575	3938	41310	9.5	100.0	5251	265.0
Mar	31	352	212420	603	37040	0	37040	0	0	115	78137	4360	37040	11.8	100.0	5813	237.0
Apr	30	440	196000	445	29110	0	29110	0	0	115	75616	4219	29110	14.5	100.0	5626	325.0
May	31	432	176000	407	21790	0	21790	0	0	115	78137	4360	21790	20.0	100.0	5813	317.0
Jun	30	486	184600	343	10700	0	10700	0	0	115	75616	4219	10700	39.4	100.0	5626	365.0
Jul	31	496	274000	552	8500	0	8500	0	0	115	78137	4360	8500	51.3	100.0	5813	381.0
Aug	31	504	268400	532	8570	0	8570	0	0	115	78137	4360	8570	50.9	100.0	5813	389.0
Sep	30	496	278600	533	9110	0	9110	0	0	115	75616	4219	9110	46.3	100.0	5626	381.0
Oct	31	480	217600	560	20140	0	20140	0	0	115	78137	4360	20140	21.6	100.0	5813	365.0
Nov	30	396	195200	493	23850	0	23850	0	0	115	75616	4219	23850	17.7	100.0	5626	281.0
Dec	31	372	219600	590	38920	0	38920	0	0	115	78137	4360	38920	11.2	100.0	5813	257.0
TOTALS	365		2672200		289510		289510		0		9220000	51336	289510			68446	
MINIMUM		352		343										10	100		237
MAXIMUM		504		603										51	100		389

MONTH	(F) SUPPL'L ENERGY USE (kwh)	SUPPL'L LOAD FACTOR (Hours)	SUPPL'L OIL USAGE (Gal)	POTENTIAL GAS RED'TION (Therms)	SUPPL'L INT'BLE GAS USE (Therms)	(G) SUPPL'L FIRM HTG SALES (Therms)	(H) GAS FOR POWER (Therms)
Jan	121863	453	0	5813	0	34657	9455
Feb	128225	484	0	5251	0	34059	8540
Mar	134263	567	0	5813	0	31227	9455
Apr	222384	370	0	5626	0	23484	9150
May	119463	377	0	5813	0	15977	9455
Jun	86984	244	0	5626	0	5074	9150
Jul	195563	514	0	5813	0	2687	9455
Aug	90263	489	0	5813	0	2757	9150
Sep	202384	531	0	5626	0	3484	9455
Oct	139463	382	0	5813	0	14327	9150
Nov	119584	426	0	5626	0	18224	9450
Dec	141463	550	0	5813	0	33107	9455
TOTALS	1702200		0	68448	0	221062	111320
MINIMUM		244					
MAXIMUM		567					

Table 34-5. Cost Data

Electric

Customer Charge	$12/month
Energy Charge	
First 15,000 kWh	6.00¢/kWh
All excess kWh	5.5¢/kWh
Fuel Adjustment Charge	1.76¢/kWh
Demand Charge	$3/kW/month
	with a 70 percent 11-month ratchet
Buyback Rate	4.00¢/kWh
Standby Rate	$1.50/kW/month

Natural Gas

Customer Charge	$12/month
Commodity	
First 10,000 Therms	$.54/therm
All excess	$.45/therm
Purchased Gas Adjustment	$.015/therm

---

for the demand charge. In addition, electric power usage must be allocated to billing blocks (Columns B and C) in accordance with the published rate. In this step, the actual utility tariff is used, with appropriate values for the Fuel Adjustment Charge (FAC) and Purchased Gas Adjustment (PGA) factors for both conventional electric and gas billings. Similar calculations are performed to compute the cost of supplemental power and fuel. The entire set of calculations are shown in Table 34-6.

The monthly total cost of operating the 115 kW module (Column I) also is computed with the cost of fuel (Column D), O&M

Table 34-6. Cost Analysis

CASE:

CAPACITY 115 kW	INT HTG 0.75 (M/Th)	0	OIL a $ 0.75 /GAL	O&M a $ 0.01087 /kWH
BLR EFF 0.75	PGA 0.015 (M/Th)	0.015	Standby 1.50 $/kW	

BUYBACK a $ 2 $ FAC 2 $

ELECTRIC COSTS 0.04 /kWH 0.0176 /kWH

MONTH	DAYS	CONV'L DEMAND (kW)	CONV'L ENERGY (kWh)	BILLING DEMAND	BLK 1 USAGE (kWh)	BLK 2 USAGE (kWh)	DEMAND CHARGE ($)	ENERGY CHARGE ($)	FAC CHG ($)	CUST CHG ($)	TOTAL POWER ($)	INST CAP (kW)	PROD'TION	REV DEMAND (kW)	REV ENERGY (kWh)	REV BLK 1 (kWh)	REV BLK 2 (kWh)	REV BLG DEMAND (kW)	REV BLK 1 (kWh)	REV BLK 2 (kWh)	SUPP'L PWR COST ($)	CONV'L GAS USE (Therms)	NON-DISPL'BLE FIRM GAS (Therms)	CONV'L TOTAL ($)	COGEN COST ($)	SAVINGS ($)
Jan	31	384	200000	384	15000	185000	1152	11075	3520	12	15759	115	78137	269	121863	15000	106863	272	15000	106863	9751	40470	0	35490	32282	3208
Feb	28	390	198500	390	15000	183500	1140	11009	3499	12	15660	115	78575	265	128275	15000	113275	272	15000	113275	10213	41310	0	35781	32888	2893
Mar	31	353	212400	353	15000	197400	1058	11757	3738	12	16566	115	78137	237	134263	15000	117263	272	15000	117263	10651	37040	0	34701	31587	3114
Apr	30	440	196000	440	15000	181000	1320	10855	3450	12	15637	115	75616	325	120304	15000	105304	325	15000	105304	9802	29110	0	30005	26960	3117
May	31	480	197600	480	15000	182600	1296	10743	3478	12	15529	115	75616	317	119463	15000	104463	317	15000	104463	9711	21790	0	26773	23555	3218
Jun	30	480	144600	480	15000	194600	1440	9128	2897	12	13477	115	75616	365	88984	15000	73984	365	15000	73984	7442	10700	0	19364	16248	3117
Jul	31	50%	274000	50%	15000	259000	1488	15145	4822	12	21467	115	78137	381	190263	15000	175263	381	15000	175263	15450	8500	0	26197	23114	3083
Aug	31	4%	264400	4%	15000	253400	1512	14837	4724	12	21085	115	78137	381	192263	15000	177263	399	15000	177263	15647	6570	0	25953	22764	3009
Sep	30	480	276000	480	15000	263000	1488	15365	4893	12	21758	115	75616	381	202304	15000	187304	381	15000	187304	15923	9110	0	26826	23797	3037
Oct	31	480	217600	480	15000	202600	1440	12043	3830	12	17325	115	78137	365	139463	15000	124463	365	15000	124463	11307	20140	0	27602	24336	3218
Nov	30	3%	195200	3%	15000	180200	1180	10811	3436	12	15447	115	78137	281	119504	15000	104504	281	15000	104504	9612	23850	0	27449	24332	3117
Dec	31	372	219600	372	15000	204600	1116	12153	3865	12	17146	115	78137	257	141443	15000	126443	272	15000	126443	11174	39920	0	34156	32984	3172
TOTALS	365		2622200	180000	180000	2442200	15638	145121	44151	144	207054		920000		1702200	180000	1522200		180000	1522200	136303	289510	0	352276	314895	37382

(A) (B) (C)

FUEL COSTS

	DISPL'BLE FIRM GAS (Therms)	INT'BLE GAS (Therms)	CONV'L GAS COST ($)	SUPP'L FIRM SALES (Therms)	SUPP'L INT SALES (Therms)	SUPP'L GAS COST ($)	CONV'L OIL USE (Gal)	OIL COST ($)	SUPP'L OIL (Gal)	SUPP'L OIL COST ($)	SUPP'L FUEL ($)	CONV'L FUEL ($)	SUPP'L FUEL ($)
	40470	0	19731	34657	0	13078	0	0	0	0	13078	19731	13078
	41310	0	20121	36059	0	13730	0	0	0	0	13730	20121	13730
	37040	0	18136	31227	0	11443	0	0	0	0	11443	18136	11443
	29110	0	14448	23484	0	7803	0	0	0	0	7803	14448	7803
	21790	0	5808	15977	0	4392	0	0	0	0	4392	5808	4392
	10700	0	4730	5074	0	1539	0	0	0	0	1539	4730	1539
	8500	0	4768	2687	0	1503	0	0	0	0	1503	4768	1503
	6570	0	5868	2757	0	1504	0	0	0	0	1504	5868	1504
	9110	0	10277	3446	0	1515	0	0	0	0	1515	10277	1515
	20140	0	12002	14327	0	3625	0	0	0	0	3625	12002	3625
	23850	0	19010	18224	0	5437	0	0	0	0	5437	19010	5437
	39920	0		33107	0	12257	0	0	0	0	12257		12257
	289510	0	145222	221042	0	78046	0	0	0	0	78046	145222	78046

COGENERATION SYSTEM COSTS

(D) (E) (F) (H) (G) (I)

TOT COST FIRM GAS (Therms)	TOTAL FIRM USE (Therms)	FUEL PWR (Therms)	COST GAS POWER	RUN COST ($)	STBY O&M COST ($)	STBY O&M & ADM COST ($)	PWR SALES ($)	TOTAL ON-SITE ($)
21424	44111	9455	8344	849	173	85	0	9452
21650	44599	9540	7920	767	173	85	0	8945
19829	40681	9455	8344	849	173	85	0	9452
16087	32634	9150	8204	822	173	85	0	9283
12738	25431	9150	5997	841	173	85	0	9452
7526	14424	9455	5655	841	173	85	0	7067
6558	12141	9455	5086	822	173	85	0	6161
6590	12211	9150	5272	841	173	85	0	6193
6787	12634	9150	6767	822	173	85	0	6351
11970	23781	9455	8344	849	173	85	0	9452
13441	27374	9150	8204	822	173	85	0	9283
20703	42561	9455	8344	849	173	85	0	9452
111320			87455	10000	2070	1020	0	100545

(Column E), electric standby charges (Column F), administration and insurance (Column H) itemized separately. Revenues, if any, from power sales also are included (Column G).

Monthly incremental costs are used to compute both supplemental electric power costs and the cost for the natural gas used as a power plant fuel.

The product of this set of calculations is an estimate of the before tax, annual, operating cost saving resulting from the installation of a 115 kW module. Each major cost component also is totaled and summarized (Table 34-7).

Table 34-7. Annual Cost Summary

Conventional Costs	
Purchased Power	$207,054
Heating Gas	145,220
Fuel Oil	0
TOTAL	$352,276
Cogeneration Costs	
Supplemental Power	$136,303
Supplemental Heating Gas	78,046
Supplemental Heating Oil	0
Fuel for Power	87,455
O&M Costs	10,000
Standby Charges	2,070
Administration & Insurance	1,020
TOTAL	$314,894
Operating Cost Reduction	$37,382

The next set of calculations at this stage of the analysis is a cash flow projection wherein financing costs, tax implications and inflation are considered. Basic assumptions are shown in Table 34-8. At this point, the total installed cost of the system must be estimated and a simple payback is computed. In addition, anticipated cost escalators and the module owner's financing details and tax position can be input to produce projected cash flows and an Internal Rate of Return (IRR).

Table 34-8. Cash Flow Assumptions

Escalation Rates	(%/Yr)	Capital Costs	($)
Purchased Power	7.0	Equipment	72,753
Heating Gas	5.0	Installation	25,000
Oil	0.0	Total	97,753
Powerplant Gas	5.0		
O&M	5.0	*Financing*	
Power Sales	4.0	Equity	100%
General Inflation	5.0	Term	0 years
		Interest	0%
Tax Rates	*(%)*	*Tax Parameters*	
Federal	46.0	Tax Life	5 years
State	4.0	ITC	8%
Local	0.1		
Property	0.1	*Economic Life*	15 years

The above calculations are readily computerized on most microprocessors. The resulting cash flow projections are shown in Table 34-9. Once the initial model has been developed, alternatives can be examined quickly and the sensitivity of the economic analysis explored. For example, the impact of differing inflation rates can be examined, as was done to prepare Figure 34-5.

Key parameters shown in Table 34-9 are the total annual savings

Table 34-9. Cash Flows

	1	2	3	4	5	6	7	8	9	10	11	12	13	14	15
YEAR	1	2	3	4	5	6	7	8	9	10	11	12	13	14	15
DEBT	0	0	0	0	0	0	0	0	0	0	0	0	0	0	0
BOOK VALUE	97753	83090	61564	41056	20528	0	0	0	0	0	0	0	0	0	0
CONV PWR	207054	221548	237056	253650	271406	290404	310732	332484	355758	380661	407307	435618	466326	498968	533896
CONV GAS	145222	152483	160108	168113	176519	185345	194612	204342	214560	225288	236552	248380	260798	273838	287530
CONV OIL	0	0	0	0	0	0	0	0	0	0	0	0	0	0	0
(A) CONV OPER COST	352276	374031	397164	421763	447924	475749	505344	536826	570317	605948	643859	684198	727124	772807	821426
PURCH PWR	136303	145845	156054	166977	178666	191172	204555	218873	234194	250598	268129	286898	306981	328470	351463
SUP OIL HTG	0	0	0	0	0	0	0	0	0	0	0	0	0	0	0
(B) SUP GAS HTG	78046	81949	86046	90348	94866	99609	104590	109819	115310	121076	127129	133486	140160	147168	154526
GAS PWR	87455	87455	87455	87455	87455	87455	87455	87455	87455	87455	87455	87455	87455	87455	87455
O & M	10000	10500	11025	11576	12155	12763	13401	14071	14775	15513	16289	17103	17959	18856	19799
STANDBY	2070	2215	2370	2536	2713	2903	3107	3324	3557	3806	4072	4357	4662	4988	5338
O&M & ADM	1020	1071	1125	1181	1240	1302	1367	1435	1507	1582	1661	1745	1832	1923	2020
PWR SALES	0	0	0	0	0	0	0	0	0	0	0	0	0	0	0
COGEN OPER COST	314895	329034	344075	360074	377095	395205	414474	434978	456798	480020	504736	531044	559049	588961	620601
NET SAVING	37382	44997	53089	61689	70829	80544	90870	101848	113519	125928	139122	153153	168075	183945	200826
ACRS	14663	21506	20528	20528	20528	0	0	0	0	0	0	0	0	0	0
OWNER TAXES	98	83	62	41	21	0	0	0	0	0	0	0	0	0	0
INTEREST	3738	4499	5309	6169	7082	8054	9087	10185	11352	12593	13912	15315	16808	18395	20083
BEFORE TAX PROFIT	18883	18909	27191	34951	43198	72490	81783	91663	102167	113335	125210	137838	151268	165551	180743
INCOME TAXES	9104	9117	13110	16851	20827	34950	39431	44195	49259	54643	60369	66457	72932	79819	87143
AFTER TAX PROFIT	9779	9792	14081	18100	22370	37539	42352	47469	52908	58692	64841	71381	78335	85732	93600
ITC	7820	0	0	0	0	0	0	0	0	0	0	0	0	0	0
ACRS	14663	21506	20528	20528	20528	0	0	0	0	0	0	0	0	0	0
PRINCIPAL	0	0	0	0	0	0	0	0	0	0	0	0	0	0	0
NET CASH TO OWNER	-6549	31298	34609	38628	42898	37539	42352	47469	52908	58692	64841	71381	78335	85732	93600
CUM CASH	-65491	-34194	416	39043	81942	119681	161834	209302	262211	320903	385744	457125	535460	621192	714792

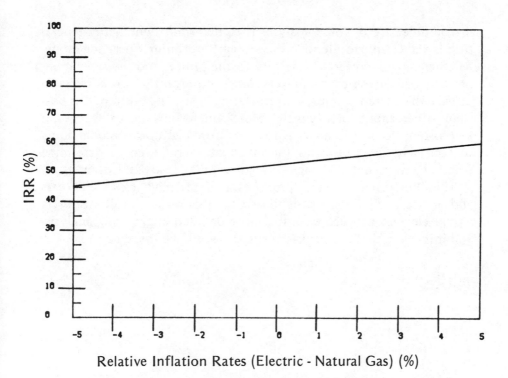

Relative Inflation Rates (Electric - Natural Gas) (%)

Figure 34-5. IRR vs. Energy Inflation Rates

(Row C), which is calculated as the difference between the projected cost of utility service without cogeneration (Row A) less the costs with the cogeneration system (Row B). The income tax liability (Row H) is computed after the net profit is reduced by charges for ACRS (Row D), local property taxes (Row E) and interest charges (Row F) for a before-tax margin or profit (Row G). The after-tax operating cost saving is shown (Row I).

The resulting net cash to the cogeneration system owner (Row M) is the after-tax profit less principal payments on debt service (Row L) or, in the first year, the equity payment plus the Investment Tax Credit (Row J) and previously deducted ACRS charges (Row K). The cumulative net cash to the owner also is computed (Row N). The net cash to the owner is used to calculate the project's IRR.

The objective of these analyses is to determine whether cogeneration is viable and to identify the system's optimum operating mode, whether or not power will be sold to the utility, and the type, number and size of the prime movers. Most importantly, it is possible to project the system's impact on operating costs. In addition, it is possible to determine how sensitive these conclusions are to changes in key assumptions. The analysis requires from eight professional hours to four days and provides the information required to determine whether a more detailed engineering feasibility study is warranted.

This illustrative example produces a simple payback of 2.6 years and an IRR of 56.0 percent. Based on this return, most end users would elect to proceed with the more detailed engineering analysis, and it is likely that the packaged system would be installed.

Chapter 35

Case 2:
Industrial,
Gas Turbine Application

This chapter illustrates a different technique for use in assessing economics during the initial walkthrough. It also considers a gas turbine as the prime mover and highlights some of the unique characteristics of that technology.

WALKTHROUGH

The facility consists of a chemical manufacturing facility that operates various production processes for 24 hours a day, 365 days a year. In addition, various chemicals and feedstocks must be constantly maintained at elevated temperatures, requiring continuous heating. In general, because of the year-round need for heat, the chemical industry has shown a strong interest in cogeneration. During the initial site visit, the facility's central boiler plant (which was technically compatible with cogeneration) natural gas service and electric service were examined. In addition, 12 months of utility billing and fuel oil use data were collected. This information and data are summarized in Tables 35-1 through 35-3.

The energy use data shown in Table 35-2 indicate several characteristics favorable to cogeneration feasibility. First, the local utility

Table 35-1. Site Characteristics

— Three dual fuel boilers operating at an average annual efficiency of 78.1%

— 10,000 therms of natural gas (or oil equivalent) required for purposes which cannot be replaced by recoverable heat

— Steam required at 150 psig, condensate returned at approximately 190°F to 200°F

— No. 6 oil used for alternate boiler fuel

— Interruptible gas price tied to monthly fuel oil price postings

— Incoming electric service at 13.8 kV

— 13.8 kV substation located near existing boiler plant

— Limited space for cogeneration equipment

Table 35-2. Baseline Electrical Requirements

MONTH	DAYS	CONV'L DEMAND (kW)	CONV'L ENERGY (KWH)	CONV'L LOAD FACT (HOURS)
Aug	28	4176	2116800	507
Sep	30	4128	2342400	567
Oct	32	4080	2481600	608
Nov	30	3840	2212800	576
Dec	30	3600	2025600	563
Jan	32	3744	2160000	577
Feb	29	3888	2304000	593
Mar	29	3840	2294400	598
Apr	33	3840	2414400	629
May	30	3696	2217600	600
Jun	32	4032	2390400	593
Jul	29	4009	1737600	433
TOTALS	364		26697600	
MAX		4176		629
MIN		3600		433

Table 35-3. Baseline Fuel Requirements

MONTH	DAYS	CONV'L GAS USE (THERMS)	INT'BLE GAS (THERMS)	CONV'L OIL USE (GAL)	CONV'L Steam (M lb)	CONV'L Efficiency (%)
Aug	28	233430	223430	35700	16136	59.8
Sep	30	290610	280610	504	17589	64.1
Oct	32	320820	310820	6132	20891	67.0
Nov	30	308910	298910	33096	23652	69.6
Dec	30	110850	100850	217350	31028	74.6
Jan	32	51640	41640	300132	40374	84.2
Feb	29	226620	216620	171906	38534	83.3
Mar	29	146110	136110	224112	40833	88.7
Apr	33	378940	368940	21924	34500	88.1
May	30	344030	334030	30954	29360	79.2
Jun	32	262010	252010	41664	24342	79.4
Jul	29	137660	127660	25200	14265	88.5
TOTALS	364	2811630	2691630	1108674	331504	78.1
MAX						88.7
MIN						59.8

has an 11-month trailing ratchet of 80 percent; however, the minimum annual billing demand of 3600 kW is 86 percent of the peak annual demand of 4176 kW. This ratio would indicate that the facility has fairly constant year-round billing demands. The assumption was confirmed by a review of historic billing data.

Second, the electric load factor, which is defined as the total monthly kilowatt hours divided by the monthly billing demand, ranged from 433 hours per month to 629 hours per month. With an average load factor of 570 hours per month, based on a nominal maximum of 730 hours, the site's load factor indicates that usage is fairly constant throughout the month.

The boiler efficiency varies considerably, ranging from an apparent high of 88 percent during winter months to a low of 60 percent during summer months. The high efficiencies should be viewed with caution, in that allocations of fuel deliveries to determine fuel usage can introduce some error. Overall, a review of steam data as maintained by the plant engineers indicates an annual efficiency of approximately 78 percent.

Based on the information that is immediately available at the site, it can be concluded that cogeneration is viable from a technical perspective. The next step is to determine whether or not cogeneration is economically viable. This procedure is illustrated by Figure 35-1 and described below.

This procedure differs from that used in the first hypothetical case and is included to illustrate alternative analytic procedures.

Item "A," the lowest monthly thermal load that the plant will supply, can be determined from Table 35-3. July steam output is almost 14,300 thousand pounds (M lb), while August output is 16,100, and an average of 15,200 M lb was used. In fact, the plant shuts down for two weeks during July. Assuming 1,000 Btu per pound of steam, this is equivalent to a requirement of 15,200 mmBtu. However, not all of this steam can be provided by the cogeneration system, as approximately 10,000 therms of fuel cannot be displaced. At an efficiency of 78.1 percent, this fuel requirement is approximately 800 mmBtu, so the lowest net thermal load is 14,400 mmBtu per month.

In order to calculate item "B," the corresponding monthly cogeneration system power output, it is necessary to estimate the heat available for each kilowatt-hour produced. To do this, it is necessary to identify the type of prime mover. Because of the requirement for high-pressure steam, 150 psig, a gas turbine was initially selected. Table 35-4 provides approximations that can be used to determine the amount of heat available. In this case, based on the site's electrical requirements, a turbine in the size range 1,000 kW to 5,000 kW was chosen. The calculations could be repeated easily with alternative choices.

Monthly generation of electricity is estimated to be 2,060,000 kilowatt-hours. The annual generation, "C," is estimated as 24,700,-000 kilowatt-hours. In order to estimate the plant capacity, the monthly generation is divided by the hours of operation per month. A gas turbine is capable of operating at very high availabilities approaching 8,400 hours or more per year, or 700 hours per month (reciprocating engines are capable of operating from 7,500 hours to 8,000 hours per year). Dividing the monthly output of 2,060,000 kilowatt-hours by 700 hours per month results in 2940 kW of capacity (item "D").

In this size range, gas turbines are available at installed costs of $800 per kilowatt; and the estimated plant cost, item "E," is $2,350,000.

Based on 12 months of electric billing information, the average cost of purchased power, item "F," was computed to be 8.5¢ per kilowatt-hour. The electrical cost savings per year, item "G," were then equal to the product of the plant's annual output of 24,700,000 kilowatt-hours and the $.085 per kilowatt-hour cost of purchased power, or $2,100,000 per year.

The next set of calculations addresses the cogeneration system operating costs. In this case, it was found that gas, item "H," is available for cogeneration at an average cost of $4.50 per mmBtu. However, in order to estimate the amount of fuel required, it is necessary to make some assumptions about the prime mover's efficiency. Table 35-5 provides one set of assumptions.

The annual natural gas usage, item "I," is estimated to be 329,000 mmBtu. Based on the projected usage, the annual fuel cost, item "J," is computed to be $1,480,000. Annual incremental operating and maintenance costs, item "K," at 0.5¢ per kilowatt-hour for a gas turbine is $124,000.

Item "L," is the amount of recovered thermal energy and is computed from the minimum monthly usage as 172,800 mmBtu. The savings due to this recovered heat, item "M," can now be estimated as $972,000 per year.

Summing the above cost items results in a total annual cost of plant operation, item "N," of $1,604,000; $1,480,000 for power plant fuel plus $124,000 for maintenance. Annual gross savings, item "O," are the sum of electric savings of $2,110,000 and fuel savings of $972,000, for a total of $3,072,000. Net savings, item "P," is the difference in these two numbers, or $1,478,000 per year. The simple payback period, item "Q," can then be computed and is estimated to be 1.6 years.

A. The lowest monthly thermal load
    (to be supplied by the cogeneration   _14,400_ MMBtu/month
    plant)

B. Corresponding monthly generation
    of electricity

$$B = \frac{A}{HRF} \times 10^6 = \frac{14,400 \times 10^6}{7000} = \underline{2,060,000} \text{ kWh/month}$$

    (HRF = heat recovery factor for selected prime mover - see table)

C. Annual generation of electricity

$$C = 12 \times B \qquad\qquad \underline{24,700,000} \text{ kWh/year}$$

D. Installed electrical capacity of
    the cogeneration plant $= 2,060,000 \div 700 =$

$$D = \frac{B}{\text{operating hours per month}} \qquad \underline{2,940} \text{ kW}$$

E. Cost of the cogeneration plant

$$E = D \times \text{cost/kW (installed capacity)} \qquad \$\underline{2,350,000}$$
$$= 2,940 \times 800$$
    (use \$1,200 per KW unless better estimates are available)

F. Average cost of electricity (utility
    data, or from electric bills inlcuding   _.085_ \$/kWh
    demand charge)

G. Savings per year on electricity

$$G = C \times F = 24,700,000 \times .085 = \underline{2,110,000} \text{ \$/year}$$

    Note: If part of generated electricity is to be sold back to the
           utility at buy-back rates, a more detailed analysis of "G"
           is needed.

H. Cost of natural gas   _4.50_ \$/MMBtu

I. Fuel consumption per year

$$I = \frac{C}{\text{kWh/MMBtu of gas (see table)}} \qquad \underline{329,000} \text{ MMBtu/year}$$

    (use values from table in "Generated Electrical Energy per Unit of
    Fuel" section)

$$I = \frac{24,700,000 \text{ kWh}}{75 \text{ mmBtu/kWh}}$$

Figure 35-1. Preliminary Evaluation Form for
Cogeneration Applications

J. Annual cogeneration fuel cost

$$J = I \times H = 329,000 \times 4.50 \qquad \underline{1,480,000} \text{ \$/year}$$

K. Annual maintenance ~~and labor~~ cost
    for the cogeneration plant

~~K = $0.010/kWh x C (for engine)~~ $\underline{124,000}$ \$/year
    K = \$0.005/kWh x C (for turbine)
    $= \$0.005/kWh \times 24,700,000$

L. Recovered thermal energy per year

$$L = 12 \times A = 12 \times 14,400 \qquad \underline{172,800} \text{ MMBtu/year}$$

M. Savings on heat, normally generated
    by 80% efficient boilers or equivalent

$$M = \frac{L}{0.8} \times H$$

$$= \frac{172,800}{.8} \times 4.50 = \qquad \underline{972,000} \text{ \$/year}$$

N. Annual operating cost of the
    cogeneration plant

$$N = J + K$$
$$= 1,480,000 + 124,000 \qquad \underline{1,604,000} \text{ \$/year}$$

O. Annual gross savings

$$O = G + M$$
$$= 2,110,000 + 972,000 \qquad \underline{3,082,000} \text{ \$/year}$$

P. Annual net savings *

$$P = O - N$$
$$= 3,082,000 - 1,604,000 \qquad \underline{1,478,000} \text{ \$/year}$$

Q. Simple payback period

$$Q = \frac{E}{P} = \frac{2,350,000}{1,478,000} \qquad \underline{1.6} \text{ years} **$$

   * Does not include tax benefits from investment tax credits and
       accelerated depreciation

** *Does not include taxes and debt service.*

Figure 35-1
(Continued)

Table 35-4. Heat Recovery Factors

— 5,500 Btu/kWh for reciprocating engines

— 9,500 Btu/kWh for gas turbines smaller than 1,000 kW

— 7,000 Btu/kWh for gas turbines between 1,000 kW and 5,000 kW

— 5,000 Btu/kWh for gas turbines greater than 5,000 kW

Table 35-5. Prime Mover Efficiency

Reciprocating Engines

— 88 kWh generated per mmBtu

Gas Turbines

— 60 kWh per mmBtu for turbines less than 1,000 kW

— 75 kWh per mmBtu for turbines between 1,000 and 5,000 kW

— 90 kWh per mmBtu for turbines greater than 5,000 kW

This analysis did not consider debt service or tax benefits and liabilities. However, based on this analysis, it was concluded that a cogeneration system was potentially viable at this site, and that a more detailed analysis was warranted.

PRELIMINARY ECONOMIC SCREENING

In order to conduct a more comprehensive analysis for the site, it was necessary to examine a specific turbine, and the particular prime mover's characteristics are summarized in Table 35-6. Gas turbine capacity is very sensitive to both ambient temperature and altitude, and it is necessary to rate the machine for actual site conditions. Manufacturers can provide the necessary rating curves. Gas turbines also require natural gas pressures ranging from 125 psig to over 600 psig. Again, manufacturers can provide information on pressure requirements and on the required compressor power. If no information is available, a 3 percent loss for low-pressure fuel and a 6 percent loss for high-pressure fuel may be used.

Table 35-6. Turbine Characteristics

Capacity:	3400 kW at the site
Compressor Losses:	105 kW
Heat Rate:	12,600 Btu/kWh (HHV)
Recoverable Heat:	6,500 Btu/kWh at 150 psig
Availability:	8664 hours per year
Duct Burner Capacity:	6,600 Btu/kWh
Duct Burner Efficiency:	91.4%
Alternate Fuel:	No. 2 oil at 140,000 Btu/-gallon

Gas turbine exhaust is high in oxygen, and it is possible to fire a fuel directly in the turbine's exhaust. As shown in Table 35-6, the supplemental capacity and the combustion efficiency for the specific machine should be used. Lacking any specific information, the supplemental capacity may be set equal to the recovered heat rate and the efficiency set equal to 92 percent. These values are an approximation; higher capacity is available; however, costs are increased.

The actual monthly energy calculations are shown in Table 35-7. The significant variations required in these calculations are to adjust the turbine capactiy (Column A), and the resulting supplemental demand (Column B), for the peak monthly temperature; and the electrical output which is based on the average capacity (Column C), and the resulting requirement for supplemental kilowatt-hours (Column D), for the average monthly temperature. In addition, the turbine electrical output (Column E) must be decreased to allow for fuel compressor losses.

Finally, it is necessary to adjust the efficiency used to produce supplemental steam (Column F) to show the increased performance of the duct burners as opposed to a conventional boiler. The overall efficiency is the average of supplemental duct burner efficiency and conventional boiler efficiency weighted by the amount of steam produced by each.

Table 35-8 summarizes the relevant cost data. In this case, both conventional power and supplemental power are purchased under the same rate. Gas is available either under a demand commodity rate or through direct purchase. Operating and maintenance costs were based on the vendor's quotation of four mils per kilowatt-hour. Standby charges were reduced to take advantage of the site capability to shed 500 kW of load during periods of equipment outage.

Based on the energy calculations of Table 35-5, as shown in Table 35-9, the system would produce annual savings of $1,612,140. This is considerably higher than was estimated using the simpler screening technique and accounts for cost savings resulting from increased efficiency of duct burners, availability of contract gas in the volumes required for the gas turbine and more precise incremental cost calculations.

Table 35-10 summarizes the cash flow analysis. The cogeneration system, with an installed cost of $3,190,000, including a 10 percent allowance for contingency, results in a simple payback of two years and an Internal Rate of Return of 46.5 percent. Again, the screening indicates that the project is economically viable and could proceed to detailed engineering analysis. At this point, however, the owner must decide whether the system will be owned by the corporation itself or a third-party developer.

Table 35-7. Energy Calculations

CASE: Industrial

CAPACITY		
HEAT RATE @ 3400 (KW)	Compressor Parasitics	105 kW
HEAT RATE @ 12600 BTU/KWH		0 kW
ALT HTG FUEL 150000 BTU/GAL		

TURBINE WITH SUPPLEMENTAL FIRING
SUP'TLE EFFICIENCY @ 0.914
HOURS PER YEAR @ 8664
HEAT RECOVERY @ 6500 BTU/KWH
BLR EFFICIENCY @ 0.781

SUPP'L CAPACITY @ Alt Turbine Fuel @
GAS HHV @

6600 BTU/KWH
140000 BTU/GAL
BTU/CF

Table 1

MONTH	DAYS	CONV'L DEMAND (KW)	CONV'L ENERGY (KWH)	CONV'L LOAD FACTOR (HOURS)	CONV'L GAS USE (THERMS)	NON-DISPL'BLE FIRM GAS (THERMS)	DISPL'BLE FIRM GAS (THERMS)	INT'BLE GAS (THERMS)	CONV'L OIL USE (GAL)	COGEN Fire Cap (KW)	COGEN Ave Cap (KW)	COGEN PRODU'TION (KWH)	Total Fuel for Power (Therms)	Alt Fuel Requir't (Therms)	REC'V HEAT AVAILABLE (THERMS)	TOTAL DISPL'BLE FUEL REQ'D (THERMS)	AVAIL/REQ'D HEAT (%)
Aug	28	4176	2116800	507	233430	10000	10000	224330	35700	2590	2670	1723680	224105		116773	274980	42.2
Sep	30	4128	2342400	567	290610	10000	10000	280610	504	2660	2740	1897200	244665		128386	281366	45.6
Oct	32	4080	2461600	608	320620	10000	10000	310820	6132	2780	2830	2092800	272096		141372	320018	44.2
Nov	30	3840	2212800	576	308910	10000	10000	299910	33896	2840	2920	2026600	263515		136763	348554	39.2
Dec	30	3600	2025600	563	110850	10000	10000	100850	217350	2950	3010	2021680	262875	262875	134273	426875	31.9
Jan	32	3744	2160000	577	51640	10000	10000	41640	300132	2990	3040	2254080	293065	293065	151848	491838	30.9
Feb	27	3808	2304000	593	226620	10000	10000	216620	171906	2980	3040	2042760	265590	265590	137629	474679	29.0
Mar	27	3840	2294400	598	146110	10000	10000	136110	224112	2920	2940	1973160	256541		133039	472278	28.2
Apr	29	3840	2414400	629	370940	10000	10000	368940	21924	2820	2890	2205720	286777		149917	401826	27.1
May	33	3696	2217600	600	344030	10000	10000	334030	30954	2730	2810	1947600	253218		131658	380461	34.6
Jun	30	4032	2370400	593	262010	10000	10000	252010	41164	2640	2770	2008320	261112		135948	314506	43.2
Jul	29	4009	1737600	433	137660	10000	10000	127660	25200	2660	2660	1594320	207286		107890	165460	65.2
TOTALS	364		24697600		2811630	120000		2691630	1108674			23798320	3092843	821530	1608494	4354441	
MINIMUM		3600		433													28
MAXIMUM		4176		627													65

Table 2

USED/AVAIL HEAT (%)	POTENTIAL FUEL SAVING (THERMS)	SUPP'L DEMAND (KW)	SUPP'L ENERGY USE (KWH)	SUPP'L LOAD FACTOR (HOURS)	Baseload Blr Fuel (Therms)	SUPP'L CAP'TY/ REQ'D THERMAL (%)	EFFECTIVE THERMAL EFFICIENCY (%)	SUPP'L OIL USAGE (GAL)	POTENTIAL GAS RED'TION (%)	Int'ble Gas Power (THERMS)	SUPP'L INT'BLE GAS USE (THERMS)	SUPP'L FIRM HTG SALES (THERMS)	GAS FOR POWER (THERMS)	OIL FOR POWER (GAL)
100.0	149506	1671	393120	232	86037	274	82.4	0	125706	224105	92631	10000	224105	0
100.0	164374	1573	445200	283	92183	505	80.9	0	164438	244665	112546	10000	244665	0
100.0	181000	1405	388800	277	98328	339	82.0	0	176912	272096	127549	10000	272096	0
100.0	175099	1085	186000	171	92183	165	84.3	0	153035	263515	135156	10000	263515	0
100.0	174472	755	3720	5	92183	83	86.5	91184			91163	10000	262875	187768
100.0	194413	859	-94080	0	98328	75	86.8	153430			37469	10000	293065	209332
100.0	176208	1013	261240	258	89110	64	86.6	49087			195373	10000	265590	189707
100.0	170332	1025	321240	313	89110	61	86.5	99854			129902	10000	254541	0
100.0	190660	1125	208680	185	101401	133	85.0	0	176044	286771	177251	10000	286777	0
100.0	165563	1071	270000	252	92183	107	85.6	0	147927	253218	169810	10000	253218	0
100.0	174056	1497	382080	255	98328	315	82.1	0	144280	261112	100567	10000	261112	0
100.0	138132	1454	143280	99	27151		78.1	0	121332	207286	6328	10000	207286	0
	2054815		2707280				68.3	393555	1211275	2271309	1368664	120000	3092843	586007
100		755		313			87.2							
100		1671												

Table 35-8. Cost Data

Electric

Customer Charge:	$205 per month
Demand Charge:	$10.60 per kilowatt
Energy Charge:	
First 300 hours use of demand:	$.075 per kilowatt-hour
All excess:	$.065 per kilowatt-hour
Standby:	$3.95 per kilowatt
Avoided Cost:	$.0525 per kilowatt-hour

Natural Gas

Demand Charge:	$.104 per therm per month
Commodity Charge:	$.490 per therm
Contract Gas:	$.405 per therm

Operations & Maintenance $.004 per kilowatt-hour

Fuel Oil

No. 2 fuel oil:	$.85 per gallon
No. 6 fuel oil:	$.74 per gallon

Table 35-9. Cost Analysis

CASE Industrial
CAPACITY 3400 kW
BLR EFF 0.781

Load Shedding			
INT HTG @ s		500 kW	
PGA @ s		0.4687 /TH	
		-0.025 /TH	

BUTBACK @ s	0.0525 /KWH
FAC @ s	0.3044 /KWH
Standby Capacity	1975 kW

Htg Oil	150000 Btu/Gal
Htg Oil	0.74 $/Gal
Per Oil	0.85 $/Gal
Per Oil	140000 Btu/Gal

OLM	0.004 $/KWH
Compressor	105 kW
Parasitics	0 kW

Cost Chg	205 $/mo
Demand @	10.60 $/kW
Ratchet @	0.60
Minimum	100 kW

ELECTRIC COSTS

MONTH	DAYS	CONV'L DEMAND (kW)	CONV'L ENERGY (KWH)	BILLING DEMAND (kW)	BLK 1 USAGE (KWH)	BLK 2 USAGE (KWH)	DEMAND CHARGE ($)	ENERGY CHARGE ($)	FAC CHG ($)	COST CHG ($)	TOTAL POWER ($)	Average Cost ($/kWh)	Max CAP (kW)	PROD'TON (KWH)	REV. DEMAND (kW)	REV. ENERGY (KWH)	REV BLG DEMAND (kW)	REV. BLK 1 (KWH)	REV. BLK 2 (KWH)	SUPP'L PWR COST ($)	Cost Chg Ratchet @ Minimum	Average Per Cost ($/kWh)
Aug	28	4176	2116800	4176	1252800	864000	44266	150120	9314	205	203905	0.096	2590	1723680	1691	393120	1691	393120		4343		0.126
Sep	30	4128	2342400	4128	1238400	1104000	43757	164640	10307	205	218908	0.093	2660	1897200	1573	445200	1573	445200	0	52228		0.117
Oct	32	4080	2481600	4080	1224000	1257600	43248	173544	10919	205	227916	0.092	2780	2092800	1405	398600	1405	398600	0	45949		0.118
Nov	30	3840	2212800	3840	1152000	1040800	40704	155352	9736	205	205997	0.093	2840	2092600	1085	184000	1085	184000	0	26474		0.142
Dec	30	3600	2025600	3600	1080000	945600	38160	142464	8913	205	189742	0.094	2750	2021680	1015	3770	1015	3770	0	11255		3.026
Jan	32	3744	2160000	3744	1123200	1036800	39656	151632	9504	205	201027	0.093	2990	2254080	659	-94080	659	-94080	0	10960		0
Feb	29	3688	2304000	3688	1164400	1337600	42213	161424	10138	205	212779	0.092	2980	2042760	1013	261240	1015	261240	0	31702		0.121
Mar	29	3840	2294400	3840	1152000	1142400	40704	160656	10095	205	211660	0.092	2920	1931640	1025	307560	1025	307560	0	36439		0.113
Apr	33	3840	2414400	3840	1152000	1262400	40704	160456	10623	205	219908	0.091	2920	2205720	1125	208680	1125	208680	13740	28699		0.138
May	30	3696	2217600	3696	1108800	1108800	39178	155232	9757	205	204372	0.092	2730	1947600	1071	270000	1071	270000	0	32996		0.122
Jun	32	4032	2390400	4032	1209600	1180800	42739	167472	10518	205	220934	0.092	2640	2003320	1497	382080	1497	382080	0	46610		0.121
Jul	29	4009	1737600	4009	1202700	534900	42695	124971	7645	205	175317	0.101	2580	1594320	1534	143280	1534	143280	0	27842		0.194
TOTALS	364		26697600		14061900	12635700	496854	1875963	117469	2460	2492746	0.093		23788320		2909620		2909280	13740	400317		0.136

Energy Blk Size		
Blk 1 Rate	0.0750 $/KWH	
Blk 2 Rate	0.0650 $/KWH	

300 Hours Use	
Blk 1 Rate	0.0750 $/KWH
Blk 2 Rate	0.0650 $/KWH

Rate for Supplemental Power	205 $/mo
	10.60 $/kW
Cost Chg	0.60 %
Demand @ Ratchet @ Minimum	0 kW

Energy Blk	
Blk 1 Rate	
Blk 2 Rate	

300 Hours Use Oil	
0.0750 $/KWH	
0.0650 $/KWH	

Gas Per @ Cap Cost	0.4050 2720000
Demand	0.1040 $/th/Mo
Commodity	0.4900 $/th

FUEL COSTS

MONTH	NON-DISPL BLE FIRM GAS (THERMS)	CONV'L GAS USE (THERMS)	DISPL'BLE FIRM GAS (THERMS)	INT'BLE GAS (THERMS)	CONV'L GAS COST ($)	SUPP'L FIRM SALES (THERMS)	SUPP'L INT SALES (THERMS)	SUPP'L GAS COST ($)	CONV'L OIL USE (GAL)	OIL COST ($)	SUPP'L OIL (GAL)	SUPP'L OIL COST ($)	CONV'L FUEL ($)	SUPP'L FUEL ($)	Fire Gas Power (THERMS)	TOTAL FIRM USE (THERMS)	TOT COST FIRM GAS ($)	Int Gas Power (Therms)	COST GAS POWER ($)
Aug	10000	233430	0	223430	110404	10000	92600	49089	35700	26618	0	0	136822	49089	0	10000	5690	224105	90762
Sep	10000	290610	0	280610	137203	10000	112511	58420	504	373	0	0	137576	58420	0	10000	5690	244665	99899
Oct	10006	320820	0	310820	151361	10000	127554	65470	4132	4538	0	0	155899	65470	0	10000	5690	272096	110199
Nov	10000	308910	0	298910	145779	10000	135100	69007	330%	24491	0	0	170270	69007	0	10000	5690	263515	106723
Dec	10000	110850	0	100850	52955	10000	91016		217350	160839	9184	67476	247374	115622	0	10000	5690		0
Jan	10000	51640	0	41640	25205	10000	37446	23249	300132	222098	153430	113538	247303	136787	0	10000	5690		0
Feb	10000	226620	0	216620	107213	10000	195341	97240	171906	127210	49087	36324	234423	133564	0	10000	5690		0
Mar	10000	146110	0	136110	69400	10000	122933	63304	224112	165943	99854	73892	235323	137196	0	10000	5690	256541	103899
Apr	10000	378940	0	368940	178600	10000	177217	88746	21924	16224	0	0	194824	88746	0	10000	5690	286277	116145
May	10000	344030	0	334030	162239	10000	169777	85259	27906	0	0	165145	85259	0	10000	5690	253218	102553
Jun	10000	262010	0	252010	123799	10000	100595	52836	41664	30831	0	0	154630	52836	0	10000	5690	261112	105750
Jul	10000	137660	0	127660	65520	10000	6151	8573	75200	18648	0	0	84168	8573	0	10000	5690	207286	83951
TOTALS		2811630		2691630	1329757	120000	1368261	709539	1108674	820419	393554	291230	2150176	1000769	0	120000	68280		919882

(Continued)

Table 35-9. Continued

$/th
$

COGENERATION SYSTEM COSTS

OIL POWER (GAL)	COST OIL POWER ($)	COST FUEL POWER ($)	O&M COST ($)	STBY COST ($)	PWR SALES ($)	Owning & Admin ($)	TOTAL ON-SITE ($)	CONV'L TOTAL ($)	COGEN COST ($)	SAVINGS ($)
0	0	90762	6895	7801	0	2267	107725	340727	206157	134570
0	0	99897	7569	7801	0	2267	117556	356484	228204	128280
0	0	110199	8371	7801	0	2267	128638	383815	240077	143738
0	0	106723	8107	7801	0	2267	124899	376268	220380	155888
187768	159603	159603	8008	7801	0	2267	177758	403536	304836	98700
209332	177932	177932	9016	7801	0	2267	192077	448330	339824	108507
189707	161251	161251	8171	7801	-4939	2267	179690	447402	344756	102646
0	0	103899	7893	7801	0	2267	121860	446983	295495	151489
0	0	116145	8823	7801	0	2267	135036	414812	252450	162332
0	0	102553	7790	7801	0	2267	120411	389517	238646	150851
0	0	105750	8033	7801	0	2267	123852	375564	223098	152466
0	0	83951	6377	7801	0	2267	100396	259405	136811	122674
586807	498786	1418668	95153	93615	-4939	27200	1628697	4642922	3030782	1612140

COST SUMMARY

CONV PWR	2492746
CONV GAS	1329757
CONV OIL	820419
TOTAL CONV	4642922
SUP PWR	400317
SUP GAS	709539
SUP OIL	291230
O & M	919082
GAS PWR	95153
STDBY	93615
PWR SALES	-4939
OIL PWR	498786
O&M & ADM	27200
TOTAL COG	3030782
OPER MARG	1612140

Table 35-10. Cash Flow Analysis

ESCALATION RATES	(%/YEAR)
PURCH PWR	7.0
HTG GAS	5.0
OIL	5.0
PWR FUEL	5.0
O & M	5.0
PWR SALES	2.5
GEN INFL	5.0

PAYBACK	2.0 YEARS
IRR	46.5 %

TAX RATES	(%)
FEDERAL	46.0
STATE	6.0
LOCAL	0.0
PROPERTY	0.0
TAX LIFE	5 YEARS
ITC	8.0

RATE OF RETURN	
ECON LIFE	15 YEARS
DISC RATE	15 %

COSTS	($)
Cap Cost	2900000
Interest	0
Cont'gcy	290000
TOTAL	3190000

DEBT	
EQUITY	100 %
TERM	0 YEARS
INTEREST	0 %

SHARED SAVING	(%)
TO SITE OWNER	0
TO PLANT OWNER	100

Industrial:

CONV PWR	2492746
CONV GAS	1332757
CONV OIL	820419
TOTAL CONVENTIONAL	4642922
SUPP'L PWR	400317
SUPP'L GAS	709539
SUPP'L OIL	291230
FUEL POWER	1418668
O & M	95153
STANDBY POWER	93615
POWER SALES	-4939
OWN & ADMIN	27200
TOTAL COGEN	3030782
INITIAL MARGIN	1612140

YEAR	1	2	3	4	5	6	7	8	9	10	11	12	13	14	15
DEBT	0	0	0	0	0	0	0	0	0	0	0	0	0	0	0
BOOK VALUE	3190000	2711500	2009700	1339800	669900	0	0	0	0	0	0	0	0	0	0
CONV PWR	2492746	2667738	2853945	3053721	3267482	3498205	3740940	4002805	4283002	4582812	4903609	5246861	5614142	6007131	6427631
CONV GAS	1332757	1396245	1464057	1539340	1616328	1697144	1782002	1871102	1964457	2062890	2166034	2274336	2388053	2507455	2632828
CONV OIL	820419	861440	904512	949738	997224	1047086	1099440	1154412	1212133	1272739	1336376	1403195	1473355	1547022	1624374
CONV OPER COST	4642922	4924723	5224514	5542819	5881034	6240435	6622391	7028319	7459791	7918441	8406019	8924392	9475549	10061609	10684632
DISC'T COST	4642922	4282542	3950483	3644493	3362500	3102599	2863038	2642205	2436819	2250915	2077839	1918238	1771048	1635293	1510073
CUM'TIVE DISC'T COST	4642922	8925464	12875947	16520440	19882940	22985539	25848578	28490783	30929402	33180317	35256156	37176394	38947442	40582734	42092807
PURCH PWR	2492746	428339	456323	490406	524734	561465	600768	642822	687819	735966	787484	842608	901591	964702	1032231
GAS PWR	1329757	1489601	1564081	1642286	1724400	1810620	1901151	1996208	2096019	2200820	2310861	2426404	2547774	2675110	2808866
SUP GAS HTG	820419	745016	782267	821380	862449	905572	950850	998393	1048312	1100728	1155764	1213552	1274230	1337942	1404839
SUP OIL HTG		305592	321081	337135	353992	371691	390276	409790	430279	451793	474383	498102	523007	549158	576615
O & M		99911	104906	110151	115659	121442	127514	133890	140584	147614	154994	162744	170881	179425	188396
STANDBY		100168	107180	114682	122710	131300	140491	150325	160848	172107	184155	197046	210839	225598	241389
PWR SALES		-5062	-5319	-5452	-5558	-5728	-5871	-6018	-6168		-6322	-6480	-6642	-6808	-6979
OWN & ADMIN		28560	29988	31487	33062	34715	36451	38273	40187	42196	44306	46521	48847	51290	53854
COGEN OPER COST	4642922	3202449	3373015	3552846	3742657	3942393	4153228	4375572	4610066	4857393	5118269	5393456	5683762	5990032	6313170
NET SAVING	0	1722474	1851699	1989972	2136576	2298042	2469153	2652747	2849725	3061048	3287749	3530934	3791787	4071577	4371663

(Continued)

Guide to Natural Gas Cogeneration

Table 35-10. Continued

	Y1	Y2	Y3	Y4	Y5	Y6	Y7	Y8	Y9	Y10	Y11	Y12	Y13	Y14	Y15
USER SHARE	0.00	0.00	0.00	0.00	0.00	0.00	0.00	0.00	0.00	0.00	0.00	0.00	0.00	0.00	0.00
USER PAYMENTS	478500	701800	649900	649900	649900	0	0	0	0	0	0	0	0	0	0
DEPRECIATION	0	0	0	0	0	0	0	0	0	0	0	0	0	0	0
OWNER TAXES	0	0	0	0	0	0	0	0	0	0	0	0	0	0	0
INTEREST															
BEFORE TAX MARGIN	-478500	1020674	1181599	1320072	1468876	2298042	2469153	2652747	2849725	3061048	3287749	3530934	3791787	4071577	4371663
TAXES	-248820	530750	614431	688438	763712	1194982	1283960	1379429	1481857	1591745	1709630	1836086	1971729	2117220	2273265
AFTER TAX MARGIN	-229680	489923	567167	633635	704965	1103060	1185193	1273319	1367868	1469303	1578120	1694848	1820058	1954357	2098398
ITC	255200	0	0	0	0	0	0	0	0	0	0	0	0	0	0
DEPRECIATION	478500	701800	669900	669900	669900	0	0	0	0	0	0	0	0	0	0
PRINCIPAL	-2685980	1191723	1237067	1303535	1374865	0	0	0	0	0	0	0	0	0	0
NET TO OWNER	-2685980	1036281	935401	857095	786083	1103060	1185193	1273319	1367868	1469303	1578120	1694848	1820058	1954357	2098398
PRESENT VALUE	-2685980	-1694257	-257189			548416	512392	478698	447158	417668	390087	364296	340182	317638	296564
CUM CASH	-2685980		-714298	142797	928880	3524270	4709464	5982783	7350650	8819953	10378073	12092922	13912979	15867336	17945734
CUM DISC'T CASH		-164969				1577296	1989688	2468376	2915534	3333202	3732289	4087565	4427767	4745405	5041968
TOTAL UTILITY COST	1956942	4394173	4610083	4856381	5117322	5045453	5338422	5648890	5968890	6326696	6698389	7088306	7503819	7944389	8411568
DISC'T TOTAL COST	1956942	3821020	3485885	3193149	2925846	2508482	2307847	2123627	1954197	1798442	1655245	1523583	1402518	1291185	1188796
CUM DISC'T TOTAL	1956942	5777962	9263846	12456996	15382841	17891323	20199270	22322897	24277095	26075536	27730781	29254365	30656882	31968060	33136863

Section VIII

Technological Developments

SELECTED AUTHORS

Chapter 36

R&D Progress Report: Commercial Cogeneration

HENRY R. LINDEN
Gas Research Institute

After several slow years the concept of commercial cogeneration is beginning to gain momentum in the United States. Encouraged by the large potential market, several engine manufacturers and independent entrepreneurs are now offering packaged cogeneration systems in the 10- to 500-kilowatt size range. Commercial energy users have become more aware of the savings achievable by producing their own electricity and thermal energy.

Gas companies are playing an important role by responding to this market pull with marketing support, gas pricing incentives, and in some cases project financing. There is a potential gas demand of almost one trillion cubic feet, of which one-half would be entirely new load for the gas industry. Cogeneration offers something for everyone—even electric utilities, who benefit from additional electrical capacity that could defer new power plant construction.

The gas industry's commercial cogeneration R&D effort, through GRI, is starting to have an impact. Two advanced cogeneration components developed as part of a 70-kW cogeneration system have been commercialized by participating manufacturers. One of these is the Paraflow heat recovery chiller-heater, developed by Hitachi, Ltd., for GRI and Gas Energy Inc. The nominal 40-ton chiller-heater offers high performance and reduced heat-recovery cost by handling exhaust and jacket coolant heat in a single unit. Gas Energy Inc. reports the sale of nine units since its initial introduction less than a year ago.

A second new product is a 70-kW engine introduced by Waukesha Engine Division of Dresser Industries, Inc. The model VRG-330S engine features low-pressure gas induction, which allows increased turbocharging for high power output without requiring high-line-pressure gas. It's being offered individually and as part of a hot-water cogeneration package being marketed by Waukesha. Both of these innovative products will improve the performance and cost-saving advantages of commercial cogeneration systems, enhancing prospects for future market penetration.

MARKET TRENDS

The commercial cogeneration market is still in its infancy in spite of the proliferation of system packagers and suppliers. Many of these are venture businesses without a fully developed marketing and support infrastructure. Furthermore, the systems offered are targeted at the "cream" of the market—end-use applications with a large and continuous demand for hot water. The major issues remain the same as those that formed the basis of our cogeneration R&D strategy over three years ago: namely, the need for reliable, cost-effective systems that can provide a variety of services including electricity, refrigeration, heating, cooling, dehumidification, and high-quality steam. The bottom line is the need for advanced system components that improve performance and reduce costs. Our goal is to expand the market penetration of cogeneration in terms of both geographical markets and end-use applications.

Fortunately, the cogeneration industry enjoys a supportive and stable regulatory environment that is essential to future growth. The Public Utilities Regulatory Policies Act (PURPA) of 1978, providing for purchase of backup power and sale of surplus power by cogenerators at reasonable rates, has been consistently upheld in the courts.

However, new ratemaking policies that are emerging in the electric utility industry are having mixed effects on the prospects for commercial cogeneration. In an attempt to recover costs in a way that more closely reflects peak generating capacity requirements, some utilities have adopted the strategy of securing revenue increases by raising their demand charge—the portion of a commercial or industrial customer's bill based on its highest energy demand during some specified period of time. The demand charge, which can be as high as $20 per kilowatt per month, is calculated the same whether the customer draws the peak demand continuously or only for a short period during the entire month.

The implication for gas-fired cogeneration installations is that even a temporary shutdown for maintenance or system failure will result in a large demand charge that negates some of the cost saving that the system is expected to provide. To avoid the demand charge, near-100-percent reliability or a backup system is required. The impacts of occasional outages can also be minimized through the use of multiple cogeneration modules instead of a single unit or through temporary shedding of nonessential electrical loads.

On a more positive note, demand-weighted ratemaking may actually help expand the cogeneration market. Specifically, by imposing a heavy cost to daytime energy consumers, demand charges increase potential cost savings for "part-time" cogenerators, allowing them to provide acceptable two- to four-year paybacks in only 10 to 12 hours of daily operation. New applications could thus become attractive, including office buildings and retail shopping centers. An additional plus is that demand charges brighten the economic advantages of gas cooling, another high-priority area within GRI's R&D program.

To accelerate the growth of commercial cogeneration, our strategy calls for continued development of innovative systems offering energy service capabilities that are not available from early market entries. These systems address market opportunities that are achievable today but which have been neglected by others due to their more difficult technical challenges. In particular, the use of absorption cooling as a summer heat load has received major emphasis. This allows installation of 3 to 4 times more electrical capacity than systems producing only hot water.

We've established a set of program targets for improvements based on advanced system component technology. In terms of performance, we're aiming at improvements in engine technology that would provide an increase in electrical generating efficiency from 28 percent today to 32 percent in 1990 and 35 percent by 2000. Overall system efficiency increases only a modest amount—from 75 percent today to 78 percent in 2000—due to a corresponding decline in thermal output. We also see potential for significant cost savings associated with innovations in heat recovery equipment, control systems, and grid-connect equipment. Installed system cost, presently about $1200 to $1600 per kilowatt, should be cut to about $1000 in 1990 and lower in later years, including waste-heat recovery. Operating and maintenance costs at about 1.5 cents per kilowatt-hour today should be cut in half as service intervals are extended. However, advanced component developments will do more than just reduce costs—they will also provide a needed boost in system reliability. We're projecting an increase in average system availability from 96 percent today to 99 percent by the end of the decade.

PROGRAM UPDATE

Field tests continue to be the focal point of program activities, serving as a critical milestone in both system and component developments. With the growth and maturation of GRI's cogeneration effort over the past three years, several projects have reached or are about to enter field tests.

- GRI's *first cogeneration field test,* involving a 500-kW packaged system for hospitals, was recently completed at a Houston site in cooperation with Entex, Inc., and the Hospital Corporation of America (HCA). During 7000 operating hours the unit demonstrated 75 percent overall efficiency, saving the hospital $11,000 per month. System availability ran at 98 percent during the second half of the 12-month test after start-up problems were resolved. Martin Cogeneration Systems, the developer, estimates installed cost at $1300 per kilowatt for the package including a 150-ton absorption chiller for space conditioning.

Martin is now discussing several additional installations with HCA while marketing the system to hospitals in targeted geographical markets.

- A hospital in Roseburg, Oregon, has been selected as the first of three sites for tests of an *advanced heat recovery unit and a cogeneration control system* in a 310-kW package. The heat recovery unit, developed by Solar Turbines Incorporated, uses extended-surface heat transfer in a compact design that reduces the package "footprint" by 40 percent, allowing an associated saving in the cost of the package enclosure. Honeywell Incorporated designed the advanced control software, which includes economic decisionmaking to maximize savings, data collection, and system-diagnostic capability. In particular, this technology will provide electric demand charge savings by improving system availability.

- A *70-kW package*, which incorporates the Hitachi chiller and Waukesha engine developments mentioned earlier, has begun a field test at a Beverly Enterprises nursing home in Winter Park, Florida. In addition to meeting a large portion of electricity needs, the unit will meet total common-area space conditioning requirements with 40 tons of cooling and 400,000 Btu of heating. The system concept is now being redesigned and downsized to a 40-kW version suitable for restaurants that is scheduled for field testing.

- Kimmel-Motz Refrigeration Corporation is installing a *supermarket cogeneration system* at a Los Angeles location of a major grocery chain in the first phase of a three-unit field test. This system is unlike any other developed for GRI because its 110-hp Cummins engine powers four refrigerant compressors that supply cooling to refrigerated cases throughout the store. Electricity can also be produced for store lighting during cool weather when the refrigeration demand is low. Recovered heat from the engine can be used to subcool refrigerant, heat the store, or regenerate a desiccant dehumidifier. A 3.4-year payback on the system is projected for the Los Angeles site based on an annual $18,000 saving. Future sites will include a store in the southeast and northeast regions, where paybacks may be less than one year.

- In addition to these projects, GRI is conducting numerous projects related to *cogeneration components and gas engines.* Engine R&D is fertile ground since most existing designs do not take advantage of recent advances in engine combustion technology. Performance-improving options being explored include "lean-burn" and "fast-burn" combustion, high-durability valve trains, direct injection of fuel and air, and gas turbine applications. In other component-related work, heat recovery and system control work is complemented by research on low-cost grid-connection equipment.

SMALL COGENERATION SYSTEM TECHNOLOGY

Although most of GRI's efforts have been directed toward cogeneration systems larger than 40 kilowatts, cost-cutting improvements could make 5- to 30-kW systems economically attractive. The challenge increases as capacity decreases since some fixed costs—particularly maintenance—remain relatively constant while operating cost savings shrink. An additional problem is the high variability of energy demand that occurs in light-commercial and residential applications.

Nevertheless, a study conducted for GRI by Arthur D. Little, Inc., revealed a large potential market for such systems. Even if only 10 percent of this market were penetrated, over 40,000 light-commercial cogeneration units with an average 15-kW capacity could be put into service. Systems of this size might be marketed as cogeneration modules that could be combined to meet commercial loads while reducing the incidence of unscheduled downtime. Residential systems of 3- to 5-kW capacity were also found to represent a large market potential but with much more stringent cost and maintenance requirements.

To guide future research in small cogeneration R&D, Arthur D. Little defined cost and performance requirements for small systems and identified key technology issues. Particular importance was given to developing small engines with 2000-hour service intervals for light-commercial and 4000-hour intervals for residential systems. (Current service requirements for small engines range from 500 to 1000 hours.) We are, in fact, already advancing small gas engine technology for residential heat pump applications, and results will be applied to small cogeneration systems as well.

The GRI research program represents a significant effort and balanced approach with an excellent chance of accelerating the use of gas-fired cogeneration systems in the United States. The interest we've encountered on the part of system packagers and manufacturers, end-users, and gas industry field test participants shows a momentum that will almost certainly result in more cogeneration installations each year for many years to come.

Chapter 37

Integrating Heating-Cooling-Hot Water Cogeneration Packages into the Health Care Sector

R. L. RENTZ
Mueller Associates, Inc.

W. H. DOLAN
Gas Research Institute

Various packaged cogeneration systems have to date been successfully developed and commercially applied. Few, however, in the size range under 500 kWe have employed space heating and cooling as a system feature. This chapter discusses the engineering integration of cogeneration packages which supply space conditioning energy as well as domestic hot water into health care facilities or facilities with similar space conditioning requirements.* Specifically discussed is a demonstration installation for a recently developed prototype cogeneration system. Also presented are various general observations regarding other potential engineering-integration options for similar packages for both retrofit and new health care facilities.

---

*The work discussed in this paper was sponsored by the Gas Research Institute (GRI).

INTRODUCTION

Several market and application studies have identified health care facilities (e.g., nursing homes) as promising applications for cogeneration technology. This is principally because of this sector's sustained demand for both electric and thermal energy, virtually seven days per week. For example, many health care facilities independently prepare all of their meals and process all of their laundering via an in-house kitchen and laundry (typically operational 24 hours/day). When combined with the hot water usage for personal, cleaning, and sanitizing purposes, a high and sustained rate of hot water consumption, on a per capita basis, results.

Additionally, adequate space conditioning of patient rooms and day-use spaces must be provided for comfort and safety. Most health care facilities also include various medical and administrative capabilities. As a result of the combined loads, these facilities typically consume significant quantities of energy. The relative mix of energy sources (electric versus other) is of course dependent on equipment selection, geographic location, and other factors.

The near term configurations for heat engine driven packaged cogeneration systems for health care facility applications are principally of two types. The first is a "conventional" hot water package. This type of system recovers waste heat rejected by the engine in order to heat water for domestic uses. The second type of system would assume an "HVAC" (heating, ventilating, and air conditioning) orientation. This type of system converts engine recovered waste heat to a form suitable for space conditioning (i.e., heating and/or cooling of room air). The conventional hot water type of packaged cogeneration systems have seen much commercial activity and have been successfully applied to health care facility applications. However, the HVAC oriented cogeneration system concept has not yet been commercially applied to this sector.

An interesting potential advantage of HVAC oriented packaged cogeneration systems over conventional systems is that they can truly provide total energy services. HVAC oriented systems are able to serve the four principal energy requirements for typical facilities:

- space cooling

- space heating
- hot water heating
- electric power

PACKAGED COGENERATION SYSTEM INTEGRATION: DEMONSTRATION PROJECT

In order to demonstrate the feasibility of HVAC oriented packaged cogeneration systems for health care facility applications, the Gas Research Institute sponsored the development and assembly of a system for field testing. The developed cogeneration system utilized proven hardware which had been "packaged" as a system to ultimately exploit cost advantages realizable through assembly-line production and subsequent product offering as a complete system. The purposes of the field test are to:

- demonstrate the feasibility of the technology
- demonstrate the integration of the technology into a building
- accumulate operating data for subsequent review and refinement of the technology

Cogeneration System Characteristics

The HVAC oriented packaged cogeneration system which was developed, assembled, and utilized in the demonstration project is illustrated schematically in Figure 37-1. The figure also defines the principal components of the completely integrated cogeneration system consisting of an engine/generator, and absorption chiller/heater, and associated piping and controls. An interesting characteristic of the absorption chiller/heater used in the system is its ability to simultaneously and directly accept and utilize both the engine's gaseous exhaust and engine jacket coolant. This feature negates the usual requirement for an intermediate heat exchanger. Salient characteristics of the principal components of the cogeneration system package are as follows:

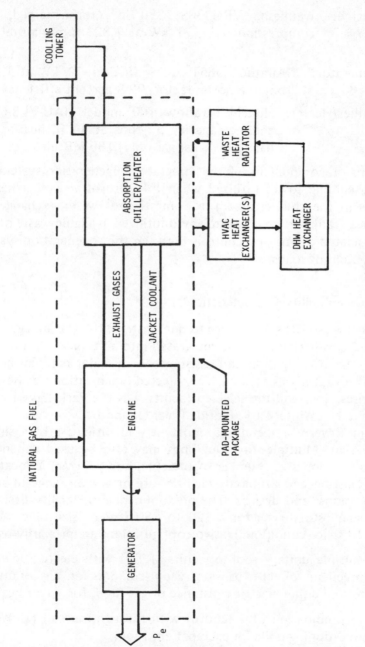

Figure 37-1. HVAC Oriented Packaged Cogeneration System

- Engine: Waukesha VRG330S, 330 in^3, turbocharged, 8:1 compression ratio, 71 kW at 1,823 rpm, natural gas fuel

- Generator: Marathon 365T AC induction; 71 kW at 1,823 rpm, 0.8 power factor, 92.8 percent efficiency

- Chiller/Heater: Hitachi Paraflow 100 model HAU-FES-60N, chilling capacity 32 tons at 44°F, heating capacity (exhaust gas only) 180 MBtuh

Other major components of the integrated cogeneration system include a cooling tower (utilized when the absorption chiller/heater operates in the chilling mode), a domestic hot water exchanger, a waste heat radiator (to dissipate any unutilized available waste heat), and associated piping and controls between the cogeneration system and the building loads.

Health Care Facility Characteristics

Since it was GRI's preference to demonstrate the feasibility of the packaged cogeneration system integrated into a "typical" health care facility, the participation and assistance of a major national health care organization was secured. The selected organization was Beverly Enterprises, Inc. and the selected facility was the Park Lake Health Care Center in Winter Park, Florida (near Orlando).

Within Beverly's operations, most new facilities are of a single-story design. Multiple-story buildings may also be used depending principally upon size (number of patients) and geographic location. Older (existing and acquired) facilities are of a wide range of sizes, configurations, and designs. The predominant new facility design is of the single-story type for a 120 to 180 patient size range. With respect to space conditioning, two configurations are primarily used.

1. Multiple unitary roof-top units (RTU) with electric-powered forced-air, direct-expansion, vapor-compression air conditioning and either electric resistance or gas-fired, forced-air heat.

2. A combination of 1 (above) with through-the-wall (TTW) air conditioning units for each patient room.

The facility selected for the demonstration project is representative of configuration 2 although a configuration 1 facility could have served equally as well.

Figure 37-2 is the roof plan for the facility selected for the demonstration. It illustrates the lay-out of the HVAC and domestic hot water (DHW) systems of interest to integration with the subject cogeneration system. With respect to space cooling and heating, Figure 37-2 illustrates the TTW air conditioning units typically employed for each patient room.

It also shows various unitary RTUs (the actual number varies with application). The RTUs are used to space condition the entire core area of the building which houses the administrative offices, laundry, kitchen, and day-use/dining-room space.

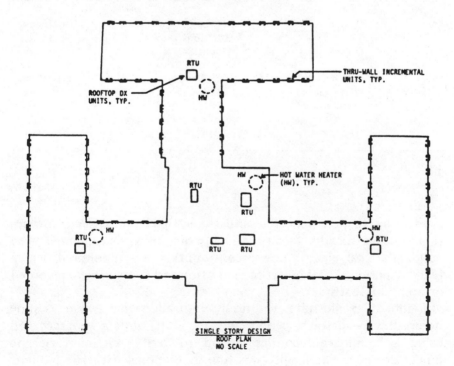

Figure 37-2. Representative Roof Plan for a Health Care Facility

RTUs are also used to space condition centrally located nurse's station areas on each wing of the building. With respect to DHW systems, Figure 37-2 also illustrates the location of the facility's water heaters. Note a dedicated water heater for each patient wing and a centrally located water heater used principally to serve the kitchen and laundry. Table 37-1 defines the pertinent characteristics of this 180 patient bed, 48,590 ft^2 health care facility.

Table 37-1. HVAC and DHW Systems Installed at Park Lake Health Care Center

System	Description	Total Installed Capacity	Energy Source
HVAC	RTUs	41 tons/87 kW	electricity
	TTWs	92.5 tons/266 kW	electricity
	Other	2 tons/64 kW	electricity
DHW	Core Area Boiler	725,000 Btuh	natural gas
	Wing Heaters	144 kW	electricity
	Miscellaneous booster heaters	77 kW	electricity

In order to determine the actual electricity and DHW consumption profiles for the facility, the building was instrumented. The building's total electric power consumption was monitored over a five-day period. Total power consumption and demand were recorded on an hourly basis.

Figure 37-3 illustrates the results. An additional two days (one before the five-day test period, and one after) of data were recorded using a 15-minute recording period to verify minimum electric power demand. The facility was fully-operational with 167 patients during the test period. The three electric hot water heaters and the gas-fired boiler were separately instrumented for one 24-hour test period. The resulting data indicated an average DHW load for these systems of 163,300 Btuh (note that Orlando has an average nearly constant, yearly ground water temperature of 70°F).

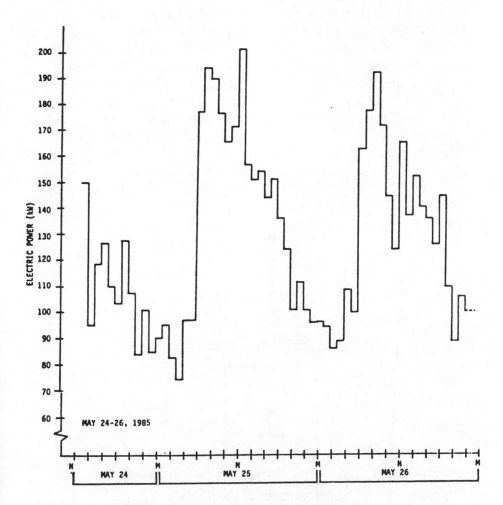

Figure 37-3. Total Electric Power Consumption and
Average Demand, Selected Three Day Period,
Park Lake Health Care Center

Demonstration Cogeneration System Integration

The cogeneration system was designed for this field test to integrate with the existing facility's HVAC and DHW systems in a fashion to ensure full back-up service by the existing systems. At the owner's request, the cogeneration package and supporting equipment were placed within and outside of an existing general storage building.

Figure 37-4 illustrates the lay-out of the equipment. The allotted space within the storage building for the cogeneration equipment is about 400 ft^2. Note the cooling tower, radiator, DHW heat exchanger, water storage tank (1,000 gallon), and capacitors. The only other required equipment and materials are associated with the piping to and interfacing with (and control of) the utility transformer and the HVAC, and DHW equipment.

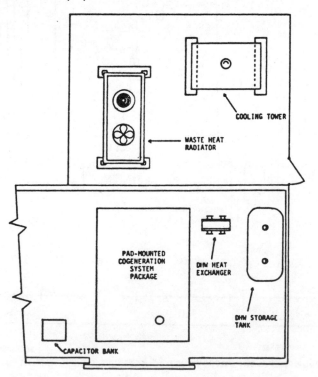

Figure 37-4. Cogeneration System Insulation Floor Plan

Various system integration decisions were made during the integration/design process. With respect to the electrical power generated by the cogeneration system, it was noted that the cogeneration system would meet the baseload electric power need of the facility without any need to sell back power to the local electric utility.

However, due to the relative technical and administrative ease of interconnection and an interest in demonstrating utility interconnect requirements, the decision was made to interconnect at the existing utility-owned transformer. With respect to HVAC integration, it was noted that the sum of the four core area RTUs matched the capacity of the absorption chiller when powered by the engine's "waste" energy. Thus it was decided to tie-into the four core area RTUs.

Finally, in order to demonstrate the potential of this system to provide truly full energy service to the facility, a tie-in to the four DHW heaters was constructed. The following summarizes the relatively simple interconnects which were required.

Electrical Integration—The electrical interconnection was made on the secondary side of the utility transformer. Two separate underground cables separate from the facility service entrance cables were used, one for the cogeneration generator and controls and the other for various cogeneration building loads and the system's capacitors. The separate cables allow for electrical isolation of the capacitors from the generator to avoid any self-excitation condition. The local electric utility required capacitors to reduce the deleterious contribution to power factor caused by the cogeneration system's induction generator. Figure 37-5 illustrates the electrical interconnection design.

HVAC Integration—The HVAC integration was performed by providing a dual-temperature water coil within each of the existing RTUs, see Figure 37-6. Thus, depending on whether the existing core space zone thermostat calls for heating or cooling, hot or cold water can be supplied by the cogeneration system to the new coil. The existing RTU fan would then blow air across the new coil thereby providing heating or cooling to the space.

For one RTU, however, it was necessary to install the new coil in the existing ductwork. This was due to the orientation of the unit on the roof and resulting space constraints. Because of the demonstration related nature of the project, full back-up of the existing RTUs was required.

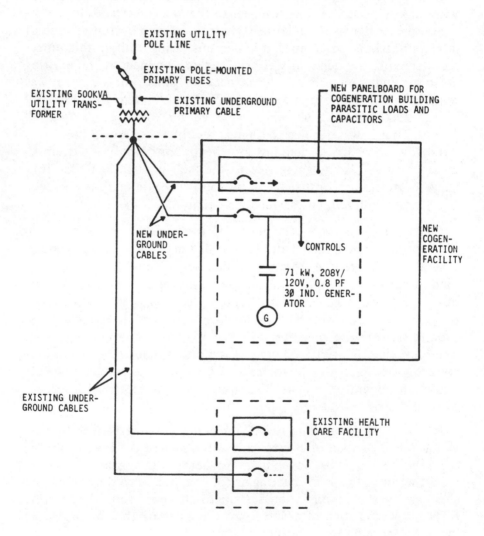

Figure 37-5. Electrical Interconnection Design

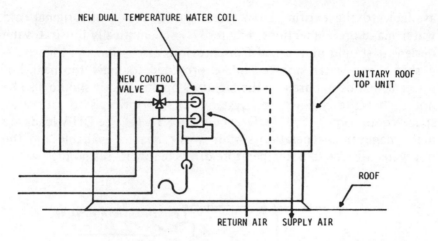

NEW DUAL TEMPERATURE WATER COIL

NEW CONTROL VALVE

UNITARY ROOF TOP UNIT

ROOF

RETURN AIR   SUPPLY AIR

Figure 37-6. Typical Installation of New Dual-Temperature Coil in Existing RTU

As a result, none of the existing equipment was removed (e.g., the existing evaporator coil). Because of the added static pressure drop on the air system caused by the new coil, higher capacity fan motors were installed to maintain the required air flow rate.

Based principally upon the low requirement for space heating in central Florida, it was decided to provide a two-pipe system for space conditioning. A computer simulation of the facility using the DOE-2 building simulation program supported this decision. The analysis showed, for the zones served by the four RTUs of interest, that coincident demands for space cooling and heating were rare and heating/cooling seasons were readily definable with definable daily swings in requirements during the transitional periods of the year.

As a result, clear and simple instructions for seasonal changeovers are possible. It should be noted that the absorption chiller/heater is capable of simultaneously providing chilled and hot water and can be configured for a four-pipe system.

DHW Integration—The DHW integration was designed for simplicity. The cogeneration system simply preheats water which is made available to the existing DHW heaters in lieu of their original cold water make-up connections. Figure 37-7 schematically illustrates the design. It should be recalled from previous discussion and Figure 37-4 that a hot water storage tank is provided to meet the peak hot water demands imposed by the facility's operations. It should also be noted that the cogeneration system's first priority is to satisfy all space conditioning loads (reference Figure 37-1). The DHW loads are met whenever cogeneration system waste heat is available and the hot water storage tank temperature drops below its set point.

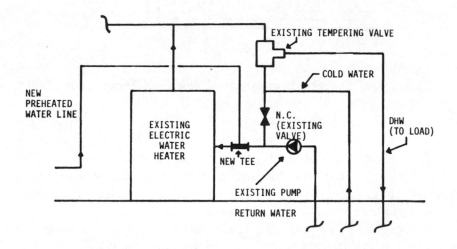

**Figure 37-7. Typical Integration of Cogeneration System
Supplied Preheated Water with Existing DHW System**

The above discussions illustrate the straightforward engineering integration of HVAC-oriented packaged cogeneration systems into an existing typical health care facility.

PACKAGED COGENERATION SYSTEM INTEGRATION:
OTHER OPTIONS/TECHNIQUES

It is not the intent of this chapter to present a treatise on the broad subject of options and techniques for integrating packaged cogeneration into existing and/or new health care facilities. However, it is important to note that moderate technical advances could further reduce the effort involved in integrating cogeneration into existing facilities. It is also observed that consideration of the integration of packaged cogeneration systems at the new building design phase could result in dramatically reduced requirements relative to a retrofit. The following general observations are offered:

- Some facilities already use piped water systems for space conditioning (e.g., a centrifugal chiller/water boiler supplying multiple fan-coil units and/or air handlers). The mechanical integration of packaged cogeneration HVAC capabilities for this situation may be no more involved than one or more piping connections.

- For the case of new facility construction, consideration of packaged cogeneration at the design phase can result in a simpler integration relative to the retrofit situation. For example, the cogeneration system may be centrally located relative to the loads it serves, thereby essentially negating any additional piping and cabling. Also, for the case of piped-water facility HVAC designs, credits may be claimed for downsizing the main HVAC plants. Other innovative techniques for reducing system requirements may become apparent. For example, it may be feasible to use a common cooling tower/DHW system for both the cogeneration system's absorption chiller and the central cooling plant.

- There may also be one or more innovative techniques for integrating packaged cogeneration HVAC capabilities with existing RTUs in an even more straightforward manner. For example, the concept of using a refrigerant-based loop between the cogeneration system's absorption chiller and the existing RTU's evaporator coil has been proposed. This idea, in fact, is currently being investigated by the Gas Research Institute. Figure 37-8 illustrates the basic concept.

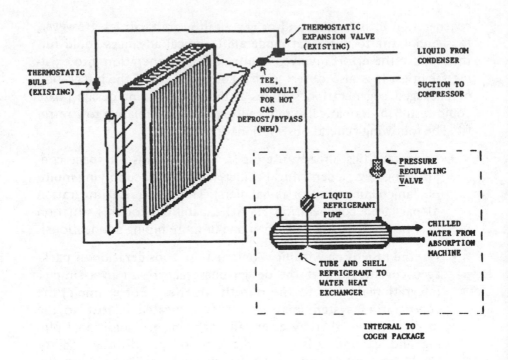

Figure 37-8. Conceptual Schematic of Refrigerant Loop Integration: Cogeneration System with Existing Unitary RTU

CONCLUSIONS

- It has been demonstrated that the engineering integration of HVAC-oriented packaged cogeneration systems into health care facilities may be accomplished in a straightforward manner.

- The consideration of packaged cogeneration systems at the building design stage may offer advantages.

- Additional advances and refinements of both cogeneration systems and their integrative ability are possible and could lead to reduced system and tie-in complexity of installed systems.

Chapter 38

A New Approach to Cogeneration in the Supermarket Industry

D. H. WALKER, S. J. HYNEK, B. N. BARCK,
D. L. FISCHBACH, R. D. TETREAULT
N. B. LONGO
Foster-Miller, Inc.

Supermarkets offer an attractive application for cogeneration because of their large needs for shaft power, electricity, and some thermal loads. Because of the unique characteristics of supermarket energy requirements, a special cogeneration system has been designed for this application. A natural gas-fueled engine is used to directly drive four refrigeration compressors through a multi-output transmission. Hot water is generated through engine heat recovery and is used to drive an absorption chiller to provide refrigerant subcooling.

Space heating can also be supplied using recovered engine heat and reclaimed heat from the refrigeration system. The system is also equipped with an induction motor/generator that can act as a backup to the gas engine or be used to generate electricity during periods of low refrigeration compressor power requirements. Economic analysis of this system for five metropolitan regions has shown favorable paybacks. A supermarket cogeneration system has been constructed and will be field tested in a supermarket in Los Angeles.

BACKGROUND

Supermarkets have several characteristics that make them attractive for cogeneration application. First, they represent one of the most intensive energy users in the commercial sector, typically consuming two million kWh annually per store. Secondly, the general trend in the supermarket industry is toward the construction and operation of larger stores which are often combined with enlarged drug and dry goods retailing. Accompanied with this store expansion is an increased consumption of energy for lighting, space heating, and air conditioning.

The energy use breakdown for a typical supermarket is given in Table 38-1. As can be seen, the largest single use is refrigeration compressors, which operate on a virtually continuous basis. Space heating requirements for supermarkets are also high because of the large amount of refrigerated fixtures employed. While most of the space heating load is met through the use of refrigeration reject heat (referred to as heat reclaim), a substantial portion must still be met by purchased fuel.

Supermarket air conditioning also has unique characteristics; again, because of the large amount of refrigeration employed within the store. Air conditioning loads typically have a very low sensible to latent heat ratio, and store air conditioning is often operated for dehumidification purposes only. In such situations space heating is needed for conditioned air reheat to maintain an acceptable dry bulb temperature within the store.

It has become apparent that retail supermarkets represent an excellent opportunity for packaged natural gas-fueled cogeneration systems. The uniqueness of the application, however, necessitates development of a cogeneration package specific to supermarkets. To provide for the large shaft power requirements of a supermarket for refrigeration, the cogeneration package presented here has been designed to directly drive refrigeration compressors with the shaft

Table 38-1. Energy Use in a Typical Supermarket

Use	Fraction (%)
Refrigeration System (54 percent)	
— Low temperature compressors	21
— Medium temperature compressors	18
— Case fans and lights	9
— Case anti-sweat heaters	6
Lighting (25 percent)	
— Sales area ceiling lights	19
— Back room office, sign lights	6
Space Conditioning System (17 percent)	
— Heating (electric)	8
— Air handler blower	5
— Air conditioning	4
Miscellaneous (4 percent)	4
TOTAL	100

output of the natural gas-fueled engine. The advantages of this approach compared to a cogeneration system producing electricity are the following:

- By combining the cogeneration and refrigeration systems, a reduction in initial capital expense can be obtained when compared to the cost of separate cogeneration and combined refrigeration systems.

- The amount of machine room space required is less than that needed for separate systems.

- The power losses associated with the generator and compressor motor efficiencies are eliminated.

- The use of open-shaft rather than semihermetic compressors also reduces power requirements for refrigeration.

SYSTEM DESCRIPTION

The supermarket cogeneration system presented in this chapter is capable of providing all refrigeration, space heating, water heating, and some electricity for lighting that a typical supermarket requires. Figure 38-1 shows an artist's conception of the supermarket cogeneration system. Four refrigeration compressors and an induction motor/generator are driven through a gearbox by a gas-fueled internal combustion engine. All components except the condenser are mounted on two skids to facilitate installation in existing machine rooms and to reduce field installation costs.

The compressor skid houses four open-shaft type refrigeration compressors capable of providing refrigeration at any desired suction condition. All four compressors are connected to the gearbox through hydraulic clutches. Also on the compressor skid, connected to the gearbox, is the aforementioned 125 hp induction motor/generator.

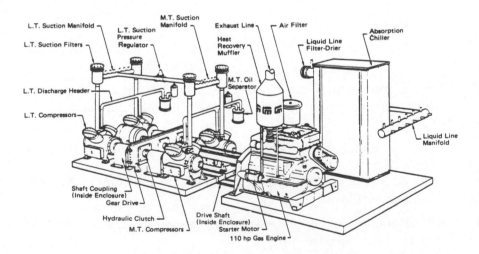

Figure 38-1. Artist's Conception of Prototype Gas-Fired
Supermarket Cogeneration System

The motor/generator can either be driven by the gas engine to produce electricity, or can be used to directly drive the compressors during periods of engine maintenance or gas outage. Refrigerant suction manifolds, oil separators, and suction filters have been arranged to provide easy access to the compressors, generator, and gearbox.

A heavy duty, 110 hp gas engine is mounted on the engine skid. The skid itself is configured to accept gas engines of several manufacturers, depending upon buyer preference and availability of local service support.

The engine's reject heat is recovered in the form of hot water circulated through engine jacket cooling water and exhaust gas heat exchangers. The recovered hot water is used primarily to drive an absorption chiller (also mounted on the engine skid), which increases refrigeration capacity by subcooling the liquid refrigerant enroute to the display cases.

The remaining heat contained in the hot water leaving the absorption chiller can be employed for building space heating, if required. By employing the cogen system hot water, and also using heat reclaimed from the discharge of the compressors (a common practice in supermarkets), all space heating requirements of the store can be met in most areas of the country.

In some locations during severe winter weather, some auxiliary heating may be required to meet space heating demand. Potable hot water heating is also accomplished by desuperheating the discharge gas of the refrigeration compressors.

This cogeneration system has two distinctly different modes of operation depending upon the need for space heating. The two modes are referred to as the summer and winter modes. During the summer, in meeting the refrigeration demand, the engine rejects more heat than can be used by the absorption chiller for refrigerant subcooling. In this situation it is wise to limit engine power output to that required by the compressors, because without a way to use the additional reject heat, the efficiency (and thus cost) advantage of cogenerating are not realized.

Therefore, during summer operation, system performance is optimized by adjusting engine output to match compressor power requirements. This amount varies with ambient temperature and with any load variations imposed by, for example, defrosting the evaporators. Since the engine drives the compressors directly, the speed of the engine and compressors can vary to exactly match the refrigeration load. Variable speed operation gives much better capacity control than the usual method of turning compressors on and off.

Figure 38-2 shows the energy balance for the cogeneration system operating at an ambient temperature of 95°F. From the diagram, it can be seen that the absorption chiller alone uses about two-thirds of the reject heat from the engine.

The only other summer mode use for reject heat in the present system is hot water, which is a comparatively small load compared to that required by the chiller. (In future configurations, desiccant dehumidification will also make use of reject heat.)

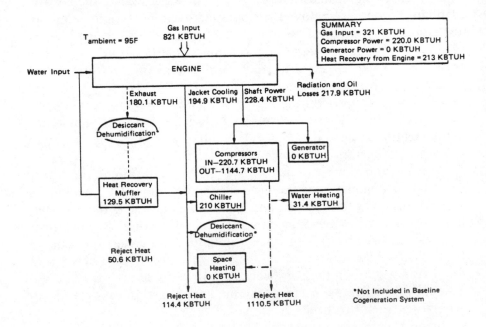

Figure 38-2. Summer Operation Energy Balance

During the winter, the store can use all the reject heat available for space heating. However, the engine provides less reject heat due to the decreased power requirements of the refrigeration compressors. Under these conditions, it makes sense to generate electricity with the surplus shaft power available, and to utilize the additional reject heat produced for space heating. To produce electricity, however, the generator must run at 1800 rpm, so compressor capacity control must be achieved by unloading cylinders, declutching compressors, or both.

Supermarkets have appreciable electric demands for other than refrigeration compressors. The electricity generated by the cogeneration system is totally consumed by the store itself. Therefore, resale to the electric utilities is not a consideration and has not been included in the economic analysis.

Figure 38-3 shows the energy balance for the system operating at an ambient temperature of 35°F. Note that space heating consumes all of the recoverable engine reject heat and part of the compressor discharge superheat.

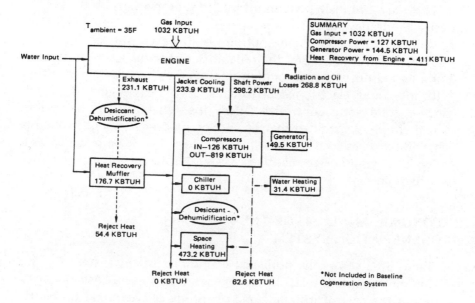

Figure 38-3. Winter Operation Energy Balance

To employ this two mode control strategy, a custom control system was designed and developed. The primary control objective was to make the system as easy to operate as existing electric-driven refrigeration systems. It accomplished this by:

- Automatically making all power and capacity adjustments for both summer and winter modes of operation.

- Monitoring operating parameters, such as engine and compressor oil temperatures and pressures, engine and compressor discharge temperatures, engine water temperature, engine shaft speed, and engine manifold pressure. It also automatically telephones service personnel when any parameter gets out of range.

- Recognizing and reacting to component failures. For example, should the engine be unable to maintain the assigned speed, the control system automatically cuts in the electric motor to help; if the engine subsequently fails to keep up, it is declutched and the load is placed on the electric motor.

There are two main benefits offered to supermarkets by the system design being developed: lower operating costs, and equal or better reliability. No energy saving option could be considered for supermarkets if it compromised the reliability of the refrigeration system. For this reason, the cogeneration package has been designed with a fully integrated electric motor to back up the gas engine. If the engine or solid state controller failed, the reliability of the system would simply degrade to that of an electric-driven system (which is what it becomes in those cases). Until then, it can be regarded as a redundant gas-and-electric-driven system, tolerant of losing either source of energy.

ECONOMIC ANALYSIS OF THE COGENERATION SYSTEM

Table 38-2 shows the projected economics of the system for five major metropolitan areas. In it, cogeneration system costs are compared to the cost of installing and operating conventional all-electric refrigeration compressor racks with heat reclaim. Presented in the

Table 38-2. Economic Benefits of Supermarket Cogeneration for Selected Locations

(First Cost Difference = $62,260)

Annual Operating Cost Differences ($)	Atlanta 1982/ 1984	Chicago 1982/ 1984	Dallas 1982/ 1984	Los Angeles 1984	New York 1982/ 1984
Gas	35.970/ 39,446	27,150/ 34,154 25,952*	30,650/ 22,186	50,110	59,560/ 36,800
Electric	(76,490)/ (90,216)	(51,410)/ (55,248)	(35,560)/ (37,790)	(80,548)	(93,380)/ (116,100)
Maintenance	12,000/ 12,000	12,000/ 12,000	12,000/ 12,000	12,000	12,000/ 12,000
Total	(28,520)/ (38,770)	(12,260)/ (9,094) (17,296)*	7,090/ (3.604)	(18,438)	(21,820)/ (67,300)
Payback (years)	2.2/1.6	5.1/6.8 3.6*	/17.3	3.4	2.9/0.9

Note: Baseline cogeneration system compared to reference conventional system, both with evaporation heat rejection.
*Number based on prime mover gas rate.

table are the calculated annual operational cost differences between the two systems which consists of:

- The increase in gas cost due to increased gas consumption by the cogeneration system

- The decrease in electric cost achieved by the cogeneration system by directly driving the refrigeration compressors and generating electricity

- The increased cost in maintenance attributable to the engine

The net annual savings achieved by the cogeneration system is the difference between the electric cost savings and the increased costs for gas and engine maintenance. The payback given represents the amount of time for the net savings to equal the cost difference between the cogeneration and conventional electric refrigeration systems. The estimated difference in first cost between the two systems is $62,260.

Local utility rates for the years 1982 and 1984 were employed for the analysis, along with regional climatic data.

The savings on electricity are roughly double the amount spent for both gas and maintenance. The paybacks are quick enough to merit serious consideration in most locations, offering returns on the order of two to three years.

PRESENT PROJECT STATUS
AND FUTURE PLANS

The cogeneration system described here has been constructed and successfully tested in the laboratory to verify system performance. The system has been delivered to the construction site of a new supermarket in Los Angeles. The cogeneration system will be instrumented and field tested for one year.

A second field test is now planned. In this second test, the supermarket cogeneration system will operate in conjunction with a desiccant system dehumidification system. The cogeneration system will supply heat to the desiccant system for regeneration air heating. Initial analysis of this combined operation for a supermarket site in

Atlanta indicates that annual savings are increased by approximately $10,000 and payback is reduced to 1.5 years.

A site in the southeast is now actively being sought. It is hoped that the second field test will begin some time in 1987.

Reprinted with permission from the Proceedings of the 21st Intersociety Energy Conversion Engineering Conference. Copyright 1986. American Chemical Society.

Chapter 39

Intelligent Cogeneration Control: A Key Factor

R. K. ASSEN
Honeywell Technology Strategy Center

What is "intelligent" or "advanced" control, and are such techniques being employed in cogeneration systems today? Recent control systems development work at Honeywell verifies that advanced control techniques are applicable to cogeneration and will enhance performance and economics without sacrificing reliability. This chapter describes an advanced control approach developed by Honeywell and the results of its implementation in a packaged cgeneration system. Other controller features are addressed briefly.

COGENERATION CONTROLS TODAY

Advanced control is a term applied to a variety of control system techniques such as multivariable, optimal, and heuristic control. These features typically are not found in today's cogeneration controllers for two reasons:

- The on/off, baseloaded applications currently employed do not require them;

- Considerable control and control systems development exper-
tise/tools are needed to develop and implement them.

Control requirements are directly related to system and application
complexity. Today's cogeneration system applications typically require only simple on/off control and equipment protection via discrete control and nonintegrated control loops. However, as the
packaged cogeneration industry matures, complex control requirements for systems with interdependent interactions are becoming
more common.

Advanced control systems become essential in providing the following features:

- Integrated control implementation:

  — Stable operation at nominal operating point (regulatory controls), start-up/shut-down, and disturbance rejection while
  meeting stringent equipment operating specifications;

  — Electrical/thermal load following (ELF/TLF), servo controls
  during mode switching, and thermal prioritization/distribu-
  tion to multiple loads;

- System operation optimization:

  — Integration of cogeneration and HVAC systems and optimal
  performance and savings for nonbaseloaded system applications;

  — A companion, flexible controller for standard cogeneration
  packages that will help enlarge the cogeneration application
  market base.

Today's cogeneration system application design requirements are
achievable; engine loop, absorption AC supply, and heating/cooling
load temperatures can be maintained. However, when any actuator
(the AC valve, heat valve, etc.) affects conditions throughout the
cogeneration system, simple application needs become complex, interdependent control implementation requirements, and it is here
that advanced control is essential.

COGENERATION CONTROLS
DEVELOPMENT PROGRAM

In 1984 Honeywell began a program under GRI contract to develop an advanced controller to meet the range of cogeneration industry needs. The cogeneration control subsystem (CCS-2C) that evolved, and that is currently undergoing field evaluation, possesses the following features: advanced control, trend-based diagnostics, data archival/information, and flexibility via tailoring, a start-up package, and modular software design. The CCS-2C controller and two other models comprise a controls family that address the spectrum of packaged cogeneration market needs. The CCS-2B, a lower cost version of the advanced controller, possesses a streamlined subset of the CCS-2C features. The CCS-2A, the third model, is a tailorable remote monitoring system.

ADVANCED CONTROLS
DEVELOPMENT PROCESS

An advanced controls development process was applied to development of the CCS-2C's supervisory and implementation control strategies. Supervisory control is the decision-making process used to determine the best strategy (e.g., lowest cost or highest performance) for operating a given system. Control implementation is the processing of sensor inputs and sending of command signals, both discrete and proportional, to actuators.

A systems approach was used in developing an effective cogeneration controls system, an approach that considers interactions within the controlled system and its environment. This required expertise in control system design, cogeneration and energy system design/integration, and control system implementation expertise, along with control development tools. The interactive methodology proceeded from system needs definition to control strategy development, then to target system simulation, and finally to control strategy verification, all leading to controller evaluation in the field. The steps (depicted in Figure 39-1) are described below.

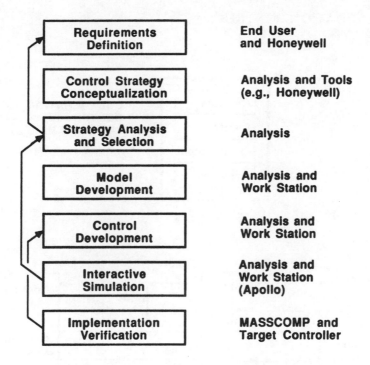

Requirements Definition	**End User and Honeywell**
Control Strategy Conceptualization	**Analysis and Tools (e.g., Honeywell)**
Strategy Analysis and Selection	**Analysis**
Model Development	**Analysis and Work Station**
Control Development	**Analysis and Work Station**
Interactive Simulation	**Analysis and Work Station (Apollo)**
Implementation Verification	**MASSCOMP and Target Controller**

Figure 39-1. Control Development Process

Requirements Definition—The target cogeneration package system selected for control is depicted in Figure 39-2. Requirements are a function of cogeneration component operational specifications, building HVAC interface, and desired performance requirements (implementation and supervisory). The primary requirements were:

- Operate cogeneration components within specifications, particularly engine coolant (e.g., maintain engine inlet (T-E) at 190°F under all conditions and enclosure ambient at less than 120°F);

- Deliver cogenerated energy (chilled water, hot water, electricity) within specified ranges and at the desired set points (e.g., chilled water (T-C) and heat load supply (T-H) at 44° and 140°F, respectively);

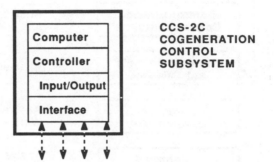

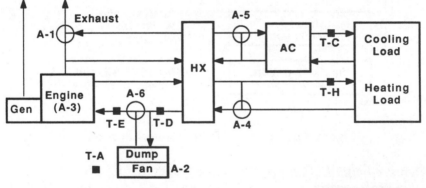

File T60244-1028MDF

Control Requirements

- Engine Inlet (T-E) = Design
- Head Load Supply (T-H) < Design
- Cool Load Supply (T-C) > Design
- Enclosure Air (T-A) at Design

- Thermal to Priority Load (%)
- Minimize Thermal Dumping
- Component Constraints
- Electrical Export, as Selected

Figure 39-2. Target Cogeneration System

- Constrained by the above, provide integrated control of actuators to produce stable, effective operation under both steady-state and transitional conditions (e.g., maintain engine inlet temperature (T-E), regardless of loads, by integrated control of the silencer, the dump valve, and the fan);

- Based on external inputs, select the operating levels and control set points for the most cost-effective total cogeneration system operation (e.g., load prioritization, thermal load following via engine throttling, and the "economic/least-cost" mode);

- Implement control with the following proportional actuators (depicted in the target system shown in Figure 39-2): A-1, exhaust HX damper; A-2, fan; A-3, engine throttle; A-4, heat valve; A-5, cool valve; and A-6, dump valve.

Control Strategy Conceptualization, Analysis, and Selection— These requirements and cogeneration/HVAC operation and performance were evaluated to help prioritize wants/needs. Supervisory and implementation control strategies, from simple to advanced, were conceptualized and analyzed for applicability. An engineering and computer analysis was conducted to evaluate design and off-design performance. Approaches that appeared complex or inherently unstable were eliminated; the most promising were selected for further analysis.

Model Development— Models were then developed on an engineering work station for each of the cogeneration and HVAC components, the building and its loads, and the sensors/actuators/piping. These models are integrated to produce an emulation of the composite cogeneration system that is used for checking out and modifying alternate control strategies. An executive program links the modular cogeneration system models to the alternate control strategy algorithms.

Interactive Simulation/Control Development— The control development effort used simplified models of the cogeneration system and included evaluation of multivariable control schemes, interacactions of controlled loops, various PI implementations, and the impact of dynamics. Use of various computer-based control design tools provided a clearer understanding of cogeneration system operational interactions. Requirements were continually scrutinized to ensure development of simple, but fully responsive, controls strategies.

Control Implementation Verification—The most promising control algorithms and detailed process models of cogeneration components were then implemented and tested in the target controller hardware and on a real-time work station, respectively. The computer work station was then hard-wired, via input/output boards, to the controller hardware. Once enabled, the cogeneration model/emulation sends its state variables (such as temperatures) to the controller, which then calculates and returns control signals to the emulation. This iterative process continues, with the target controller directly controlling the cogeneration system/emulation in real time. Further control modifications are identified and control algorithms in the target hardware/operating system are verified via this control verification process. Field checkout issues and problems are drastically reduced.

ADVANCED CONTROL
DEVELOPMENT RESULTS

The results of the CCS-2C control development process are significant. The target cogeneration system control requirements—which ultimately were similar to multivariable, interactive process control requirements—were satisfied with integrated control algorithms. Control strategies for following the thermal load and for distributing thermal energy to the heating and cooling load were developed. A performance optimizing algorithm was developed for overall cogeneration system operation (the economic mode and heating/cooling priority selection). The composite effect of these results is a controlled cogeneration system that operates reliably at maximum economic performance for the target applications yet still meets the cogeneration and HVAC operating constraints.

Integrated Control—Attaining effective, stable, integrated control of a cogeneration system becomes exceedingly complex as the number of proportionally controlled points increases. Today's typical cogeneration system may have one or two single-loop PI controllers; the target system has six interactive PI loops, including the engine throttle. The keys to the CCS-2C's success in producing effective control are:

- PI Control—Proportional/integral/reset control implementation.

- Control Set Point Supervision—Since the target system's control loops are interdependent, the relationship between set points is critical. Extensive analysis led to defining control loop set points based on a single reference set point (such as engine inlet temperature), with a predetermined offset from the reference.

- Auctioneering Feedback Control (AFC)—The AFC strategy uses multiple sensors within a single control algorithm for the target actuator. AFC control is stable and inherently "bumpless," an extremely difficult and critical feature for a nonmultivariable controller. The TLF engine throttling, the heating and cooling valve, and the dump fan are all controlled via an AFC algorithm.

Figure 39-3 is an example of integrated control effected to maintain the engine inlet temperature (T-E). Figure 39-4 depicts dual loop control to maintain both T-E and heating loop (T-H) temperatures. Both examples include a dramatic heat load reduction at time = 1 minute when the engine is fixed at full load.

Advanced Control (Load Following and Prioritization)—Following the thermal load and prioritization of thermal energy distribution are advanced control concepts not employed in cogeneration systems today. TLF is the throttling of the engine to a level such that engine thermal output equals the total thermal load of the building and, by definition, results in no dumping of thermal or use of auxiliary fuel to meet the heating/cooling loads. Load prioritization is the delivery of thermal energy to the heating or cooling load with the most costly auxiliary system; the remainder will go to the secondary load. The CCS-2C controller implements these advanced control algorithms while ensuring that the cogeneration system runs at design

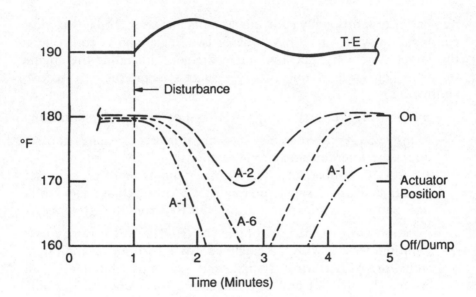

Figure 39-3. Integrated Control

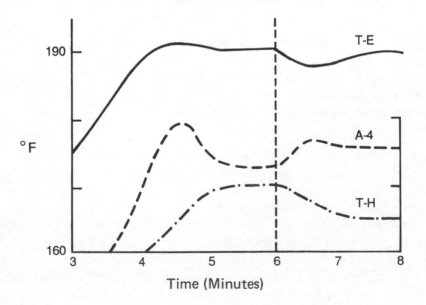

Figure 39-4. Dual Loop Control

temperature, utilizing the AFC strategy and set point supervision described above.

Figure 39-5, an example of advanced control, depicts the engine throttling back to keep the dump systems from wasting heat in the TLF mode. A disturbance (dramatic heat load reduction) was created at time=1 minute.

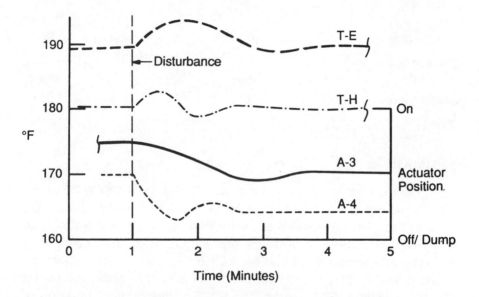

Figure 39-5. Advanced Thermal Load Following Control

System Control Optimization—This control feature of the CCS-2C is supervision of the above-described control implementation process. Decisions are made continually in real time and ultimately are based on economics, resulting in mode selection or set point selection for implementation by the controller. Mode selection enables a load following control algorithm (TLF, ELF, or LIMit the electrical demand) or sets the engine throttle to full ON, OFF, or a FIXed output. The load prioritization algorithm supervises changes to control set points for the heating and cooling valve AFC control implementation.

Table 39-1 depicts the controller mode selection for least-cost system operation, given the noted conditions.

Table 39-1. Economic Mode Supervisory Control

	Case 1	Case 2	Case 3	Case 4	Case 5
Gas cost ($/Btu)	Low	Low	Low	Low	Med.
Electrical usage (cents/kWh)	Low	Med.	N/A	High	Low
Demand charge ($/kW)	Low	Low	High	N/A	Low
Selected mode	TLF	ELF	ON	ON	OFF

SUMMARY

Advanced controls such as those developed for the CCS-2C have a significant value to the cogeneration industry today and in the future. Such control enhances performance and economics without sacrificing reliability; it enhances application flexibility and thereby enlarges the potential application base for current cogeneration systems; it makes possible the migration of advanced control into less complex systems; and it paves the way for design and operation of more intricate cogeneration systems and applications in the future.

Development of advanced control features requires control system design and implementation expertise, control development tools (software and hardware), and an intimate knowledge of cogeneration, HVAC, and energy management systems operation and interplay. For this reason, control system development companies will continue to figure prominently in cogeneration controls.

Reprinted with permission from the Proceedings of the 21st Intersociety Energy Conversion Engineering Conference. Copyright 1986. American Chemical Society.

Chapter 40

Cogeneration Prime Mover Advancements— "The Heart of the Systems"

ALLEN D. WELLS
LAWRENCE J. KOSTRZEWA
Gas Research Institute

The implementation of cogeneration has been accelerating rapidly in recent years as a result of the substantial energy savings potential of this technology. Although most of the installed capacity has been in multimegawatt industrial systems, a growing number of commercial and light industrial energy users have applied systems of less than 1mWe capacity to take advantage of cogeneration efficiency in their buildings. These early-entry installations have generally been in areas where the price differential between electricity and natural gas has made the comparatively high capital costs justifiable. Cogeneration in the commercial and light industrial sectors could have an ultimate potential on the same order as the industrial sector, but broad penetration requires the research and development of advanced energy conversion technology.

In this market sector and size range, provision of heating, cooling and electricity are merely incidental to the main line of business being pursued. HVAC equipment employed are standard catalog items, with proven reliability and service support, automatically controlled, and requiring little or no user attention. For cogeneration to be widely applied in the commercial and light industrial sectors, the systems will need to satisfy similar constraints. Packaged cogeneration systems, standardized and pretested with autonomous controls, are an appropriate and achievable means of addressing commercial and light industrial energy requirements. Packaged cogeneration systems of 15-650 kW electric capacity with varying degrees of component integration and standardization are being offered by an increasing number of suppliers.

Operating costs and operating savings accruing from a heating/cooling cogeneration system are illustrated in Figure 40-1. The values shown are representative of a 300 kWe system, but are expressed in $/kWe/year. A host of assumptions go into this representation, the most important being electric rates ($8/kW/mo. demand, 6¢ kWh energy, $1/kW/mo. standby), gas rates ($5/mmBtu), installed cost ($1600/kWe), maintenance cost (1¢/kWh), operating load factor (75 percent), and capital financing (15 year mortgage-type loan at 12 percent). On a simple payback basis (without financing), the system would return its initial cost in 7.8 years.

This particular set of assumptions (including financing) results in a first year before tax operating loss of $31/kWe of capacity, but a 0.4¢/kWh increase in electric energy rates, a 40¢/mmBtu decrease in gas rates, a $211 decrease in installed cost, or increase in load factor to 94 percent would make the system break even. This verifies the observation that implementation of commercial cogeneration systems is most active in areas with higher electricity prices, in applications with high load factors or where the scale economies of 500–2000 kWe systems can be taken, when the added cost of an absorption chiller and related components can be avoided, or in applications where the cost of a standby power system can be taken as a capital credit.

That same $31/kWe loss could also be addressed by a decrease in fuel consumption to 10,700 Btu/kWhe (HHV) from a current level of

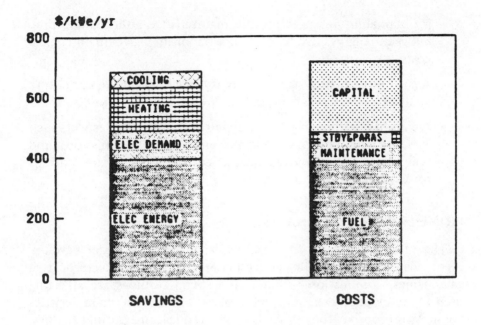

Figure 40-1. Cogeneration Operating Savings and Costs

11,500 Btu/kWhe, a 0.5¢/kWh decrease in maintenance cost, or an increase of recovered heat from the assumed 44 percent of input fuel to 50 percent. It is clear, then, that the application of advanced engine, heat recovery, and system technology has the potential to improve the attractiveness of natural gas cogeneration, particularly in areas and for applications that are currently marginal. Needed improvements and their impacts on the items in Figure 40-1 can be broken down as follows:

- Efficiency—increased shaft efficiency decreases fuel operating costs, while increased efficiency of heat recovery increases cooling and heating savings.

- Capital cost—reduced equipment and installation costs directly reduce annual capital charges.

- Maintainability—longer routine maintenance intervals and time between engine overhaul reduce labor-intensive maintenance costs.

- Availability—high system availability must be achieved to ensure credit for electric demand savings.

- Flexibility—although not translated as a cost item, aspects such as space requirements, noise and environmental emissions, and temperature of recovered heat affect the range of applications for which cogeneration is suited.

PRIME MOVER ALTERNATIVES

The various prime mover types available today or under development possess the attributes above in varying degrees. Reciprocating gas engines, combustion turbines, and steam turbines are all being used today as cogeneration prime movers where their characteristics are most appropriate for the needs of the application. Stirling engines, Wankel engines, and fuel cells are now in the development phase, and will likely be seen in cogeneration systems in the future. Table 40-1 compares the characteristics of these engines in the 25–1000 kW size range, and Figure 40-2 compares the economics of their operation meeting a fixed 1 mmBtu/h thermal load, 6570 hr/yr.

In sizes below 1000 kW, although current production reciprocating spark ignition engines fueled by natural gas are not ideal in all respects, they are expected to be the overwhelming choice for at least the next five years and probably longer. The reasons for this are varied:

- Highest electrical efficiency of any production prime mover.

- Technology is well developed with over 90 years of experience.

- Many well-tooled engine manufacturers exist, resulting in competitive pricing, considerably lower than alternatives.

- Internal combustion engines provide high overall efficiencies due to the high percentage of rejected heat which can be easily recovered.

Table 40-1. Natural Gas Prime Mover Characteristics

Prime Mover Type	Electric Efficiency, %HHV	Overall Efficiency, %HHV	Recovered Heat Temperature, °F	Engine Cost, $/kW	Maintenance Interval, Hrs	Maintenance Cost, cents/kWh
Heavy Duty Reciprocating Engine	28	75	210*	250	750	1.0
Advanced Reciprocating Engine	32	75	240*	150	3000	0.7
Simple Cycle Combustion Turbine	20	75	350	400	3000	0.5
Recuperated Combustion Turbine	30	65	200	600	3000	0.5
Steam Turbine	15	85	350	300	4000	0.2
Stirling Engine	36	80	140	?	?	?
Wankel Engine	26	75	210	150	3000	0.7
Phosphoric Acid Fuel Cell	36	80	180	not appl.	4000	0.5
Solid Oxide Fuel Cell	45	80	400	not appl.	4000	0.5

*Temperatures 40°F higher can be attained if oil cooler is separately cooled.

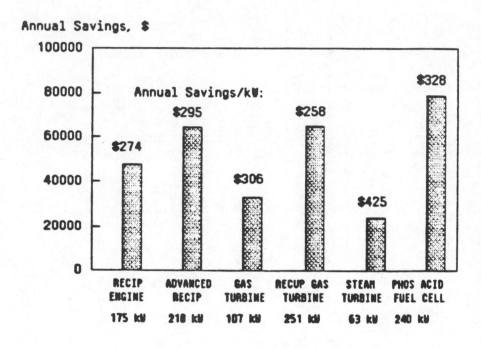

Figure 40-2. Impacts of Prime Mover Characteristics on
Operating Economics

- Life and reliability of gas-fueled, spark-ignition engines is good
  and is constantly being improved.

The reciprocating engine is, though, a complex machine requiring
considerable periodic maintenance and has the drawback of two
waste heat streams (coolant and exhaust), which complicates the
system somewhat.

The Gas Research Institute (GRI) is strongly committed to advancing natural gas cogeneration technology and is working with
major system and component manufacturers to accomplish these
needed improvements.

Advances in prime mover technology will promote penetration of
cogeneration into a much broader segment of the market. GRI's
strategy emphasizes reciprocating engines as the prime mover of
choice for gas-fueled commercial cogeneration because they are

currently the most efficient and lowest cost machine in the 50–1000 kWe range. Also, their liquid-fueled counterparts are already receiving substantial private and government support. GRI's efforts focus on research that is unique to gas-fueled engines, yet which allows gas to take advantage of evolutions in diesel engine technology. This strategy will enable the gas industry to reap the benefits of other useful developments.

Advances in reciprocating gas engine ignition systems, special low ash engine oils formulated for natural gas engines, automatic valve lash adjustors, and long life spark plugs have substantially increased the potential for increased intervals between servicing of natural gas engines. GRI-sponsored R&D programs currently in progress have goals of 4000-hour service intervals and 24,000 hours between top-end overhauls. The feasibility of 4000-hour service intervals has been demonstrated in lab tests by Thermo Electron Corp. and field testing started mid-1986 on two new 75 kW cogeneration packages with plans to implement the 4000-hour service interval.

Realization of the 24,000-hour goal between top-end overhaul is somewhat longer range, and requires breakthroughs in materials. The primary goal of a GRI-sponsored R&D program with Southwest Research Institute is to find a solution to the exhaust valve and valve seat wear problem, which exists to some degree on all high output gas engines. Elimination of the valve and valve seat wear, or a significant reduction, will allow increased service intervals for valve lash adjustment on those gas engines which do not use automatic valve lash adjustors (hydraulic tappets) and will also allow increased intervals between top-end overhauls.

Since valve lash adjustment is the most frequent and most demanding service need on many gas engines, a solution to the valve wear problem is seen as a major first step in realization of the goal of 4000 hours between scheduled service intervals. Experience to date and the potential of such concepts as automatic valve lash adjustment, synthetic oils, onboard centrifugal oil cleaners and automatic oil addition make even longer service intervals conceivable. These could be combined with diagnostics such as overall system efficiency and oil analysis, to tell the engine user when service is needed rather than servicing the engine at regularly scheduled intervals.

Waukesha Engine Division of Dresser Industries is performing R&D for GRI on a lean-burning gas engine that incorporates a prechamber for ignition of the oxygen-rich main chamber fuel/air mixture. Compared with existing engines, these improvements should yield higher efficiency and a reduction in emissions. Targets for the Waukesha project are a 32 percent (HHV) electric efficiency and reduced heat rejection to engine jacket coolant.

Ricardo Consulting Engineers are performing theoretical and laboratory R&D on a subcontract to Waukesha Engine Division. Work to date includes computer simulation of gas engine lean operation with a prechamber, constant volume combustion bomb experiments to provide experimental verification of the model results and most recently, implementation into a single cylinder test engine. Results will be transferred to a multicylinder engine by Waukesha.

Another reciprocating gas engine project, with Caterpillar Tractor Co., also employs lean combustion but in a "fast-burn" system. This design is more difficult than the prechamber concept because the lean mixture is ignited in a compact, highly turbulent open chamber. Early plans show natural gas/air mixing upstream of the turbocharger virtually eliminates cylinder-to-cylinder air/fuel variation. This will also allow the engine to use gas supply pressures below one psi without requiring a fuel-gas compressor. The jacket coolant temperature may be pushed as high as 270°F, and electric efficiency is targeted at 33 percent (HHV). Prototype experimentation is currently underway.

In a much longer term effort, Eaton Corporation, a major manufacturer of engine components, is investigating methods of high-pressure fuel injection and ignition during the engine power stroke. This would permit the use of natural gas in a "diesel" cycle, which allows high compression ratios and is thus more efficient than a spark-ignition engine, Initial work will focus on determining injection pressures and ignition mechanisms, as well as examining the trade-off between central compression versus unit injectors. Early results appear promising, and will be confirmed in single cylinder engine tests shortly.

GRI's R&D program will also be looking at applying the injection/-ignition system to engines incorporating ceramic materials that are still in the development stage. This could push electric efficiencies still higher and, eventually, reduce or even eliminate the need for cooling the engine. This would make the system smaller, less complex, and easier to maintain.

Garrett Corporation is developing a 50 kWe gas turbine based cogeneration prime mover for cogeneration application. This state-of-the art system will be highly recuperated for high electrical efficiency and through such advanced concepts as air bearings, a shaft speed alternator, and catalytic combustion will provide many features to make it ideally suited for the varied demands of the cogeneration market. These features include high reliability, low maintenance requirements, low noise and vibration, low fuel gas pressure, low NO_x emissions, load following capabilities, and good part-load efficiency.

The phosphoric acid fuel cell, subject of a major development area at GRI in itself, is another "prime mover" technology for cogeneration being advanced by United Technologies Corporation. In a fuel cell, natural gas which has been reformed to hydrogen reacts electrochemically with air to produce DC power and heat in a solid state (no moving parts) process. Unlike heat engines, fuel cells have no Carnot limits on conversion efficiency, although practical size and cost constraints limit phosphoric acid fuel cells (phosphoric acid refers to the electrolyte employed) to an electric efficiency of 36 percent (HHV). Currently completing a 46-unit field test, this technology should begin to be commercially available in the late 1980s.

Since the prime mover is the "heart" of a cogeneration system, the cost and performance improvements projected will have a significant effect on the cost-effectiveness of natural gas cogeneration and will help to preserve cogeneration as a long-term energy and cost-saving option.

Reprinted with permission from the Proceedings of the 21st Intersociety Energy Conversion Engineering Conference. Copyright 1986. American Chemical Society.

Chapter 41

Closing the Grid-Interconnect Gap

J. R. IVERSON
Onan Corporation

The activities involved in assessing the performance requirements for interconnecting small (40 kW to 800 kW) gas-fired engine-driven cogeneration packages to the electric utility grid and a selection of related key technical issues are presented in this chapter. Results of the efforts described will be used to evaluate alternative methods and technologies to achieve the level of performance required for reliable, safe, and effective parallel operation of small cogenerators with utility systems.

Although cogeneration is not a new technology nor parallel operation of dispersed generators with the electric utility a new concept, no standards for interconnection have been developed due to the wide variation in utility circuit configurations with which cogenerators may be connected. Faced with a wide diversity of potential requirements and a certain degree of unfamiliarity with utility system operational and protection practices, manufacturers of small packaged cogeneration systems are faced with a knowledge/information gap in developing cost-effective interconnect equipment, an element of the package that is not of primary interest to them, but, nevertheless very important.

As shown in Figure 41-1, the interconnect contains several functions key to the utility and on-site load interfaces.

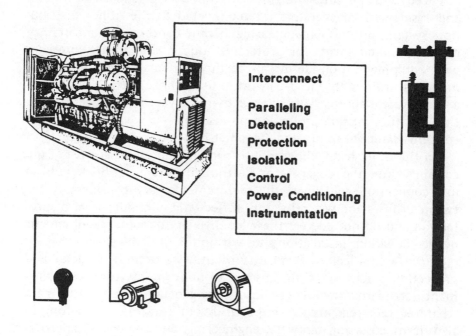

Figure 41-1. Interconnect Functions

Although most of the installed capacity has been in multimegawatt industrial systems, a growing number of commercial and institutional energy users have applied systems of less than 1 mWe capacity to take advantage of cogeneration efficiency in their buildings. Cogeneration in the commercial and institutional sector could have an ultimate potential on the same order as the industrial sector, but broad penetration requires the research and development of advanced energy conversion technology.

The Gas Research Institute (GRI) has a very active R&D program for packaged cogeneration systems and advanced component technology. This project effort resulted from a recognition that the cost of utility interconnection (hardware and approval process) becomes an impediment to developing these small system applications to their fullest market potential.

Faced with potentially significant numbers of small commercial and residential cogenerators interconnected to the utility distribution system, utilities experience significant concern and uncertainty regarding requirements for protective functions in utility interconnect equipment. Considering all of the related responsibilities and pressures utilities are faced with; it is no surprise the utilities follow a conservative approach using their own conventional power generation and transmission protection philosophies to cover their primary concern—safety of utility and customer personnel.

On the other hand, the grid interconnect represents an area of uncertainty for the cogenerator. Although several utilities publish interconnection guidelines, the requirements vary considerably between utilities and generally only reflect the minimum level of protection and do not address the wide diversity of actual requirements needed at various installation sites within the utility system.

Faced with a lack of familiarity of utility system operational and protection practices and the absence of uniform interconnect performance standards, small cogenerators are generally relegated to meet whatever requirements are imposed by utilities, often resulting in significant costs for hardware, engineering, and the overall approval process. In the size range of interest, the interconnect represents a gap between utility and cogenerator interest and a significant cost barrier. GRI contracted this project effort with the following objectives in order to close this gap:

(1) Develop performance standard requirements for both synchronous and induction systems based on a detailed examination of grid-interconnect issues.

(2) Identify advanced technology options for generator systems (including protective devices) meeting the performance-based requirements at reduced costs and/or when higher safety and reliability than present commercially available systems.

The objectives require grid-interconnect issues be examined, performance-based requirements be generated, existing grid-interconnect hardware be evaluated in terms of the performance-based requirements, and advanced technology concepts be identified to reduce cost and/or increase reliability.

UTILITY/COGENERATOR PARTNERSHIP

The primary justification for small cogeneration installations is the increased energy conversion efficiencies associated with satisfying both on-site electric and heat loads from a single fuel energy source. Small cogeneration installations typically are complicated by the fact that on-site electric load profiles are not totally consistent with thermal load needs (variable power to heat ratio) as shown in Figure 41-2. If generation capacity is set to match thermal needs, the cogenerator needs the utility as a partner to provide a source of load during periods of low on-site electrical consumption and as a supply of power during periods of high on-site electrical consumption. The need for parallel operation with the utility is critical and the interconnection interface is very important.

In all cases, the cogenerator recognizes his responsibility to maintain quality of electric service to nearby customers, cooperate in utility relaying to clear faults, and avoid the hazards of feeding a dead utility line. Additionally, the cogenerator is clearly interested in protecting his own generation equipment. Protective relaying is mutually beneficial, but care must be taken that the cost is not disproportionately high and that nuisance outages do not unreasonably disrupt the cogenerator's facility or cause peak demand charges to be incurred. Clear, consistent, and timely communication between the cogenerator and the utility on interconnection requirements is important to effectively implement cogeneration.

UTILITY CONCERNS

In order for cogenerators to effectively communicate with utilities and discharge their responsibilities in this partnership, they need to be acutely aware of the multitude of utility concerns and capable of responding positively to the related issues. Review of utility

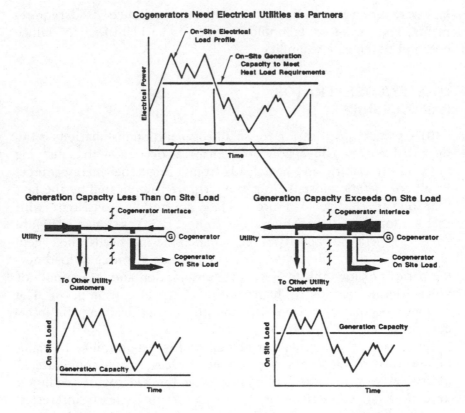

**Figure 41-2. Impact of On-Site Power to Heat Ratio
on Power Flow at Utility Interface**

interconnection guidelines and other studies/reports reveals the following major utility concerns:

(1) *Safety*—Utilities are dealing with one of the most useful but inherently dangerous commodities, electricity. Therefore, protection to customers, utility personnel, and the general public is of paramount concern. Interconnected cogenerators, being another potential source of power even during a utility system interruption, can significantly complicate efforts to achieve this objective.

(2) *Power Quality*—Maintaining acceptable power quality in utility systems is essential to providing power to customer loads cost effectively, and for sensitive electrical loads, for the proper operation of customer equipment. Power quality is measured by voltage level, frequency, power factor, and harmonic content.

(3) *Equipment Protection*—Utilities have a heavy capital investment in equipment to meet the objectives above resulting in a final objective—that of protecting all of this equipment within anticipated ranges of load, fault current, and voltage.

(4) *Fault Detection and Isolation*—Utility systems are subject to a variety of temporary and permanent faults, principally short circuits, grounds, and open conductors. The entire protection scheme is configured to detect these faults quickly, isolating faulted circuits to the minimum extent possible, limiting subsequent equipment damage and disturbance to the remainder of the system.

UTILITY SYSTEM OVERVIEW

To gain appreciation for these concerns, it is important to achieve some level of understanding for utility system configurations and protection practices.

System Configuration—As represented in Figure 41-3, typical utility systems are configured to transmit power from a central generating station (where power is generated at voltages up to 20 kV) over many miles (at voltages from 69 kV to 765 kV) to locations where the power is distributed to load consuming customers (larger users at primary distribution voltages from 4 kV to 34.5 kV and smaller customers at secondary distribution voltages up to 15 kV).

The vast majority of cogenerators will be interconnected to radial distribution circuits (power from a single generation source). Few will be connected to network systems used to supply power to high load density areas from multiple power sources. Therefore, emphasis is placed on radial systems with special requirements noted for network applications.

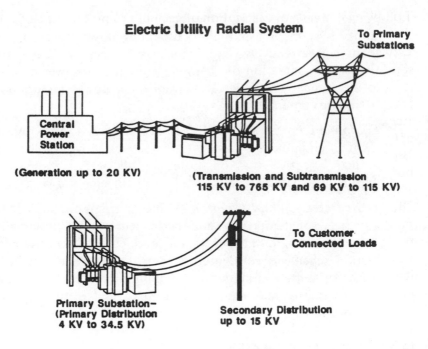

Electric Utility Radial System

To Primary
Substations

Central
Power
Station

(Generation up to 20 KV)

(Transmission and Subtransmission
115 KV to 765 KV and 69 KV to 115 KV)

To Customer
Connected Loads

Primary Substation—
(Primary Distribution
4 KV to 34.5 KV)

Secondary Distribution
up to 15 KV

Figure 41-3. Electric Utility System Configuration

Protection Practices—Utilities employ a number of devices including circuit breakers, reclosers, sectionalizers, and fuses throughout the system to provide selectively coordinated protection against faults. These devices, their ratings and the operating characteristics are carefully chosen on the basis of detailed fault current studies to detect faults, interrupt temporary faults and isolate permanent faults as close to fault locations as practical. The primary objective is to minimize disturbance to the remainder of the system and subsequent equipment or property damage as well as secure safety of customers, public, and utility personnel.

Since the majority of faults on utility systems are transient or temporary in nature (faults that are either self-extinguishing or are interrupted and do not reoccur after interruption), reclosing is a commonly applied practice in utility systems. This practice creates an area of great concern for utilities and cogenerators.

Following is a general description of each of the protective devices:

Circuit Breakers—Circuit breakers are the devices on the power system with the highest interrupting capability and voltage ratings. The control intelligence of a power circuit breaker is an external accessory. This intelligence may be provided by a simple control switch, an overcurrent relay, or by an entire relay scheme that will supply tripping and closing signals to the breaker.

Breakers are generally applied at generation, transmission and distribution substation facilities. Several different types are used employing different construction and arc interrupting media depending on the current interrupting capability required and the voltage of the circuit. Breakers are used to open or close circuits under all anticipated fault conditions either manually or automatically. Breakers are used for reclosing with supplemental external control logic in circuits of high voltage and/or high available fault currents with operating characteristics similar to those described below for reclosers.

Reclosers—The automatic circuit recloser is a self-contained device with intelligence to sense overcurrents, to time and interrupt fault currents and to reenergize the line by reclosing automatically, restoring service after a momentary outage. If a fault is permanent, the recloser locks open after a preset number of operations, usually three or four, and isolates the faulted section from the rest of the system. Interrupting capability and voltage ratings are not as high as for circuit breakers. They are, therefore, usually applied on distribution systems where the available fault current is lower.

For overhead distribution systems, approximately 80 to 90 percent of all system faults are temporary in nature and most last only a few cycles to a few seconds. If the line can be tripped open momentarily, a subsequent reclosing very likely will be successful. By then, the cause of the fault would be gone. The recloser eliminates prolonged outages on distribution systems due to temporary faults or transient overcurrent conditions.

A recloser responds to fault current with an interruption as shown in the typical recloser operating sequence of Figure 41-4 which represents a four-shot recloser. Reclosers can be applied with from one to

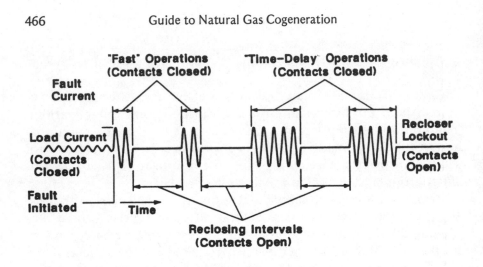

Figure 41-4. Typical Recloser Operating Sequence

four reclosing operations generally with a combination of instantaneous and time delayed operations.

In this sequence, the initial two responses are instantaneous allowing fault current to flow for only two cycles to interrupt temporary faults. If, however, the fault is permanent in nature, the recloser will continue to try to clear the fault using the longer time delay operations. If the fault is not cleared, the recloser will ultimately interrupt flow of fault current and lock out to isolate the section of the circuit down-stream from the recloser.

A recloser has a time/current control operating characteristic that operates the recloser in very short periods of time for high levels of current and longer periods of time for low levels of current. The recloser will not operate with levels of current below the minimum sensitivity or pick-up point of the recloser.

To further minimize the extent of service outages, sectionalizers and fuses are used in a coordinated fashion with circuit breakers and reclosers to isolate circuits containing permanent faults. Descriptions of these devices follows:

Sectionalizers—The sectionalizer must be used in conjunction with an interrupting device such as a recloser or circuit breaker. The sectionalizer is not capable of interrupting fault current, but merely capable of sensing the interrupting operations of an upstream interrupting device such as a recloser. Figure 41-5 illustrates the coordinated operation of recloser and sectionalizer. A sectionalizer counts the number of reclosing interrupting operations. Following a preset number of counts, the sectionalizer opens while the recloser is open isolating the fault, allowing the recloser to once again reclose after the sectionalizer is opened resuming the flow of normal load current to down-stream circuits.

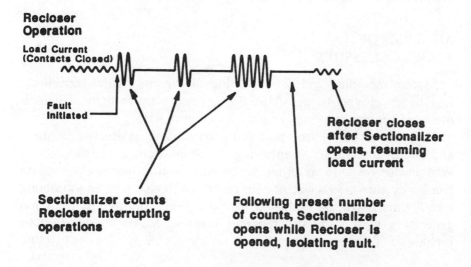

Recloser Operation

Load Current (Contacts Closed)

Fault Initiated

Recloser closes after Sectionalizer opens, resuming load current

Sectionalizer counts Recloser interrupting operations

Following preset number of counts, Sectionalizer opens while Recloser is opened, isolating fault.

Figure 41-5. Recloser/Sectionalizer Coordination

In the sequence shown, the recloser opens and the sectionalizer counts one. The recloser then closes back in. If the fault was temporary, it will have been cleared without any prolonged outage. If the recloser is set for four operations, and the sectionalizer set for three operations, the sectionalizer will lock open after three recloser operations for a permanent fault.

Fuses—Power fuses consist of a fusible element and an arc-extinguishing means. The fusible element is made of silver, tin, copper, aluminum, or some alloy. The fusible element is designed to carry rated current continuously without damage to the element. The element melts open with a specific time-current characteristic on overloads or short circuits.

With proper method of arc extinction and element shape, current limiting action can be achieved opening the circuit before the current can reach the crest of the first half-cycle or short-circuit current, limiting the maximum fault current. Fuses are only expected to operate on permanent faults and are coordinated with reclosers such that reclosers interrupt temporary faults prior to fuse operation.

ANALYSIS OF
TECHNICAL ISSUES

System modeling and analysis using analog and digital techniques will be employed to ascertain the necessary performance requirements of the interconnection shown in Figure 41-6.

Major emphasis will be placed on the sensing, protective controls and circuit interrupter, identifying the desired characteristics to prevent and/or respond to abnormal conditions of voltage, current and frequency for various circuit configurations likely to be encountered. Requirements will be evaluated for both synchronous and induction generators within the size range of interest. Major technical issues include the following:

Islanding—Islanding is one of the greatest utility concerns and one of the more difficult to prevent. Islanding occurs when the utility system is separated (usually as a result of a fault condition) from the circuit to which a cogenerator is interconnected and the cogenerator has adequate capacity to carry the load connected to the isolated circuit. This condition is of particular concern for synchronous generators capable of self-excitation. Although less concern exists for induction generators, islanding can occur for a long enough duration with the right combination of power factor correction capacitors and connected load. Generally, though, the greater concern for induction generators following utility separation is rapid development of overvoltages.

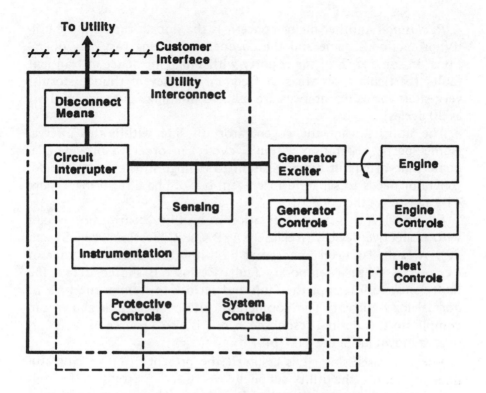

Figure 41-6. Interconnection Elements

The electric utility is concerned with islanding when other customers' loads are being supplied by a cogenerator especially if there is a possibility of going undetected by protective devices. Personnel safety and quality of service (voltage level and frequency) are the main concerns.

The basic philosophy used by many utilities to minimize the probability of islanding for small cogenerators is to ensure the cogenerator is connected to a circuit that contains a large amount of load compared to cogenerator capacity. This often requires a dedicated transformer to connect the cogenerator to a major distribution feeder.

Reclosing—Another major concern is the impact on both the utility protection scheme and the cogenerator during reclosing operations. In some areas of the country with a high incidence of transient faults (particularly in areas of frequent and severe thunderstorms) very short reclosing intervals are used to minimize outages (as short as 30 cycles).

The utilities want the cogenerator off line within this interval (some specify separation within 9 cycles) in order to provide adequate time of circuit deenergization to extinguish the fault. The cogenerator needs to separate to prevent out-of-phase reclosing. Figure 41-7 illustrates the concerns.

Figure 41-7(a) is a representation of a utility distribution circuit with protective devices (Reclosers A, B, and C; Sectionalizers S1 and S2). A fault has been imposed at F2 within the protective zone of Recloser B. With a temporary fault, Recloser B should detect the fault, open to interrupt the fault and reclose to resume the flow of normal load current. The cogenerator interconnected as shown can complicate this desired response since it is also a source of fault current as shown in Figure 41-7(b).

For this discussion, it is assumed the presence of the generator does not reduce the utility supplied level of fault current to the fault below the minimum sensitivity of Recloser B. This can, however, also be a potential problem in fault detection. As stated earlier, the cogenerator needs to separate as soon as possible from the utility system to:

- Allow a temporary fault to be extinguished during the recloser open interval

- Allow the recloser/sectionalizer to isolate a permanent fault

- Prevent cogenerator damage due to out-of-phase paralleling when the recloser closes following an initial interruption

Since the cogenerator is likely to be connected to the utility system where the connected load is large relative to generator capacity, the combination of the fault and connected load will cause a rapid decrease in frequency and/or voltage as reflected in Figure 41-7(c). After a short period of time (as little as 2 cycles) the generator will

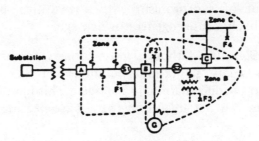

(a) Distribution Circuit with Fault F2 Imposed within Recloser B Protective Zone

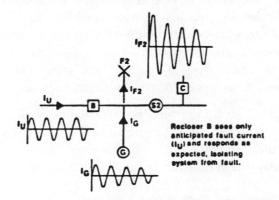

(b) Fault Current Flow from Utility and Cogenerator

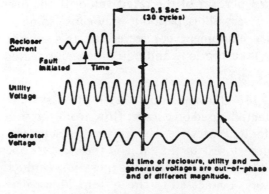

(c) Out-of-Phase Reclosing

Figure 41-7. Fault Reclosing Sequence

have gone out of synchronism with the utility and subsequent reclosing will likely cause cogenerator damage.

Harmonic Currents—Although most concern for harmonics exists for inverter systems, saturation and imperfect flux distribution in generators also produce harmonic voltages. Third and other odd harmonic voltages can be large enough to cause heating from circulating currents.

Third harmonic currents (and multiples called triplen harmonics) originating at the same three-phase source have the same relative phase angle independently of the phase in which they are flowing.

If the generators are connected directly to the distribution system, triplen harmonic currents may flow in the distribution feeder if the generator and utility system neutrals are grounded. One way of avoiding this is to use a transformer to isolate zero sequence currents.

Overvoltage—Distribution feeders are usually grounded only at the main substation. When the substation breaker opens on a single line to ground fault, overvoltages occur if a generator remains connected through a distribution transformer connected in delta on the primary side.

Voltage sensing may solve the problem, but if the generator is large enough, the voltage between the unfaulted phases and ground may reach 173 percent of the rated value. If the overvoltage is present even for a very short period it may cause some damage to arresters and other equipment connected to this circuit. Conventional relaying may take too long and equipment connected line-to-neutral may be damaged.

Nuisance Fuse Blowing—Most distribution systems are radial and they are protected based on current flow from the source (substation) toward the fault. Protective devices are time-current coordinated so that the device closest to the fault is the first to operate and isolate the fault. With fault supporting cogeneration (induction or synchronous generators) connected to the system, abnormal conditions such as faults, will have short-circuit current contributions from the substation and from the additional dispersed sources which may affect the coordination.

An example of distribution feeder protection is shown in Figure 41-8. For a temporary fault downline from the lateral fuse, the recloser back at the substation will interrupt the circuit allowing the fault to be cleared and then reclosing to restore service after a minimal interruption. For permanent faults, the lateral fuse is expected to confine service interruption to the smallest area possible. The recloser is set to operate faster than the fuse for the initial fault and slower than the fuse after an unsuccessful reclose has been attempted by the substation device. Typical time-current curves (TCC) for this example are included in Figure 41-9.

A cogenerator with the ability of contributing short-circuit current can cause the lateral fuse to blow before the recloser operates. This causes service interruption to an area which, although small, would not be affected if proper coordination was maintained. Investigators have found this problem more prevalent on systems where the dispersed generation capacity exceeds 1000 kW.[1]

Resonant Overvoltages—Resonant overvoltages can result from the isolation of a synchronous generator with some capacitance (power factor capacitors or line capacitance) during a single line to ground fault. Some investigators have found that conventional sensing is not fast enough to avoid equipment damage. Therefore, surge arresters may be needed to protect against this condition.[1]

Grounding—Power system grounding involves intentional connection between the phase conductors and ground. In general, there are four types of grounding on a distribution system:

(1) Ungrounded

(2) Solidly grounded

(3) Resistance grounded

(4) Reactance grounded

The type of grounding on a distribution circuit has a direct effect on system performance in terms of ground fault current, harmonic currents, and overvoltages.

For illustration purposes, only solidly grounded and ungrounded systems will be considered here.

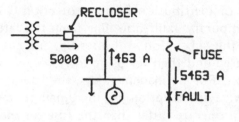

**Figure 41-8. Fault Current Supplied by
Utility System and Cogeneration**

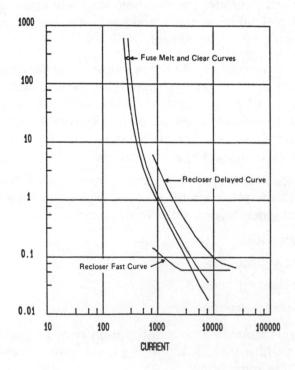

Figure 41-9. Time-Current Curves for Figure 41-8

Ground Fault Currents—A common utility practice is to have the grounded neutral of the substation transformer as their only ground. This practice permits ground fault current to flow radially from the substation to the fault point, making the coordination of protective devices an easier task. Having more than one grounding point on the feeder may permit the flow of ground fault current in more than one direction as shown in Figure 41-10.

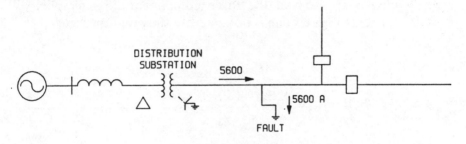

(a) System with Only One Grounding Source

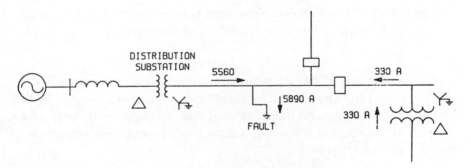

(b) System with Two Grounding Sources

**Figure 41-10. Flow of Ground Fault Current on
Distribution Feeder**

After effect of having more than one grounding source depends on the transformer connection. Transformers connected grounded-wye/delta act as a zero sequence current source for ground faults on the grounded-wye side regardless of whatever is connected to the delta side (generator, load, etc.). An example of this is shown in Figure 41-11. Here, the main substation transformer is connected delta/grounded-wye. The other transformers on the feeder have different connections to illustrate the effect of them on the ground currents. The subscripts 1, 2, and 0 refer to positive, negative, and zero sequence quantities. At any point in the circuit, the total current flowing is the sum of the three components. For simplicity, only the currents whose value is not zero are shown in the figure.

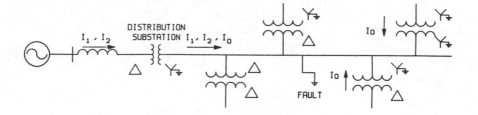

Figure 41-11. Flow of Zero-Sequence Currents on Faulted Feeder with Various Types of Transformer Connections

Another effect of having more than one grounding source, is the increase in total ground fault short-circuit current. Figure 41-10 shows the total ground fault current for a system with only one grounding source and the same system having more than one ground current source.

PROTECTION CHOICES FOR UNDESIRABLE CONDITIONS

As can be seen from the above discussion, there is no shortage of interconnection issues to be concerned about, especially considering the list is not complete. Other issues include the need for manual, visible isolating means; impact of cogeneration available fault current on utility protective device interrupting ratings; impact of cogenerators on feeder capacity; potential motoring of cogenerators; impact of utility available fault current and voltage surges on interconnect equipment and the cogenerator; impact of multiple feeders at the cogenerator; VAR/power factor control; potential of nuisance cogenerator outages as a result of protection sensitivity; etc. Detailed analysis continues to evaluate current practices for responding to the various conditions and evaluating the performance of hardware presently commercially available. Alternative methods will be evaluated including variations of present voltage, current, frequency, harmonic and sequence component detection, and investigation of alternate protection technologies including digital or microprocessor techniques.

REFERENCES

1. "Protection of Electric Distribution Systems with Dispersed Storage and Generation (DSG) Devices," ORNL/CON-123, prepared by McGraw-Edison Company for Oak Ridge National Laboratory, Oak Ridge, TN, September 1983.
2. John R. Parsons, Jr., "Cogeneration Application of Induction Generators," IEEE Trans. on Industry Applications 20, No. 3, pp. 497-503, May/June 1984.
3. "A Guide to Interconnection Requirements in New York State," NYSERDA Report 85-2, prepared by New York State Energy Research and Development Authority by Self Reliance, Inc., August 1985.
4. Andrew Politis and Verlin J. Warnock, "Cogeneration Relay and Control Problems at the Customer-Utility Interface," Proceedings of the American Power Conference, Volume 44, 1982.

Reprinted with permission from the Proceedings of the 21st Intersociety Energy Conversion Engineering Conference. Copyright 1986. American Chemical Society.

Appendices

Appendices

Appendix A

Natural Gas Supply Availability

GEORGE H. LAWRENCE
American Gas Association

There are currently two basic problems facing the natural gas industry
with respect to supply. The first is *current excess supply*. It is causing
low spot prices, pressure for more carriage, an advantage for transportation gas over system supplies, and shut in gas which—with collapsing
oil prices—is discouraging drilling activity. Also, these same low spot
prices, along with pressure for transportation gas over system gas (for
both pipelines and distributors) are adding to the huge take or pay
problem, to the point that it is approaching a genuine crisis. Such a
crisis is not just one for the pipelines directly involved, but for their
LDCs and all consumers as well.

The second problem involves *concern with inadequate future
supply*. This is an old problem. It is one which, even with the excess
supplies of recent years, the industry has never been able to completely shed. The current lack of drilling has revitalized the fear of
future gas shortages which is discouraging potential new gas users
and depressing demand. While gas supply pessimism has been a chron-
ic problem faced by gas marketers, the current low rig count is
breathing new enthusiasm into negative supply assessments, and we
now have clear indications that these arguments are not being lost on
state regulators and potential customers.

No one can predict with exact certainty what the conventional supply situation will be in five years at current low levels of drilling activity. Our analyses in Appendix A indicate that there need not be any gas shortages.

No one can say with absolute certainty what the level of gas prone exploration activity will be "beyond the bubble" when supply and demand is in balance—particularly now that field price controls are off for new exploration and drilling. A.G.A. has reason to believe that market forces will work, and drilling in gas-prone areas will increase as demand increases relative to supply. This is especially true when most new market growth "beyond the bubble" is not expected to compete just with residual fuel oil, but with higher priced alternatives. Therefore, beyond the bubble, the huge natural gas resource base should be developed by more drilling that is increasingly gas prone. This is a likely future, although it can't be predicted with certainty.

What *is* certain is the industry's current problems: lack of demand, oversupply and growing take-or-pay liabilities. The current rash of negativism about supply works against their solution. With this as an introduction, the discussion will now focus on some revealing data.

GAS SUPPLY SHORT-TERM

The Bubble

A.G.A.'s current estimate of excess production capability (the bubble) is about 3 Tcf/yr. See Figure A-1. The Energy Information Administration (EIA) of the Department of Energy, estimates a six-month bubble (for gas under contract only to interstate pipelines) at 2.3 Tcf for the period July 1 to December 31, 1986. This is up from 1.8 Tcf for the last six months of 1985. EIA's six-month number includes some Canadian gas and may include some double counting; however, this EIA number is significant. It confirms the A.G.A. numbers and shows that in 1986, the bubble is not decreasing *and* that the bubble is moving almost exclusively to system gas.

- A.G.A. published its latest bubble analysis in early 1986. The purpose was to point out that; (1) at current levels of

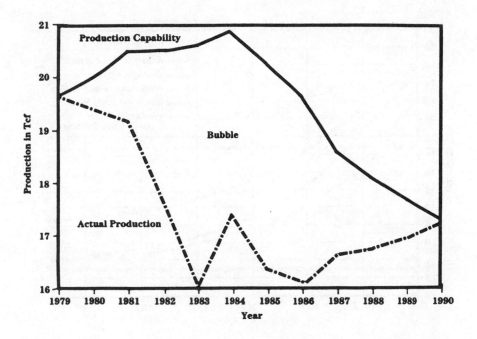

Figure A-1. Production Capability versus Production
1979–1990

exploration and drilling activity, the bubble would indeed diminish by the winter of 1989/90 to the point of near nonexistence, and; (2) that this was a clear signal that supply and demand would come into balance and that there should be steps taken toward regeneration of exploration and drilling activity.

Extended Bubble

In a recent analysis, ENRON Corporation evaluated the "extending bubble." Figure A-2 presents the ENRON assessment of how the gas bubble could last until 1991 if gas demand remains at current, depressed levels. A combination of excess production capability from producing gas wells (currently 3 Tcf), plus additional production

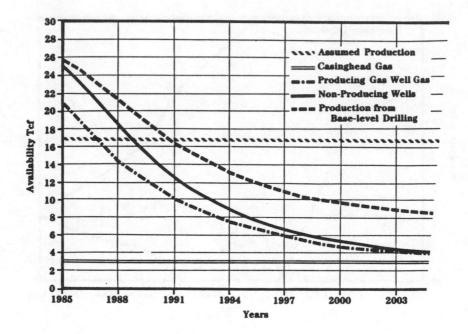

Figure A-2. Availability Analysis
17 Tcf Production Lower 48 States

capability from nonproducing reserves (both committed and uncommitted), and new reserves added at basement level drilling (6 Tcf/yr
additions) should maintain domestic production capability at a minimum of 17 Tcf through 1991. Not even the most persistently gloomy
forecasts predict less than 6 Tcf/yr of new reserve additions over this
period.

Beyond the Bubble

As the bubble contracts, several additional sources of gas will
become available to ensure that no shortage occurs during the
period it takes the exploration and production sector to turn around.
Under a scenario with higher energy prices ($25/bbl crude oil) these
incremental sources of gas could bring approximately 1.65 to 2.45
Tcf/yr to the marketplace within 12 months, as needed. Under an

extremely low priced scenario—that is, if present $15/bbl prices continue to 1990—gas that could be brought to market within 12 months from these sources could be as low as 1.15 Tcf/yr. This low price scenario is discussed at the end of this appendix. Table A-1 presents estimates of the incremental short-term gas supply potential under a high world oil price scenario ($25/bbl).

Table A-1. Possible Incremental Short-Term Gas Supply Potential at High World Oil Prices (in Bcf)

	Within 12 Months
Uncommitted Nonproducing Reserves	500–1,000
Canadian Gas[1]	300
Accelerated Infill Drilling	100–400
Mexican Gas	200
LNG[2]	50
Total Additional Supply Available	1,650–2,450

[1]Volume is in addition to the 700 Bcf anticipated to be imported in 1986. By 1990, a number of Canadian gas import projects currently pending regulatory approval should be concluded, increasing additional short-term potential from this source from 800 Bcf/yr to 1,300 Bcf/yr.

[2]Volume is well below the physical limitations of the established terminals. This 50 Bcf could be doubled given adequate time to revitalize existing terminals.

- *Uncommitted, Nonproducing Gas Reserves*

  These were estimated by EIA to be 9.0 Tcf at year-end 1984. Production from these reserves when developed and hooked-up to the transmission network could supply between 0.5–1.0 Tcf annually. This gas is not part of the A.G.A. bubble estimate. The low level of production relative to the size of the reserves reflects a time lag in bringing all this gas to market.

- *Canadian Gas*

  U.S. companies currently have contracts with Canadian pipelines and producers to import up to 1.9 Tcf/yr. Considering seasonal and physical deliverability constraints, it is estimated that 1.5 Tcf/yr of Canadian gas could be physically imported into the U.S. (with some minor additional pipeline construction). Given that current Canadian imports are about 700 Bcf/yr (1986 estimate), 0.8 Tcf/yr of additional gas could be made available to U.S. purchasers and up to an additional 1.3 Tcf/yr with major new construction.

- *Accelerated Infill Drilling*

  It has been a point of some discussion as to how much infill drilling will increase reserves, but all will agree that infill drilling will increase production capability, and all will agree that when the bubble is gone there will be more infill drilling.

  Furthermore, no doubt exists that several of the largest gas fields in the country remain amenable to infill drilling to increase production capability, and in some cases, proved reserves. Nationwide accelerated infill drilling could quickly increase production capability by 0.1–0.4 Tcf/yr.

- *Mexican Gas*

  The political uncertainty about Mexican imports is fully understood, but it is also known that Mexico has huge associated gas reserves. The potential *is* there, and the export of gas does *not* divert gas from Mexican industry. Large volumes of gas are being reinjected now and the gas-to-oil ratios are increasing. Before suspension of shipments due to pricing considerations, Mexican gas imports to the U.S. averaged 300 Mcf/day (100 Bcf/yr). This amount could be doubled to 200 Bcf/yr with existing pipeline capacity.

- *LNG*

  There are four LNG terminals which have been operational and one or more of them could become operational again. Collectively, these terminals have the capacity to handle 770 to 900 Bcf per year. In the event that additional gas supplies were needed in the U.S., 50 Bcf could be anticipated from LNG.

NATURAL GAS RESOURCE ASSESSMENT

The conventional natural gas resource base is equivalent to about *50 years* of gas supply at the current production rate. (See Figure A-3.) The resource base for near-term, currently producing *un*conventional sources (i.e., coal seam gas, Devonian shale gas, western tight sand gas and enhanced gas recovery from coproduction of gas and brine in watered-out gas fields, etc.) is two to three times the size of the conventional resource base. The pace at which these resources are developed (both conventional and unconventional) will depend on the economics of gas exploration and production, and the application of new technologies.

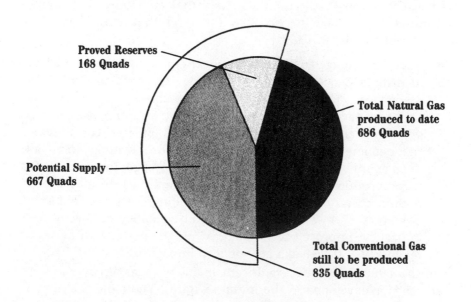

Proved Reserves 168 Quads

Total Natural Gas produced to date 686 Quads

Potential Supply 667 Quads

Total Conventional Gas still to be produced 835 Quads

Figure A-3. U.S. Natural Gas Conventional Resources for the Lower 48 States

The long-term gas supply issue should therefore focus on the timing of resource development, not on the size of the resource base. This resource development issue is a matter of economics—are future gas prices, relative to the costs of production, sufficient to permit aggressive gas-prone exploration and production programs.

- According to a new study by the Potential Gas Agency for GRI, at an average cost of $3 per Mcf (1984 $), recoverable potential lower-48 resources are 535 Tcf—about 83 percent of the total lower-48 potential. Adding the 535 Tcf potential to 160 Tcf of proved reserves (163 Tcf in 1984) equals 695 Tcf. If divided by 17 Tcf/yr of production it results in *40 years* of supply at finding costs of $3/Mcf or less. In today's market, $3 per Mcf is about twice the current spot price *but* it is also just half of what was commonly projected as minimum field prices just a few years ago. The Potential Gas Agency's findings are in general agreement with the ENRON study that concludes that finding costs today have returned to the level that existed in 1974 (about $2/Mcf in 1985 $).

- National average finding costs appear closely related to the level of drilling occurring (see Figure A-4). As more difficult (e.g., deeper water) areas are explored, the increased cost of exploration and development tends to be offset by improved technology. Further as the intensity of drilling increases, costs rise. As rigs become available during slack periods drilling costs fall. Thus supply and demand for attractive acreage, for rigs, and for other products and services, tends to be a critical influence on finding costs. When examining gas finding costs over extended periods, the upward pressure exerted by geology has been offset in part by improved technology; while the boom/bust cycles of the industry have had a pronounced impact on cost.

In the 1980–85 period, significant major new gas discoveries were made and/or developed in the lower-48 states. These indicate that at the levels of activity that existed through the 1980–85 period, adequate supplies can be made available.

While there is confidence that the drilling levels that existed in the 1980–85 period were sufficient, current prices and takes appear

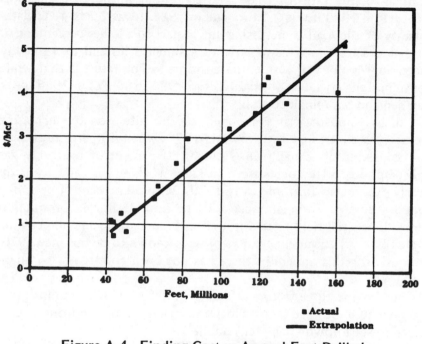

Figure A-4. Finding Cost vs Annual Feet Drilled
United States—Lower 48 States

*in*adequate to sustain either proved reserves or the current production capability level. That's why the bubble will be dissipating. Gas-well completions through May 1986 were down 27 percent relative to 1985. It is impossible to determine how much of the downturn in drilling is related to low prices and how much to a lack of buyers for long-term supplies; but overall the level of drilling is too low.

SEASONAL GAS SUPPLIES

Part of the discussion on the possibility of shortages focuses on the adequacy of seasonal and peak day deliverability. This is a question of: (1) deliverability in the field; (2) pipeline capacity constraints; and (3) storage output capacity.

In 1970, the ratio of peak sales to minimum monthly sales was 1.9 to 1; in 1980 it was 2.1 to 1 and in 1984 it was 2.7 to 1. The seasonality of the load is increasing due in part to a lower percentage of industrial sales which contribute to summer sales and to a steadily increasing number of residential customers. The result is that while consumption has fallen from 20 Tcf in 1980 to 17 Tcf in 1986, peak day demand has remained stable.

The gas industry has responded to this situation by increasing storage. As contrasted with the mid-1970s when 40 percent of peak day consumption came from storage, 50 percent of peak day gas consumption today comes from storage. Working gas capacity is up nearly 50 percent from the early 1970s. With 50 percent of peak day supplies coming from storage, a 10 percent decline in production capability in the field results in only a 5 percent decline in peak deliverability. As to pipeline capacity, it is known to be up since 1970, although for this appendix there has not been an attempt to quantify how much.

Thus, of the three factors affecting winter peak deliverability, two are favorable. Storage is up, pipeline capacity is up, and one is projected to start down—field deliverability.

There is the exception that on an historically cold day, distribution systems with poor storage capacity and/or tight pipeline capacity could have deliverability problems. However, these exceptions are not indicative of overall gas supply availability.

ECONOMICS OF SUPPLY

Both short- and long-term gas supplies will be impacted by the price of gas and the price of oil. Residual fuel oil prices are going to be a key influence on natural gas prices—and maybe in the short-term, a cap—but new gas markets will not compete with resid and will therefore diminish the importance of resid to gas pricing. However, a low oil price trajectory, at least in the short-term, implies a low natural gas price trajectory.

Low gas prices result in both less exploratory drilling as well as less gas from the transitional sources previously described. Certain transitional sources will not be materially affected by low gas prices (e.g., Canadian gas); while others will be severely impacted (e.g., LNG). Table A-2 presents estimates of the volumes available from transitional sources under a low energy price scenario.

Table A-2. Possible Incremental Short-Term Gas Supply Potential at Low World Oil Prices (in Bcf)

	Within 12 Months
Uncommitted Nonproducing Reserves	250–750
Canadian Gas[1]	800
Accelerated Infill Drilling	50–100[2]
Mexican Gas	—
LNG[3]	50
Total Additional Supply Available	1,150–1,700

[1] Volume is in addition to the 700 Bcf anticipated to be imported in 1986. By 1990 a number of Canadian gas import projects currently pending regulatory approval should be concluded, increasing additional short-term potential from this source from 800 Bcf/yr to 1,300 Bcf/yr.

[2] If the bubble extends to the 1989/90 winter period—as we expect—up to 200 Bcf/yr from this source is possible.

[3] Volume is well below the physical limitations of the established terminals.

Low energy prices mean less long-term supplies (lower exploratory drilling) and, as indicated, smaller volumes from transitional supplies. Still, even under a low price scenario, between 1.15 Tcf to 1.7 Tcf/yr of transitional supplies could be brought to market within 12 months when needed as the bubble dissipates.

CONCLUSION

In conclusion, the "transitional zone" between the dissipation of the bubble and the resumption of adequate levels of exploratory and drilling activity has a cushion of from 1.15 to 2.45 Tcf (and up to 3 Tcf/yr given new Canadian construction). When supply/demand market forces begin to operate beyond the bubble, it is anticipated that exploration activity will increase, uninhibited by field price controls which existed previously.

Given this scenario, the gas industry should not focus on shortages. Instead, the focus should be on the here and now problem of selling the oversupply of gas today and for several years to come. This is not to imply that the industry should neglect the ongoing activity needed in conventional drilling and the development of new drilling technology to support our genuine long-term supply goals.

There are others who, through pessimistic forecasts, are currently making gas supply a major issue. Concern over supply, if not constructively addressed, could seriously impair marketing efforts, thereby not only extending the bubble, but also hurting the long-term supply outlook. A contracting bubble is not a *bad* thing. With most of the domestic bubble being system supply today, any closing of the bubble will help reduce take or pay liability, the industry's biggest here and now problem. As emphasized in this appendix, an eventually balanced supply and demand does not spell shortage. Constant talk of shortages is neither a good message nor an accurate message to be sending prospective customers and regulators.

Appendix B

The Impact of Cogeneration on Gas Use

GAS DEMAND COMMITTEE
American Gas Association

EXECUTIVE SUMMARY

The results of the A.G.A. Gas Demand Committee forecast indicate that cogeneration demand has the potential to more than double by the year 2000. The cogeneration gas demand forecast in 2000 is 1.2 Tcf in the Attainable Case, 0.8 Tcf and 1.7 Tcf in the Business as Usual Case and Potential Case, respectively (see Table B-1). After steady growth in the 1985 to 1990 period, cogeneration should resume a rapid growth pattern after 1990. Tight electric utility capacity margins after 1990 will create an opportunity for more extensive use of cogeneration by industry.

Table B-1. Forecast Summary

	1985	1990			2000			2010		
		BAU	Attainable	Potential	BAU	Attainable	Potential	BAU	Attainable	Potential
Tcf	0.7	0.7	0.8	1.0	0.8	1.2	1.7	0.9	1.3	2.0
Gigawatts (GW)	10.8	10.8	12.3	15.4	12.3	18.5	26.2	13.9	20.0	30.9
# Sites	1,248	1,350	1,367	1,540	1,538	2,056	2,620	1,738	2,222	3,090

493

The forecast for 1990 is based on the view that the rapid growth in cogeneration experienced in recent years will slow down between now and 1990. The more gradual growth over the next five years is attributable to high-capacity margins on many electric utilities.

After 1990, the need for new electric generating capacity will stimulate investment in gas-fired cogeneration.

In 2010 the Attainable Case forecast is 1.3 Tcf. The BAU and Potential Cases are 0.9 Tcf and 2.0 Tcf, respectively.

The outlook for electricity supply and demand will determine just how extensive cogeneration growth will be. Cogeneration will be particularly sensitive to the outlook for new baseload capacity additions for coal and nuclear fuels, electricity imports from Canada and the infrastructure for long-distance power wheeling. The possibility of electric generating capacity shortages by the mid-1990s is considered a strong stimulus to cogeneration. In addition, the combination of higher electric rates and stable gas prices throughout the forecast period will improve the competitive outlook for natural gas cogeneration.

The economics of gas-fired cogeneration are superior to competing fuels for both large-scale industrial applications and for commercial applications. Despite the generally lower fuel costs for coal, natural gas-fired systems have lower annualized costs over a wide range of situations.

Cogeneration is the largest of the new markets for natural gas and has the potential to continue the rapid growth pattern of the 1980s. There are a number of factors that have contributed to this growth. The incentives under PURPA relating to the determination of buy-back rates for the sale of electricity by cogenerators, as well as the ability to buy power when needed from the grid, are key factors. Other important regulatory factors include the FERC policy on qualifying facilities and authorization to wheel cogenerated power.

ASSUMPTIONS AND METHODOLOGY

The market for natural gas-fired cogeneration is a function of numerous factors, both regulatory and economic. The cogeneration market is directly related to three major traditional markets: the commercial, industrial and electric generation markets. Competitive factors in these markets influence cogeneration potential as well. The economic competitiveness of cogeneration is a site-specific determination, but the same key factors are applicable to most cogeneration projects—capital costs, relative fuel prices, local electricity rates and avoided cost/buyback rate determinations.

REGULATORY FACTORS

The Public Utilities Regulatory Policy Act of 1978 (PURPA) served to both remove barriers and provide incentives for prospective cogenerators. An important incentive for cogenerators is the obligation of electric utilities to purchase electricity from a cogeneration facility based on the utility's avoided cost. The avoided cost forms the basis for the buyback rate, which is the price the cogenerator receives for electricity sold to the utility. State regulators have exercised their authority in recent years to implement the PURPA incentives in a manner that has promoted a full avoided cost basis (i.e., taking into account both avoided energy costs and avoided generating capacity costs).

For this forecasting effort, the BAU Case reflects buyback rates that are near the low end of the discretionary authority of the state regulators to determine avoided costs. For example, avoided energy costs, but not avoided capacity costs, are assumed to be included in the buyback rate. In the Attainable Case, full avoided cost treatment for buyback rates makes cogeneration projects moderately attractive. Even though full avoided costs are assumed there is the recognition that some states have already begun allowing utilities to negotiate buyback rates below full avoided cost out of concern for creating current excess generating capacity. The Potential Case forecast represents a scenario in which cogenerators receive full avoided cost treatment in the determination of buyback rates. This is coupled with active selling of power into the grid due to an assumed slow down in the addition of new baseload generating stations.

The determination of buyback rates under PURPA is an important regulatory factor, but there are others that will influence the level of natural gas-fired cogeneration in the future. One of these regulatory factors affecting cogeneration potential is the granting of qualifying facility (QF) status to applicants by FERC under PURPA. QF status means that a cogenerator is granted automatic entitlement to back-up electric power (in the event the facility is down). QF status, therefore, is very important in reducing the financial risk associated with a forced outage. Many cogeneration projects are neither owned nor operated by applicants for QF status, rather they are financed by third parties. A recent FERC ruling calls into question the ability of third-party-financed projects to obtain QF status. To reflect this condition in the forecast, the BAU Case assumes full exclusion of third-party-financed projects from QF status, the Attainable Case assumes that a case-by-case approach is adopted and the Potential Case assumes that the FERC Order is reversed. A regulatory issue of more limited applicability is the question of wheeling cogenerated electricity. In the BAU Case, wheeling or the transportation of electricity to purchasers other than the local utility, is assumed to be prohibited. Wheeling is assumed to be allowed in the Potential Case and on a limited basis in the Attainable Case.

ELECTRICITY OUTLOOK

Cogeneration will be particularly sensitive to the outlook for new baseload capacity additions (coal and nuclear), electricity imports from Canada, and long-distance wheeling capability. Because cogenerated electric power is an important source of revenue and serves to reduce the economic risk of a project, the outlook for electric supply and demand becomes a key factor for investors in cogeneration projects. From the viewpoint of an industrial customer, projected electricity price and availability will be an important factor in the decision to cogenerate and in the choice of system size.

As to the outlook for new capacity additions, A.G.A. has conducted two studies, one on coal and one on nuclear capacity, that indicate the likelihood of a capacity shortage in the mid-1990s. The basic reason for this outlook is the reluctance of electric utilities to undertake new, costly baseload construction projects in the face of large cost overruns, cancellations and construction delays experienced in recent years.

To reflect these electric capacity considerations in the forecast, the BAU Case assumes that new baseload capacity keeps up with electric demand at current reserve margins. The Attainable Case forecast assumes that new capacity keeps pace with electricity demand only at the expense of reduced reserve margins, meaning lower electric reliability and increased incentives for cogeneration. The Potential Case forecast assumes that real capacity shortages are experienced by 1995. Given the long lead times for large baseload power plants, it is assumed that there would be a period of several years in which cogeneration and other alternative sources of electricity supply would have to be relied upon. In conjunction with the electric capacity assumptions, the BAU Case assumes higher than average imports along with added long distance transmission capability, the Attainable Case assumes current levels and the Potential Case assumes lower import levels due to decreased hydropower availability.

Estimates of the potential for natural gas-fired cogeneration in the EOR market vary according to assumptions regarding heavy crude production, environmental considerations in using lease crude to generate steam and more general considerations about the ability of local electric utilities to absorb the new cogeneration capacity. One certainty is that natural gas will be priced competitively and will be available to serve this market. Based on estimates by the California Energy Commission and utilities, it is assumed in the BAU Case that gas demand in the EOR market will be 120 Bcf by 1990. The Attainable Case assumes that gas takes a higher share of the EOR market, which translates into 144 Bcf of gas demand by 1990. The Potential Case assumes that there will be clear environmental constraints on the use of lease crude and that firm gas supplies will be available. Under those conditions producers will not have to maintain dual-fuel capability for fuel supply security or standby scrubbers for emissions reduction. As a result, there will be an increase in gas demand to 228 Bcf by 1990.

Fuel cell research and development is being carried out presently. While the technology appears operable, it may be a while before capital costs can be reduced enough to be competitive with other cogeneration technologies. Accordingly, this forecast assumes that full commercialization will occur some time before the year 2000. While there will be gas-fired fuel cells in the market by 2000, this forecast assumes that significant volumes will only be achieved in the Potential Case. The potential could be 1.4 GW of capacity, or 80 Bcf of gas consumption in 2010.

RESULTS

The A.G.A. Gas Demand Committee expects that gas demand for cogeneration will grow moderately in the next few years, after recent experience of very rapid growth, and will then enter another period of rapid growth as electric generating capacity limitations begin to appear. In the Attainable Case, cogeneration gas demand is expected to nearly double to 1.2 Tcf by the year 2000 and 1.3 Tcf in 2010. The Potential Case is driven by the assumptions regarding new baseload capacity shortages in the mid-1990s and by higher buyback rates. Consequently, gas demand for cogeneration in the Potential Case would be expected to grow from 1.0 Tcf in 1990 to 1.7 Tcf in 2000. In 1990, cogeneration gas demand is expected to level off somewhat as a result of a temporary pause in the key regions, the Southwest and West. However, if all of the projects now under consideration in other regions, such as the Northeast, the Attainable and Potential Case forecasts will likely be conservative. Between 1990 and 2010, cogeneration gas demand in the Potential Case would be expected to double to 2.0 Tcf. In the BAU Case, cogeneration will grow very slowly during the forecast period. Market penetration will be based on the comparative price advantage of gas. Consequently, cogeneration gas demand would only rise to 0.9 Tcf by 2010.

To put the gas demand volumes in perspective, Table B-1 included estimates for the comparable generating capacity and number of units associated with the various forecasted levels of gas demand. In the year 2000, for example, the forecast range of 12.3 GW to 26.2 GW of generating capacity would be approximately 1.5 to 3 percent of the projected total electric generating capacity in 2000 according to the A.G.A.-TERA model base case forecast. The number of cogeneration units in 2000 would range from 1,500 to 2,600. The average size per unit would be expected to vary between cases. For example, the assumptions in the BAU Case regarding no wheeling and low buyback rates would result in the down-sizing of cogeneration equipment to levels designed to serve only on-site needs. Conversely, the Potential Case shows larger-sized units that would take advantage of the opportunity to sell electricity back to the grid at favorable buyback rates.

There may be some decline in conventional electric utility and industrial cogeneration in the Attainable and Potential Cases. However, it is very likely that the primary effect of increased cogeneration will be the delay and cancellation of new coal and nuclear baseload plants.

Table B-2. Key Demand Assumptions for Cogeneration

Market/Factor	Forecast		
	BAU	Attainable	Potential
— Avoided Cost/Buyback Rate	low buyback rates	moderately attractive rates	continued attractiveness coupled with significant reselling to grid
— Regulatory Factors			
• FUA	FUA remains	FUA remains with liberal exemptions (Repealed by 1990)	FUA repealed immediately
• Third Party financing and Qualifying Facility Status	FERC Fully excludes third party QF's	FERC ruling applied case-by case	no restrictions
• Wheeling of Cogenerated electricity	no wheeling	wheeling allowed on a limited basis	wheeling allowed
— Electricity Imports	high hydro-power output (Canada) and added long distance capability	current projections only	lower hydropower output than normal
— Electricity Capacity Additions	reserve margins remain at industry norm	new capacity additions just keep up with demand	generation capacity approaches shortage by 1995
— EOR Market Potential	120 BCF by 1990	144 BCF; gas gets 80% of steam generation (Calif. Energy Commission estimate)	switching all steam to cogen increases gas demand to 228 Bcf by 1990
			—no need for dual fuel capability & standby scrubbers
—Fuel Cell Development	capital cost remains high, commercialization delayed	commercialized by 2000, but penetrates slowly	1.4 GW (1,400 Megawatts) potential approached around 2010 (80 Bcf)

Appendix C

Cogeneration Equipment Vendors

J. A. ORLANDO, P.E.
GKCO, Incorporated

- Turbine Manufacturers
- Reciprocating Engines
- Packaged Cogeneration Systems
- Heat Recovery Boilers

Disclaimer

No claims are made regarding the completeness of this listing. To make additions or to suggest changes, write to the address below and provide all required information in your letter.

Manager Energy Systems Marketing
American Gas Association
1515 Wilson Boulevard
Arlington, VA 22209

TURBINE MANUFACTURERS

Allison Gas Turbine Operations
General Motors Corp.
P.O. Box 420
Indianapolis, IN 46206
317/242-4151

Alsthom
3, Avenue des Trois Chenes
90001 Belfort, France

Avco Lycoming Division
550 So. Main St.
Stratford, CT 06497
203/385-2000

BBC Brown, Boveri & Company, Ltd.
1460 Livingston Ave.
North Brunswick, NJ 08902
201/932-6000

Coppus Engineering Corp.
Turbine Division
344 Park Ave.
Worcester, MA 01610
617/756-8391

Dresser Clark, Div. of Dresser Industries
P.O. Box 560
Olean, NY 14760
716/375-3000

GEC Rolls-Royce Power
Generation Division of
Ruston Gas Turbines Inc.
15950 Park Row
Houston, TX 77084
713/492-0222

The Garrett Corp.
9851 Sepulveda Blvd.
Los Angeles, CA 90009
213/417-6836

General Electric Company
Turbine Marketing & Products
One River Road

Schenectady, NY 12345
518/385-0607

Ingersoll-Rand
Turbo Products Division
942 Memorial Parkway
Phillipsburg, NJ 08865
201/859-8549

John Brown Engineering Ltd.
Clydebank, Dunbartonshire,
G81 1YA
Scotland
041/962-2030

Kawasaki Heavy Ind. Ltd.
1-1 Kawasaki-cho
Akashi 673 Japan
078/923-1313

Kraftwerk Union Ag
Gas Turbine Subdivision
Hammerbacherstr 12/14
Postfach 3220
D-8520 Erlangen
West Germany
09131-18-1

Murray Turbomachinery Corp.
Burlington, IA 52601
319/753-5431

Onan Corporation
1400 73rd Ave. NE
Minneapolis, MN 55432
612/574-5000

Ruston Gas Turbines Ltd.
P.O. Box 1
Lincoln LN2 5DJ
England
0522-251212

Solar Turbines Inc.
P.O. Box 85376

GAS TURBINE MANUFACTURERS (Continued)

San Diego, CA 92138
714/238-6500

Stewart & Stevenson Services Inc.
P.O. Box 1637
Houston, TX 77001
713/868-7700

Terry Turbine Co.
P.O. Box 555
Lamberton Road
Windsor, CT 06095
203/688-6211

Turbomeca
Bordes 64320 Bizanos
France
(59) 32/84/37

Turbonetics Energy Inc.
Subsidiary of Mechanical Technology Inc.
968 Albany-Shaker Road
Latham, NY 12110
518/785-2211

Turbosystems International
Div. Vevey Manufacturing, Inc.
7 Northway Lane
Latham, NY 12110
518/783-1625

Westinghouse Electric Corp.
Power Generation Commercial Division
P.O. Box 251, Mail Stop C-180
Concordville, PA 19331
215/358-4887

RECIPROCATING ENGINES

ALCO Power Inc. (CEC)
100 Orchard St.
Auburn, NY 13021
315/253-3241

Allis-Chalmers Engine Div.
151 St. & Halsted
P.O. Box 563
Harvey, IL 60426
312/339-3300

Caterpillar Tractor Co.
Engine Division
100 N.E. Adams St.
Peoria, IL 61629
309/578-6423

Colt Industries
Fairbanks Morse Engine Div.
12253 FM, Northwoods

Houston, TX 77041
713/896-9455

Cooper Industries
1401 Sheridan Ave.
Springfield, OH 46501
513/327-4200

Cummins Engine Co., Inc.
1031 E. Rio Grande
Paso, TX 79902
915/593-8888

Detroit Diesel Allison
Division of General Motors Corp.
13400 West Outer Drive
Detroit, MI 48228
313/532-0660

Enterprise Engine & Compressor Division
Transamerica Delaval Inc.
550 85th Ave.
Oakland, CA 94621
415/577-7400

RECIPROCATING ENGINES (Continued)

Ford Motor Company
Industrial Engine Operations
3000 Schaefer Road
P.O. Box 6011
Dearborn, MI 48121
313/323-2123

Hawker-Siddeley Dynamics
Engineering Ltd.
Bridge Road East
Welwyn Garden City AL7 1LR
England
0453/885166

Ingersoll-Rand
P.O. Box 348
Broken Arrow, OK 74012
918/451-0464

International Harvester Co.
Engine Division
401 N. Michigan Ave.
Chicago, IL 60611
312/836-2000

M.A.N.-BMW Diesel GMBH
Stadtbachstrasse 1
8900 Augsburg
Federal Republic of Germany
45 8 42 10 00

Mirrlees Blackstone
Stamford, Lincolnshire
England
0780 64641

S.E.M.T-Pielstick
2 Quai De Seine
93203 St. Denis, France

Skinner Engine Co.
Box 1149
Erie, PA 16512
814/454-7103

Stewart & Stevenson
Box 1637
Houston, TX 77251
713/868-7700

Sulzer Brothers Ltd.
Diesel Engineering & Power Plant Division
8401 Winterthur, Switzerland
052/811122

Teledyne Total Power
3409 Democrat Road
Memphis, TN 38118
901/365-3600

Transamerica Delaval Inc.
Engine & Compressor Division
550 85th Ave.
Oakland, CA 94621
415/577-7400

Waukesha Engine Division
Dresser Industries, Inc.
1000 St. Paul Ave.
P.O. Box 379
Waukesha, WI 53187
414/549-2951

White Engines, Inc.
101 Eleventh St., SE
P.O. Box 6904
Canton, OH 44706
216/454-5631

PACKAGED COGENERATION SYSTEMS

Allison Energy Systems
Gas Turbine Division
P.O. Box 420
Speed Code U-1
Indianapolis, IN 46206-0420
317/242-2947

American M.A.N. Corp.
50 Broadway
New York, NY 10004
212/509-4545

American Private Power
922 S. Barrington Ave., Suite A
Los Angeles, CA 90049
213/826-1474

Cogenic Energy Systems Inc.
127 E. 64th St.
New York, NY 10021
212/772-7500

Design Systems, Inc.
1335 Midway Drive
Alpine, CA 92001
619/445-3563

Empire Generator Corp.
W190 N11260 Carnegie Drive
Germantown, WI 53022
414/255-2700

Enertech
2832 Vassar Road
Albuquerque, NM 87587
505/881-8089

Intellicon
7750 Daggat St., Suite 201
San Diego, CA 92111
619/569-7141

KW Energy Systems
75 Otis St.

Northborough, MA 01532
617/393-7855

Micro Cogen Systems, Inc.
Subsidiary of Ultrasystems, Inc.
16845 Van Karman Ave.
Irvine, CA 92714
714/863-7000

Martin Tractor Co., Inc.
1637 S.W. 42nd
Box 1698
Topeka, KS 66601
913/266-5770

Perennial Energy
Route 1, Box 645
West Plains, MO 65775
417/256-2002

Thermex Corporation
1086 N. Kraemer Place
Anaheim, CA 92806
714/630-5882

Thermo Electron Corp.
45 First Ave.
P.O. Box 459
Waltham, MA 02254
617/890-8700

Van Weld, Inc.
P.O. Box 26885
Albuquerque, NM 87125
505/243-8955

Waukesha Engine
Division of Dresser Industries
Box 379
Waukesha, WI 53187
414/547-3311

Zond Cogeneration Systems, Inc.
17752 Skypark Circle, Suite 150
Irvine, CA 92714
714/261-1030

HEAT RECOVERY BOILERS

ABCO Industries, Inc.
2675 E. Highway 80
Abilene, TX 79604

The American Schack Co., Inc.
P.O. Box 11006
Pittsburgh, PA 15237

Babcock & Wilcox Company
P.O. Box 1126, Wall St. Station
New York, NY 10268

The Bigelow Company
Lloyd & River St.
Cranford, NJ 07016

Burn-Zol, Division of New Way
Industries, Inc.
P.O. Box 109
Dover, NJ 07801

Buxton Mfg. Company
15 N. Salem St.
Dover, NJ 07801

Clayton Mfg. Company
4213 No. Temple City Blvd.
El Monte, CA 91731

Conseco, Inc.
611 North Road
Medford, WI 54451

Cyclotherm Division
Oswego Package Boiler Co.
157 E. First St.
Oswego, NY 13126

Deltak Corporation
P.O. Box 9496
Minneapolis, MN 55440

Dias, Incorporated
P.O. Box 3009
Kalamazoo, MI 49003

Eclipse, Inc.
1665 Elmwood Road
Rockford, IL 61103

Eclipse Lookout Company
P.O. Box 4756
Chattanooga, TN 37405

Econo-Therm Energy Systems
11535 K-Tel Drive
Minnetonka, MN 55343

First Thermal Systems, Inc.
Box 4756
Chattanooga, TN 37405

Green Economizers, Inc.
29425 Chagrin Blvd.
Pepper Pike, OH 44122

Indeck Power Equipment Co.
1075 Noel Ave.
Wheeling, IL 60090

Industrial Boiler Co., Inc.
221 Law St., Box 2258
Thomasville, GA 31792

The International Boiler Works Co.
5 Birch St., East
Stroudsburg, PA 18301

Johnston Boiler Company
300 Pine St.
Ferrysburg, MI 49409

E. Keeler Company
238 West St.
Williamsport, PA 17701

Kelley Company, Inc.
6720 N. Teutonia Ave.
Milwaukee, WI 53209

Kewanee Boiler Corp.
16100 Chesterfield Village Parkway
Chesterfield, MO 63107

HEAT RECOVERY BOILERS (Continued)

Midwesco Energy Systems
7720 Lehigh Ave.
Niles, IL 60648

Noranda Metal Ind., Inc.
Prospect Drive
Newtown, CT 06470

Q Dot Corporation
726 Regal Row
Dallas, TX 75247

Ray Burner Company
1301 San Jose Ave.
San Francisco, CA 94112

Riley Beaird, A Division of
U.S. Riley Corp.
P.O. Box 31115
Shreveport, LA 71130

Seattle Boiler Works, Inc.
500 S. Myrtle St.
Seattle, WA 98108

R. D. Smith International Supply Co.
215 W. 28th St.
P.O. Box 7456
Houston, TX 77008

Struthers Wells Corporation
P.O. Box 8
Warren, PA 16365

Thermal Transfer Corp.
1100 Rico Road
Monroeville, PA 15146

Trane Thermal Company
Process Division
200 Brook Road
Conshohocken, PA 19428

Vaporphase Engineering Controls
5901 Elizabeth Ave.
St. Louis, MO 63110

Henry Vogt Machine Co.
P.O. Box 1918
Louisville, KY 40201

Wabash Power Equipment Co.
444 Carpenter Ave.
Wheeling, IL 60090

Wessels Company
1625 E. Euclid Ave.
Detroit, MI 48211

Western Blower Corp.
18625 E. Railroad St.
City of Industry, CA 91748

Zurn Industries, Inc.
Energy Division
1422 East Ave.
Erie, PA 16503

Appendix D

Glossary

J. A. ORLANDO, P.E.
GKCO Incorporated

AVAILABILITY:

The ratio of the time the unit is capable of being in use to the total time.

AMBIENT TEMPERATURE:

The temperature of the air surrounding the equipment. This temperature may be expressed in degrees Celsius or Fahrenheit. Normally, ambient temperature is expressed in degrees Celsius when referring to electrical equipment. Degrees Fahrenheit is more frequently used for engines and mechanical equipment.

BACK-UP POWER:

Electric energy available from or to an electric utility during an unscheduled outage to replace energy ordinarily generated by the facility or the utility. Frequently referred to as standby power.

BASELOAD:

The minimum electric or thermal load generated or supplied continuously over a period of time. For example, the electric baseload for an industrial facility is that power

507

requirement for process, lighting, etc., which exists at that site for up to 8,760 hours a year.

A baseloaded prime mover is one which is operated at its continuous rating.

BOTTOMING-CYCLE: A cogeneration facility in which the energy input to the system is first applied to another thermal energy process; the reject heat that emerges from the process is then used for power production. For example, a steam turbine that is driven by heat recovered from a kiln.

CAPABILITY: The maximum load that a generating unit, generating station, or other electrical apparatus can carry under specified conditions for a given period of time, without exceeding approved limits of temperature and stress.

CAPACITY: The load for which a generating unit or station is rated. This capacity is applicable only under the specified conditions.

CAPACITY FACTOR: The ratio of the actual annual plant electricity output to the rated plant output.

COGENERATION: The sequential production of electrical or mechanical energy and useful thermal energy from a single energy stream.

COMBUSTOR: The mechanical component of the gas turbine in which fuel is burned to increase the temperature of the working medium.

DEMAND: The rate at which electric energy is delivered at a given instant or averaged over any designated period of time (usually fifteen minutes to one hour).

Annual Demand—The greatest of all demands which occurred in a calendar year.

Billing Demand—The demand upon which billing to a customer is based, as specified in a rate schedule or contract. It may be based on the contract year, a contract minimum, or a previous maximum and, therefore, does not necessarily equal the demand actually measured during the billing period.

DUAL FUEL SYSTEM: An engine that can switch back and forth from one fuel (e.g., gas) to another (e.g., oil) or which can simultaneously burn both fuels with no technological modification and minimal downtime.

EFFICIENCY: A ratio of the amount of input energy which is converted to useful output. Overall efficiency for a cogenerator includes the thermal as well as electric output and is expressed as the ratio of electric output plus heat recovered in Btus to fuel input in Btus.

EXHAUST HEAT RECOVERY: The process of extracting heat from the working medium leaving a prime mover and transferring it to a second fluid stream or to a product.

GRID: The system of interconnected transmission lines, substations and generating plants of one or more utilities.

GRID INTERCONNECTION: The intertie of a cogeneration plant to an electric utility's system to allow electricity flow in either direction.

HARMONICS: Waveforms whose frequencies are multiples of the fundamental (60 Hz) wave. The combination of harmonics and fundamental wave causes a nonsinusoidal, periodic wave.

Harmonics in power systems are the result of nonlinear effects. Typically, harmonics are associated with rectifiers and inverters, arc furnaces, arc welders and transformer magnetizing current. There are both voltage and current harmonics.

HEAT RATE: A measure of generating station thermal efficiency, generally expressed in Btu per net kilowatt-hour. Manufacturers will usually quote heat rate in a fuel's Lower Heating Value, while the fuel is normally sold based on its Higher Heating Value.

HEATING VALUE: The energy content in a fuel that is available as heat.

INTERCOOLER: A heat exchanger located between two compressor stages to reduce the air temperature entering the high-pressure compressor stage and thereby reducing the power required to drive the compressor.

LOAD FACTOR: The ratio of the average load supplied or required during a designated period to the peak or maximum load occurring in that period.

MAGNETIZING CURRENT: The current required to magnetize transformers, motors and other electromagnetic devices containing iron in a magnetic circuit. It is customary to speak of the lagging inductive current as a magnetizing current. See also Power Factor.

POWER: The time rate at which energy is produced or used, usually expressed in kilowatts.

Apparent—A quantity of power proportional to the mathematical product of the volts and amperes of a circuit. This product is generally designated in kilovoltamperes (kVa), and is comprised of both real and reactive power.

Reactive—The portion of "apparent power" such as the magnetizing power, that does not do work. It is commercially measured in kilovars (kVar).

Real—The energy or work-producing part of "apparent power." It is the rate of supply or energy, measured commercially in kilowatts (kW).

POWER FACTOR: Power factor is the ratio of real power measured in kilowatts to apparent power measured in kilovoltamperes for any given load.

POWER POOL: Two or more interconnected electric systems planned and operated to supply power in the most reliable and economical manner for their combined load requirements and maintenance program.

QUALIFYING FACILITY: A cogeneration facility which has been granted a "qualified" status by the FERC. To obtain the "qualified" status a facility must meet the ownership requirement (i.e., no more than 50 percent electric utility ownership) and operating efficiency standards as outlined in the FERC Order 70.

RECUPERATOR: A heat exchanger having the turbine's exhaust gas and the combustion air streams separated by a thin wall (usually of metal) through which heat is transferred by conduction.

REGENERATOR: A heat exchanger designed to transfer heat from a turbine's exhaust gases to the compressed air stream.

REHEATER: A combustor located between two turbine stages to increase the temperature of the working fluid and the power available from it.

SIMPLE CYCLE TURBINE: A turbine in which the working medium passes successively through the compressor, combustor, and turbine only.

SINGLE-SHAFT GAS TURBINE: A turbine in which all the rotating components are mechanically coupled together on a single shaft.

SUPPLEMENTAL THERMAL: The heat required when recovered engine heat is insufficient to meet user thermal needs.

SUPPLEMENTAL FIRING: The burning of fuel in the recovered heat stream (such as turbine exhaust) to raise the heat content of the stream.

SUPPLEMENTAL POWER: Electric energy supplied by an electric utility in addition to that which the facility generates itself.

THERMAL CAPACITY: The maximum amount of heat that a system can produce.

TOPPING-CYCLE: A cogeneration facility in which the energy input to the facility is first used to produce power, with the reject heat from power production then used for other purposes. An engine or gas turbine driving a generator with the exhaust heat recovered and used for space heating is an example of a topping cycle.

VOLTAGE FLICKER: Term commonly used to describe a significant fluctuation of voltage.

WHEELING: The use of transmission facilities of one electric utility system to transmit power for another system.

Appendix E

Computerized Analysis Techniques

J. A. ORLANDO, P.E.
GKCO Incorporated

The development of the cogeneration market has been accompanied by the marketing of computer software specifically designed as an aid in analyzing and/or designing cogeneration systems. The available software can be categorized as being suitable for screening or design. In general, *screening programs,* available for purchase or lease, are designed for use on microprocessors or on mainframes with access via user-owned terminals. They require varying amounts of data describing the site's energy characteristics, cogeneration and conventional equipment performance and costs, utility rate and cost data and financial information. The extent of built-in data varies considerably. The characteristic of these programs is that they are intended only for the initial screening of a cogeneration application. Vendors claim that they require anywhere from 15 minutes to several hours to exercise, once the program has been installed on the user's computer. They are not capable of use in detailed system design.

The second type of software consists of those *design programs* or software systems capable of supporting the detailed analysis of alternatives that is required during system design. These programs generally are available on either a "per run" fee or a rental basis. Among the

513

most commonly used programs are ECUBE IV, DOE 2, The Ross F. Meriwether Analysis System and AXCESS. This latter program was developed by the Edison Electric Institute.

Screening programs are catalogued and reviewed in this appendix.

□ □ □

NAME: Cogeneration Assessment Program (CAP)

TYPE: Screening

SOURCE: Insights West, Inc.
                   13293 Courtland Terrace
                   San Diego, CA 92310
                   Mr. Russel Williams
                   (619) 259-0661

DESCRIPTION: The program provides a screening of potential applications using an hour-by-hour approach. Designed for use by utility personnel, the system includes a database with process, space and water-heating energy use profiles for 25 different industrial and commercial facilities. The user has the option to override any of the built-in data. The user must input fuel and electric cost data and capital cost information.

                   The program has an option wherein the thermal and electrical loads are modified to substitute an absorption chiller for electric-driven air conditioning.

COMPUTER   An IBM PC or PC compatible with MS/DOS and 256K
REQUIREMENTS: memory. Requires 1.5 hours to install and 20 minutes to exercise.

COST: $495 for program purchase.

□ □ □

NAME: COGENOPT

TYPE: Screening

SOURCE: Decision Focus Inc.
                   4984 El Camino Real

Los Altos, CA 94022
Mr. Gary Ackerman
(415) 960-3450

DESCRIPTION: The program is an optimization model designed for use in industrial cogeneration systems. The program uses Net Present Value as the economic criteria.

Program required input includes load profiles, system sizes, equipment sizes, fuel and electric prices and capital cost data.

COMPUTER
REQUIREMENTS: An IBM PC with 256K memory. Requires 15 to 20 minutes to exercise.

COST: $6,500 for program purchase; $365/month for program lease

□ □ □

NAME: Modular Steam System Analyzer (MESA)

TYPE: Screening of steam systems

SOURCE: The MESA Co.
22 Golden Shadow Circle
The Woodlands, TX 77381
(713) 363-3133

DESCRIPTION: The program models an existing or proposed steam system. The user must input required steam temperatures, pressures and flows; boiler, generator and steam turbine characteristics; and fuel and electric costs. A file generator allows the user to apply different spreadsheet programs for detailed report generation.

COMPUTER
REQUIREMENTS: An IBM PC with 384K memory and math co-processor. The program is written in FORTRAN.

COST: $15,000 to $20,000 for program purchase.

□ □ □

NAME: PG&E Financial Analysis Program

TYPE: Screening

SOURCE: CDC Cybernet®

DESCRIPTION: The program takes user-supplied data assumptions regarding technical and cost characteristics to produce a 20-year cash flow and calculate an internal rate of return, net present value and simple payback. The user must develop required technical data, including seasonal peak, partial peak and off peak data, utility rates and fuel costs, equipment performance and costs, and financial assumptions.

The program appears to consider only steam turbine-based cogeneration.

COMPUTER REQUIREMENTS: Requires a terminal capable of communicating with the CDC Cybernet® Center.

COST: Dependent on usage.

□ □ □

NAME: COGEN

TYPE: Screening of steam systems

SOURCE: Software Systems Corp.
5766 Balcones Drive
Austin, TX 78731
Mr. M. Williams
(512) 451-8634

DESCRIPTION: The program evaluates the thermodynamic and economic performance of a cogeneration system that uses back pressure steam turbine technology.

COMPUTER REQUIREMENTS: An IBM PC or PC with 64K memory. Written in MBASIC.

COST: $495 for program purchase.

□ □ □

NAME:	Cogenerator I
TYPE:	Screening
SOURCE:	Energy Conversion Corp. 1310 Industrial Avenue Escondido, CA 92015 Mr. Don Roberts (619) 746-8390
DESCRIPTION:	The program is an interactive analytic tool for determining cogeneration feasibility. Input data include three years of previous electric utility billings and rates; fuel consumption and costs; equipment choices, performance and costs and financial information.
	Program uses a database language allowing the user to specify any parameter as a variable.
COMPUTER REQUIREMENTS:	Any PC compatible with MS/DOS, CP/M 80 or 86. Uses dBase II language.
COST:	$495 for program purchase.

□ □ □

NAME:	Cogeneration and Energy Planning Program (CEPP)
TYPE:	Screening and more detailed analysis
SOURCE:	ENCOTECH Inc. Box 174 Schenectady, NY 12301 Mr. John M. Daniels (518) 374-0924
DESCRIPTION:	The program consists of five modules including a main program which simulates plant operations, a financial model and three output programs, and is supported by an ENCOTECH developed database. The user must input load profiles, equipment sizes and performance data, cost data, financing and tax information. The program can accept part-load performance data and will output system operating data including machine loadings.

COMPUTER   Requires terminal (Diablo, Xerox or DEC) to communicate
REQUIREMENTS: with ENCOTECH Computer. Also available in an IBM
                      PC compatible version.

COST: $500 initiation fee ($100 for A.G.A. members). Time
                      sharing costs $300 to $800 per case.

□ □ □

NAME: Associated Cogeneration Evaluation (ACE)

TYPE: Screening

SOURCE: Associated Utility Services, Inc.
                      Box 650
                      Moorestown, NJ 08057
                      Mr. William Steigelmann
                      (609) 234-9200

DESCRIPTION: The program can model up to three types of prime mov-
                      ers and up to 12 utility rate structures. The program can
                      handle time varying electric and thermal loads and re-
                      quires part-load equipment performance data as part of
                      the user input.

                      The program considers inflation, fixed and variable O&M
                      costs, tax liability and computes net present value and
                      internal rate of return.

COMPUTER   An IBM PC-XT or AT with 384K memory and 8087 math
REQUIREMENTS: co-processor chip.

COST: $1,250 for program purchase. $95/rate for preprogrammed
                      utility rate structures.* $50/set for system performance
                      parameters.* $200/year annual maintenance fee.

                      *Plus $50 handling fee.

□ □ □

NAME: Cogeneration Feasibility Analysis System (CFAS)

TYPE: Screening

SOURCE: Integrated Energy Systems
                      307 N. Columbia Street

Chapel Hill, NC 27514
Mr. Daniel Koenigshofer
(919) 942-2007

DESCRIPTION: The program evaluates three types of prime movers and includes fuel switching, both with and without cogeneration. Required input includes thermal and electrical loads, equipment parameters, fuel costs, utility rate structures and operating costs. The program computes simple payback and life-cycle costs and financial analyses based on user-defined input.

COMPUTER REQUIREMENTS: Written in Microsoft® BASIC. Available for TRS-80 III and IV and IBM/PC and compatibles. Requires 15 minutes to one hour to exercise.

COST: $1,000 for program purchase. $75 per year for annual updates.

□ □ □

NAME: Dual Energy Use Systems (DEUS) Computer Program

TYPE: Screening of steam systems

SOURCE: National CSS
Marina Playa Executive Park
1333 Lawrence Expressway
Santa Clara, CA 95051
(408) 249-9500

DESCRIPTION: The program analyzes steam cogeneration from the perspective of an electric utility. It considers alternate cogeneration options for a variety of industries and includes a database with built-in default values. The program performs energy, cost and cash flow calculations.

COMPUTER REQUIREMENTS: Available through National CSS Network.

COST:

□ □ □

NAME: Optimization and Simulation of Integrated Systems
                         (OASIS)

TYPE: Screening and more detailed analyses

SOURCE: Argonne National Laboratories
                         9700 Cass Avenue
                         Argonne, IL 60439
                         Ms. Dorothy Bingamen
                         (312) 972-3978

DESCRIPTION: The program was developed by Argonne National Labora-
                         tories as an aid in analyzing and eventually designing com-
                         munity energy systems. The program accepts user-defined
                         demands and operating strategies and weather data and
                         can perform hour-by-hour simulations. Equipment type,
                         sizes and performance are input. The program computes
                         life-cycle costs as a basis for system optimization.

                         Default values are built into the program.

COMPUTER
REQUIREMENTS: An IBM 370/195, 3330.

COST:

                                    □ □ □

NAME: SYSTEMS & COGENERATION (SYSCOGEN)

TYPE: Comparative and Detailed Analysis

SOURCE: Energy Systems Engineers
                         8000 E. Girard Avenue, Suite 508
                         Denver, CO 80231
                         Mr. Don Pedreyra
                         (303) 696-6241

DESCRIPTION: The program determines actual energy consumed by
                         various types of central plant equipment in up to four
                         systems to meet the hourly energy requirements of a
                         building (or site). Hourly energy use is simulated by the
                         program from actual monthly utility use or from output
                         from another program (quikee program). The operating
                         characteristics of any type of generator (if used), chiller,

boiler, or heater as well as the characteristics of auxiliary equipment in the system are inputs to the program. The program printout is a monthly summary of the gas, auxiliary fuel, and electricity consumed and generated. The peak electric demand; the number of operating hours of each generator, chiller, and boiler; and an evaluation of thermal energy usage.

COMPUTER
REQUIREMENTS: IBM PC, XT, AT, and compatibles with 256K memory and 8087/80287 math co-processor. One 360K disk drive.

COST: $895.

□ □ □

Index

523